Deformation, Processing, and Structure

Other ASM Materials Science Seminar Books

Advances in Powder Technology
Fundamentals of Friction and Wear of Materials
Grain-Boundary Structure and Kinetics
Fatigue and Microstructure
Interfacial Segregation

Deformation, Processing, and Structure

Papers presented at the
1982 ASM Materials Science Seminar
23-24 October 1982
St. Louis, Missouri

Sponsored by the Seminar Committee of the
Materials Science Division of the
American Society for Metals

Edited by
George Krauss
AMAX Foundation Professor
Colorado School of Mines

AMERICAN SOCIETY FOR METALS
Metals Park, Ohio 44073

Library of Congress Catalog Card No.: 83-48774

ISBN: 0-87170-175-8

SAN: 204-7586

Editorial and production coordination by
Carnes Publication Services, Inc.

PRINTED IN THE UNITED STATES OF AMERICA

Preface

The papers in this book contribute an important extension to the understanding of the interrelationships among deformation, processing, and microstructure of metals and alloys. The state-of-the-art reviews examine in depth the complex interactions among deformation, strain hardening, phase transformations, and dynamic and static restoration mechanisms during thermomechanical processing. Important themes throughout the presentations are the influence of microstructure on deformation and the response of structure to deformation. An understanding of microstructure has gradually developed, but systematic characterization of microstructure coupled with thermomechanical treatment has heretofore been limited because of theoretical and experimental difficulties associated with elevated temperatures, complex thermomechanical treatments, very high strains, complex microstructures, extremes in strain rate, and/or phase transformations either concurrent with or subsequent to deformation. Although much remains to be done in defining property-processing-microstructure relationships in complex systems, the information presented in this ASM Materials Science Seminar book should both stimulate additional work and immediately contribute to improved processing, higher processing productivity and improved product quality in many materials.

The various complex interactions among deformation, processing and structure are effectively addressed in three sets of papers in this book. The first set considers strain hardening or the dynamic response of structure to deformation. The effect of very large plastic strains, typical of many metalworking and impact phenomena, on strain hardening and substructure development are developed by Hecker and Stout. Matlock et al. examine the effect of processing on multistage strain hardening of intercritically annealed steels with microstructures consisting of martensite in ferrite. Elevated-temperature solid-solution strengthening and strain hardening are theoretically developed by Kocks to complete the first set of papers.

The second set of papers deals with static and dynamic changes during hot work and various types of thermomechanical processing. Dynamic structural transitions during commercial hot working are emphasized by Roberts, and a new model for dynamic recrystallization is developed by Jonas and Sakai. The complex interactions between microalloying elements in solid solution, precipitation and static recrystallization of austenite after hot rolling of low-carbon steels are reviewed by Sellars. Starke and Williams discuss properties and structures produced by thermomechanical treatment of aluminum and titanium alloys. Sherby and Wadsworth complete this set of papers with a comprehensive discussion of the microstructural factors, processing parameters and deformation mechanisms associated with superplasticity. Also, in accord with this set of papers, Dr. M. J. Luton presented a talk in St. Louis, not reproduced here, concerning the evolution of microstructure and properties after hot work.

The last set of papers explores various aspects of the heterogeneity of deformation. Olson develops the effects of austenite transformation under stress on strain hardening and relates enhanced ductility and fracture toughness to this phenomenon. Deformation-induced thermal effects which cause strain localization are discussed by Rogers. Finally, case histories of inhomogeneities in structure and properties developed in complex austenitic stainless steel forgings are presented by Mataya and Carr.

Members of the 1982 ASM Materials Science Seminar organizing committee included M. J. Carr, R. J. Krenzer and M. Mataya, all from Rockwell International Corporation, Golden, Colorado; F. B. Fletcher, Climax Molybdenum Corporation of Michigan, Ann Arbor, Michigan; M. J. Luton, EXXON Research, Linden, New Jersey; A. T. Davenport, Republic Steel Corporation, Cleveland, Ohio; E. A. Starke, University of Virginia, Charlottesville, Virginia; and E. L. Brown, Colorado School of Mines, Golden, Colorado. Session Chairmen during the seminar in St. Louis were A. J. DeArdo, University of Pittsburgh, Pittsburgh, Pennsylvania; S. S. Hansen, Bethlehem Steel Corporation, Bethlehem, Pennsylvania; I. F. Hughes, Inland Steel Company, East Chicago, Indiana; and E. L. Brown, Colorado School of Mines, Golden, Colorado. The assistance and contributions of all of the above are warmly acknowledged.

Many ASM staff associates contributed to the conference arrangements and the publication of this volume. In particular, Lana Loar, Peg Ternovacz, Robert Uhl and Robert Stedfeld contributed much to the success of the Seminar and the published seminar proceedings. Travel grants for Drs. Roberts, Sellars and Jonas were awarded by the ASM Foundation for Education and Research.

GEORGE KRAUSS
AMAX Foundation Professor
Colorado School of Mines

Contents

The Role of Dynamically Recovered Substructure in Dynamic Recrystallization (A discussion of the paper by J. J. Jonas and T. Sakai)   231
(H. J. McQueen)

Deformation, Processing, and Structure

Strain Hardening of Heavily Cold Worked Metals

S. S. HECKER and M. G. STOUT

Materials Science and Technology Division
Los Alamos National Laboratory
Los Alamos, NM 87545

Introduction

The strain-hardening behavior of materials is very important in structural response, metalworking processes, and impact penetration problems. A precise description of strain hardening is particularly important in problems involving plastic instability. Large strains present specific complications in testing and analysis. Experiments are plagued by large geometric changes and a variety of instabilities. Hence, strain-hardening descriptions at large strains are typically inferred from uniaxial tensile tests, which are restricted to modest strains (<0.5) by plastic instability. Such data not only are inadequate, but also are often misleading. Analysis of large-strain problems is complicated by the need to account correctly for material rotations. For large-strain problems involving the behavior of isotropic materials, the Jaumann stress-rate tensor is adequate.[1] However, a number of investigators[2–5] have pointed out the shortcomings of this measure of stress when it is used with anisotropic hardening criteria such as kinematic hardening.

In this paper we will review large-strain experiments conducted by a variety of experimental techniques and will attempt to establish proper constitutive descriptions. We are particularly interested in the role of deformation geometry (or stress state) on hardening response, and we will attempt to sort out the effects of texture, microstructure and substructure. The effects of these factors are understood fairly well at small strains. Texture evolution plays a minor role at strains less than 0.3, and the variation of flow stress with material parameters, such as crystal structure, grain size and stacking-fault energy, is well documented. No satisfactory microstructural explanation exists, however, for hardening at large strains. On the basis of undisturbed dislocation interactions, one expects steady-state behavior and a saturation flow stress.[6,7] However, experiments show that hardening, rather than saturation, is more common at large strains. Hence,

1

processes such as texture development, twinning, deformation banding, shear banding, etc. play important roles.

Finally, we will cite several practical consequences of large-strain behavior — consequences related to problems of plastic instability and sheet metal forming.

Experimental Techniques

The principal experimental difficulties associated with large-strain testing are large geometric changes and a variety of instabilities which terminate experiments prematurely. These were reviewed previously.[8] A summary of techniques currently used is presented in Table 1. Torsion of very short, thin-wall tubes is the only technique that permits continuous determination of strain hardening to very large strains on reasonably homogeneously deformed specimens. All other tests require remachining or prestraining followed by a simple test. The limitation of having few decisive test techniques for determining large-strain behavior is in large part responsible for the current lack of understanding in this area.

We will cite data generated by a large number of techniques. In our own work we used the techniques of torsion on thin-wall tubes, compression with intermittent remachining, and sheet rolling followed by uniaxial tension.

Table 1. Current Large-Strain Test Techniques and Their Limitations

Test technique	Strain limits	Comments
Direct Tests		
Uniaxial tension (T)	<0.5	Plastic instability
	1 to 1.5	Necking correction
Biaxial tension (BT):		
Hydraulic bulge	<0.8	Plastic instability
Tube testing	<0.4	Plastic instability
Compression (C)	<0.7	Barreling
	3 to 4	Remachining
Torsion (TOR):		
Thin-wall tubes	<0.2	Buckling
	2 to 5	Very short tubes
Round solid bars	2 to 5	Limited by ductility
Indirect Tests		
Wire drawing plus tension (WD + T)	10	...
Strip drawing plus tension (SD + T)	4	...
Rolling plus tension (R + T)	7	...
Rolling plus plane-strain compression (R + PSC)	7	...

Macroscopic Hardening

Several phenomenological models used to describe hardening are summarized in Table 2. The first three predict continued hardening, whereas the Voce models predict saturation at a stress σ_s. Kocks' mechanistic model also predicts saturation.[6] In fact, a plot of θ versus σ presents a good graphical picture of the saturation stress. Kocks' model is based on dislocation interactions and predicts an early steady-state saturation behavior. Figures 1 and 2 demonstrate the hardening behavior of 1100 aluminum deformed under several different modes. As shown in Fig. 2, hardening does not saturate but persists to high stresses at low

**Table 2. Stress-Strain Relations Based on
Empirical (Phenomenological) and Theoretical
(Kocks) Considerations**
The Voce and Kocks Relations Predict Saturation.

Phenomenological Models:
Holloman
(parabolic) . . . . $\sigma = K\varepsilon^n$
Ludwik . . . . $\sigma = \sigma_0 + K'\varepsilon^{n'}$
Swift . . . . $\sigma = K_2(\varepsilon + \varepsilon_0)^{n_2}$
Voce . . . . $\sigma = \sigma_s - (\sigma_s - \sigma_0)\exp(-N\varepsilon)$
Modified Voce
(Hockett-Sherby) . . $\sigma = \sigma_s - (\sigma_s - \sigma_0)\exp(-N'\varepsilon^P)$

Kocks Model . . . . $\theta = \theta_0\left(1 - \dfrac{\sigma}{\sigma_s}\right)$

$$\text{where } \theta = \frac{d\sigma}{d\varepsilon}\bigg|_{\dot{\varepsilon},T}$$

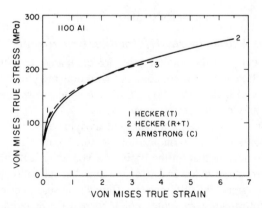

Fig. 1. Comparison of stress-strain curves as determined by tension, rolling + tension, and compression of annealed 1100 aluminum (*compression curve from Armstrong et al.*[15]).

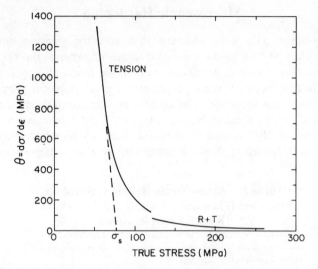

Fig. 2. Strain-hardening rate (θ) as a function of stress for the flow curves shown in Fig. 1. Extrapolation of tensile data to a meaningful saturation stress is not possible.

rates (θ). Apparently, some process intervenes at large strains (high stresses) and prevents a steady-state balance. This may be the result of deformation mode and texture development or of microstructural effects such as different deformation mechanisms, deformation banding or shear banding. A clear answer does not exist at present because no systematic studies have been conducted combining measurement of hardening, texture evolution and substructural evolution with analytical predictions.

Deformation Mode

Torsion Versus Axisymmetric Deformation Experiments

Stress-strain curves from torsion tests are compared with those from axisymmetric tests (tension, compression, and wire drawing plus tension) in Fig. 3 to 6 for a variety of metals and alloys (three fcc materials with widely different stacking-fault energies and one bcc material).

Unfortunately, in most cases torsion and axisymmetric tests were not conducted on the same material (composition, texture, grain size, etc.). Nevertheless, the results indicate that on the basis of a von Mises effective stress-strain criterion the flow curve is lower for torsion and that saturation occurs at lower stress levels in torsion. The differences in flow-stress levels between torsion and axisymmetric deformation are quite similar for the fcc materials. The effect is most dramatic for the bcc iron. In most of these comparisons torsion was con-

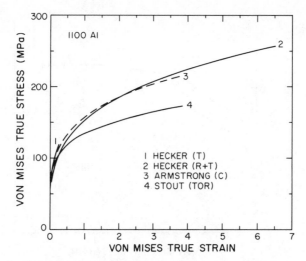

Fig. 3. Comparison of torsion with other deformation modes in annealed 1100 aluminum. Torsion was performed on rod specimens using the method of Nadai to reduce torque-twist to stress-strain. All comparisons were made on the basis of von Mises effective stress and strain.

ducted on solid rods, and some uncertainty exists concerning stress levels. However, the trends are quite clear.

Comparison of Other Deformation Modes

A careful and systematic study of the effect of stress state on hardening was conducted by the authors.[20] Thin-wall tubes of 70-30 brass were loaded under combined axial loading and internal pressure or torsion. The results were limited to moderate strain levels because of plastic instability. Stress-strain curves for a number of different stress states are shown in Fig. 7. The plane-strain stress states (axial, $\varepsilon_z = 0$; hoop, $\varepsilon_\theta = 0$; and torsion) all exhibit similar hardening — less than that in uniaxial tension. The flow curve in hoop tension is the lowest. We believe that this is a result of initial preferred orientation. The hardening rate for balanced biaxial tension appears to be slightly lower than that for uniaxial tension. Plastic instability in biaxial tension prohibits a comparison at larger strains, which would be required for a definitive answer. The lower hardening rate in plane strain agrees with earlier, more indirect experiments by Ghosh[21] and Wagoner.[22] Unfortunately, little large-strain data exists. However, the available literature data shown in Fig. 8, in conjunction with our data presented in Fig. 5, demonstrate that hardening in plane-strain deformation is less than that in axisymmetric deformation.

Razavi and Langford[27] compared axisymmetric wire drawing deformation with plane-strain (strip drawing) deformation. Their results, shown in Fig. 9, are

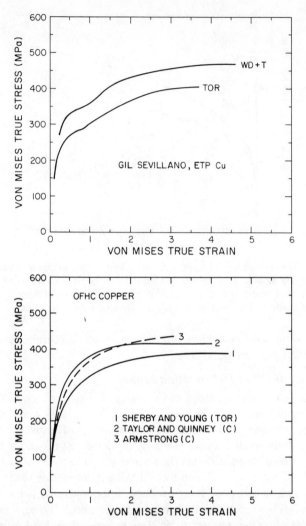

Fig. 4. (Top) Comparison of stress-strain curves for electrolytic tough-pitch copper deformed by wire drawing plus tension and by torsion (*Gil-Sevillano*[16]). (Bottom) Comparison of stress-strain curves for oxygen-free, high-conductivity copper deformed by compression (*Taylor and Quinney,*[18] *and unpublished work by Armstrong*) and by torsion (*Sherby and Young*[17]) of solid rods.

similar to the results for brass. At large strains the flow curve for strip drawing levels off whereas that for wire drawing continues to increase. The flow curve for torsion from Young and Sherby[17] is similar to that for strip drawing, albeit at lower stress levels. Similar results have been obtained on low-carbon steel. Figure 10 compares torsion with wire drawing plus tension for 1007 steel. The

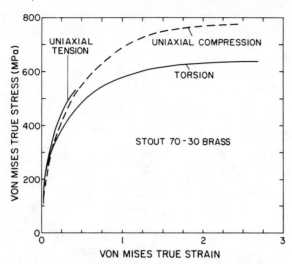

Fig. 5. Comparison of stress-strain curves for 70-30 brass for uniaxial tension, uniaxial compression, and torsion (*Stout and Hecker*[20]). Tension and torsion were carried out on identical thin-wall tubes. Compression was carried out on solid rod, which was remachined often to avoid barreling.

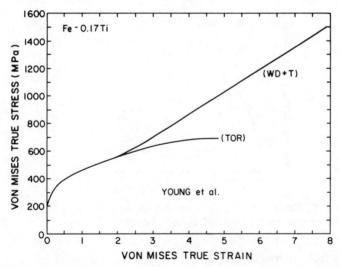

Fig. 6. Comparison of stress-strain curves for Fe-0.17%Ti deformed by torsion (solid rods) and by wire drawing plus tension.[19]

results of Ford[28] on plane-strain compression for 1008 steel are superimposed. Torsion and plane-strain hardening level off, whereas axisymmetric hardening continues at large strains.

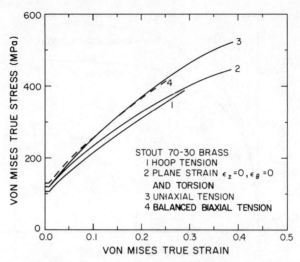

Fig. 7. Comparison of stress-strain curves for thin-wall 70-30 brass tubes.[20] Curve 2 represents the results for three different stress states: torsion, plane strain with no length change ($\varepsilon_z = 0$), and plane strain with no diameter change ($\varepsilon_\theta = 0$). Curve 1 represents uniaxial hoop tension.

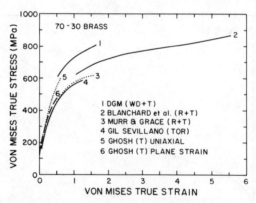

Fig. 8. Comparison of stress-strain curves for 70-30 brass for different deformation modes from Ref 21 and 23 to 26.

Changes in Deformation Mode

Most of the indirect large-strain tests fall into this category. During pre-straining, large strains are applied in a deformation mode resistant to plastic instability. The prestrain operation is then followed by a simple stress-state test (typically, uniaxial tension) to determine the flow curve. A composite stress-strain curve for 1100 aluminum determined by rolling plus tension is shown in

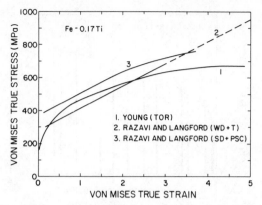

Fig. 9. Comparison of stress-strain curves for Fe-0.17%Ti deformed by torsion[19] (solid rod), wire drawing plus tension,[27] and strip drawing plus plane-strain compression.[27]

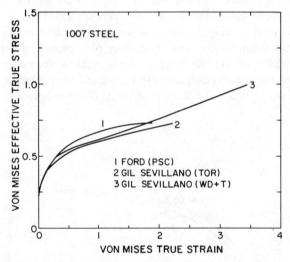

Fig. 10. Comparison of stress-strain curves for low-carbon steels: 1008 steel deformed by plane-strain compression;[28] and 1007 steel deformed by wire drawing plus tension and by torsion.[16,26]

Fig. 11. The resulting stress-strain curve can be represented as an effective flow curve by converting the rolling prestrain to a von Mises effective strain. The validity of such a curve depends, of course, on the independence of hardening from deformation mode, which has been demonstrated not to be the case. Hence, flow curves such as the one in Fig. 11 must be recognized as being complicated composites of two deformation modes.

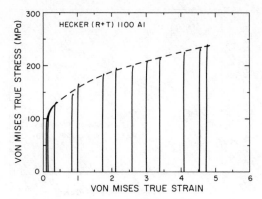

Fig. 11. Construction of a flow curve from rolling prestrain followed by uniaxial tension (R + T). The rolling thickness reduction is converted to an effective von Mises strain.

In some cases prestraining and final deformation are carried out in similar stress states. For instance, in wire drawing plus tension, both stress states are axisymmetric. Ford[28] compared the flow curves for low-carbon steel determined by plane-strain compression and by rolling plus plane-strain compression (Fig. 12). Here the stress states are very similar, and yet the rolling-plus-plane-strain compression curve is different. Ford explained this difference on the basis of redundant work, explaining that the curvature of the rolls causes some redun-

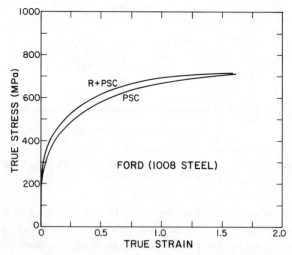

Fig. 12. Comparison of stress-strain curves for monotonic plane-strain compression with rolling prestrain followed by plane-strain compression (1008 steel[28]).

dant shearing (not contributing to thickness reduction) and extra hardening. Figure 13 shows a most dramatic effect of a change in deformation mode. Sundberg et al.[29] found that rolling plus tension in brass produced a rapidly rising flow curve, whereas rolling plus plane-strain compression resulted in immediate saturation. Unfortunately, Sundberg attached no scale to his plot; but the results are still most interesting. He noted that the plane-strain compression tests exhibited immediate shear-band formation.

Our experimental results for 70-30 brass comparing axisymmetric deformation and torsion are shown in Fig. 5. We also conducted a series of experiments prestraining in torsion followed by uniaxial tension. All specimens were thin-wall tubes. Test sections were 25.4 mm long, 12.14 mm in diameter and 0.589 mm in wall thickness. Specimens were carefully machined, annealed and electropolished before twisting. After twisting, they were unloaded, re-electropolished and strain gaged for tension testing. The resulting tensile curves are shown in Fig. 14 superimposed upon the previous torsion and compression curves. The two curves at smaller prestrains showed little uniform elongation; most of the deformation occurred in a localized neck. Hence, these flow curves are questionable. The two curves for large prestrains definitely show that significant plastic flow in tension following torsional prestraining takes much higher stresses than does continued torsion. In fact, the flow curves are very close to that observed for compression at the same von Mises strain level.

For 1100 aluminum we followed torsion with uniaxial compression. Torsional prestraining was conducted on a solid round rod. The rod was then drilled out and bored to provide a thin-wall tube 5.1 mm long, 4.8 mm in diameter and

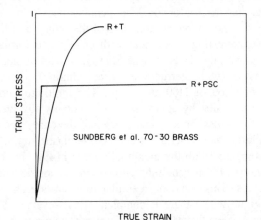

Fig. 13. Comparison of stress-strain curves for 70-30 brass determined by rolling plus tension and by rolling plus plane-strain compression. Curves are only schematic because original reference by Sundberg et al.[29] contains no units for stress or strain.

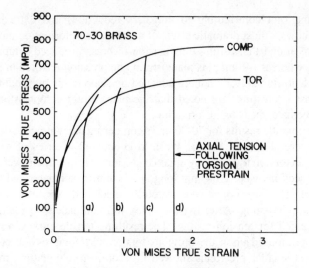

Fig. 14. Results of path-change experiments on 70-30 brass. COMP and TOR curves are from Fig. 5. Additional curves (a) through (d) represent axial tension in thin-wall tubes following torsional prestraining to von Mises strains indicated.

0.78 mm in wall thickness. The thin-wall tube was then tested in compression. The resulting stress-strain curve for a von Mises prestrain of 3.9 is shown in Fig. 15. Again, the flow curve for axisymmetric flow (this time in compression) was much higher than that for torsion. The specimen buckled at a stress very close to that observed for monotonic compression at the same von Mises strain level.

Another most interesting experiment with changing deformation mode was conducted by Armstrong, Hockett and Sherby,[15] who compared monotonic, uniaxial compression (with interruptions for remachining) with sequential, multidirectional compression of 1100 aluminum. The multidirectional compression was conducted on a cube by compressing sequentially (by identical strain increments of 7.5%) across the x, y and z faces of the cube. The flow stress was measured and plotted as a function of cumulative plastic strain. This composite flow curve is compared with the monotonic curve in Fig. 16. Initial hardening behavior is identical, but the multidirectional curve soon deviates and saturates at low stress levels. This behavior is similar to saturation observed in tension-compression fatigue where there is complete stress reversal.[30] Figure 16 also shows that when the loading is changed from monotonic to multidirectional or vice versa, the flow curves tend toward the current mode of loading. The transition from one type of hardening to the other is very gradual.

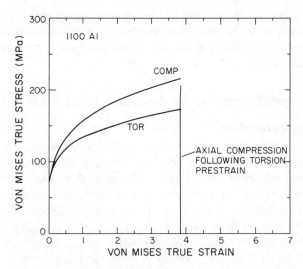

Fig. 15. Results of path-change experiment on 1100 aluminum. COMP and TOR curves are from Fig. 3. The additional curve represents axial compression following torsional prestraining.

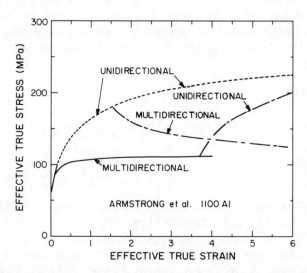

Fig. 16. Stress-strain curves for 1100 aluminum in unidirectional (dashed line) vs multidirectional (solid line) compression.[15] The dash-dot curves represent changes in deformation mode from unidirectional to multidirectional and vice versa.

The results shown in this section demonstrate that significant changes in hardening occur with changes in deformation mode. This has important consequences in many practical applications. Some of the microstructural causes of these hardening effects will be discussed below.

Crystallographic Analysis and Texture Effects

Analysis and Predictions

The most obvious effect of deformation geometry on hardening is the development of crystallographic texture. Because of the crystallographic nature of slip and the general validity of Schmid's law, one expects that, at large deformations, differences in hardening may arise from purely geometric considerations — that is, the mean inclination of the active slip planes and directions will change differently for different deformation modes. Quantitative prediction of this geometric effect has been the goal of many theoretical studies dating back to Sachs[31] in 1924 and Taylor[32] in 1938. Recently, several excellent reviews have been written on this topic.[26,33,34] Jonas et al.[35] have recently examined the crystallographic considerations necessary for comparing hardening for different deformation modes. They correctly point out that even at small strains (before significant texture development) crystallographic predictions of yield and flow differ from the macroscopic von Mises condition. For a randomly oriented fcc polycrystal one can relate the macroscopic stress (σ) to the critically resolved shear stress for slip (τ_c) and the macroscopic strain (ε) to the accumulated shear strain on all activated slip systems (γ_c) through the average Taylor factor $\overline{M}$. Specifically,

$$\sigma = \overline{M}\tau_c \tag{1}$$

$$d\varepsilon = \sum d\gamma_c/\overline{M} \tag{2}$$

$$\varepsilon = \int_0^\gamma \sum d\gamma_c/\overline{M} \tag{3}$$

The Taylor factor varies with deformation mode. Bishop and Hill[36] showed that for tension $\overline{M}_t$ is 3.06 and for torsion $\overline{M}_\tau$ is 1.65.

One can now define a crystallographic effective stress and strain criterion if one assumes that, microscopically, the strain-hardening law does not depend on deformation mode. A specific comparision of torsion and tension yields:[35]

In torsion:

$$\tau = \tau_c \cdot \overline{M}_\tau \tag{4}$$

$$d\gamma = d\gamma_c/\overline{M}_\tau \tag{5}$$

The effective stress-strain definition gives

$$\sigma^{\text{eff}} = \overline{M}_t \cdot \tau_c = (\overline{M}_t / \overline{M}_\tau) \cdot \tau \tag{6}$$

$$d\varepsilon^{\text{eff}} = d\gamma_c / \overline{M}_t = (\overline{M}_\tau / \overline{M}_t) \cdot d\gamma \tag{7}$$

For the Bishop and Hill values of $\overline{M}_t$ and $\overline{M}_\tau$ these relations become $\sigma^{\text{eff}} = 1.85 \ \tau$ and $d\varepsilon^{\text{eff}} = d\gamma/1.85$, compared with the von Mises relations $\sigma_{\text{vm}}^{\text{eff}} = 1.732 \ \tau$ and $d\varepsilon_{\text{vm}}^{\text{eff}} = d\gamma/1.732$. Therefore, the crystallographic yield and flow criterion predicts stress levels $\sim7\%$ higher than those predicted by the von Mises criterion for constant Taylor factors. It should also be noted here that the Taylor-type analysis represents an upper-bound crystallographic solution because it assumes that each grain undergoes identical deformation.

At strains greater than ~0.3, crystallographic textures develop and do so differently for different deformation modes. Hence, the Taylor factors evolve differently. Predictions of Taylor factors for different deformation modes have been made by numerous authors. Most notable are the efforts of Chin and colleagues[34,37,38] and of Gil-Sevillano, van Houtte and Aernoudt.[26,39] Most of these predictions use the Taylor formalism, and the resulting evolution of Taylor factors with deformation (taken from Ref 26) is shown in Fig. 17. For axisymmetric deformation the Taylor factor increases substantially — more so for tension than for compression. For torsion the Taylor factor decreases. One can again develop a crystallographic effective flow criterion by using these Taylor factors in Eq 6 and 7. Equation 7 now must be integrated, because the ratio $\overline{M}_\tau / \overline{M}_t$ is no longer constant.

All of the crystallographic analyses mentioned above assume random initial orientations and do not take account of changing grain shapes during deformation. The effect of changing grain shapes on crystallographic slip has recently been examined by Honneff and Mecking[40] and by Kocks and co-workers.[41,42] They found that in some deformation modes elongated grain shapes can relax some of the constraints on slip and decrease the number of necessary slip systems. Kocks and co-workers conducted a complete crystallographic analysis of torsion and predicted the evolution of $\overline{M}_\tau$. Their results are cited in abbreviated form by Jonas et al.[35] The major difference between their method (called method of relaxed constraints) and that of Gil-Sevillano et al.[26,39] (Taylor-type analysis shown in Fig. 17) is that Jonas et al. predict that $\overline{M}_\tau$ will rise after strains in excess of $\varepsilon^{\text{eff}} \cong 3$. For instance, at an effective strain of ~4 they predict $\overline{M}_\tau = 1.68$, compared with ~1.5 for Gil-Sevillano et al.

All of the above predictions are made for high-stacking-fault-energy (SFE) fcc metals which do not twin during room-temperature deformation. For low-SFE metals and alloys, deformation by twinning becomes a complicating factor. Chin et al.[43] and van Houtte[44] have developed calculational methods for incorporation of twinning into the prediction of Taylor factors. These methods are reviewed in Ref 26 and will not be discussed here.

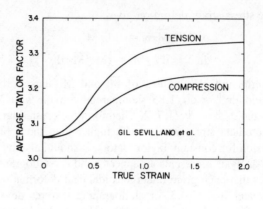

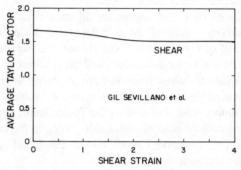

Fig. 17. Calculations of the evolution of average Taylor factors with strain using the Taylor analysis (*Gil-Sevillano et al.*[26,39]). (Top) Tension and compression. (Bottom) Shear.

Texture Experiments

A very large body of literature exists on measurements of the textures of metals and alloys. Those measurements relevant to current discussions have been reviewed by Gil-Sevillano et al.[26] In high-SFE metals (copper, nickel, aluminum) in axisymmetric extension, a strong [111] fiber texture plus a weaker [100] texture generally develop. In compression, grain rotation is in the opposite direction, resulting in a strong [110] texture. These textures are qualitatively consistent with the Bishop and Hill predictions and the changes in Taylor factor shown in Fig. 17.

In low-SFE metals and alloys the emergence of twinning as a deformation mode has the effect of increasing the [100] component of the axisymmetric extension texture. A low SFE introduces the additional complications of changes in latent hardening and deformation faulting (see Chin et al.[43] and van Houtte[44]). Experimentally for 70-30 brass the [100] fiber texture becomes much stronger than the [111] fiber components.[45] In compression a lower SFE results in a relatively weak [111] component in addition to the strong [110] texture, compared with high-SFE metals, which develop a [100] secondary component. Gil-Sevillano et al.[39] used a Taylor analysis to predict shear textures. For high-SFE metals they predicted an S1 copper-type texture consisting of two partial fiber textures:[26] (uvw) [1$\bar{1}$0] and (111) [hkl].* For low-SFE metals an S3 brass type[26] is similar but with different density distributions. Shear experiments on copper[46–48] and aluminum[47–49] generally confirm the partial fiber textures. Regenet and Stüwe[48] also found that a (100) [011] component is actually the strongest component. This is predicted by the method of relaxed constraints but not by the method of full constraints. Backofen and Hundy[50] approximated the shear texture of 70-30 brass by three ideal orientations: (11$\bar{1}$) [112], (112) [11$\bar{1}$] and (110) [001]. Williams[47] found similar results except for some of the symmetry assumed by Backofen and Hundy.

Texture Correction of Flow Curves

The torsional flow curves for copper (Fig. 4) are converted into crystallographic effective stress-strain curves in Fig. 18. The original shear stress–shear strain curves were converted into the crystallographic effective stress-strain curves by use of Eq 6 and 7 and the average Taylor factors as a function of strain shown in Fig. 17. The graphs in Fig. 18 also show several points for the crystallographic criterion for torsion using the method of relaxed constraints for torsion.[40] It is quite clear that the crystallographic criterion yields better agreement than the von Mises criterion at small to moderate strains, but overcorrects at large strains. The method of relaxed constraints gives better agreement at large strains, but the shape of the flow curve is inconsistent with experiments. Similar corrections are made for aluminum and brass in Fig. 19 and 20. The crystallographic criterion for brass is only approximate because we used the Taylor factors of Fig. 17, which were developed for high-SFE metals. These comparisons show that the texture corrections give reasonable agreement. Unfortunately, they are not conclusive because in most experiments tension tests (or tests involving other axisymmetric deformation modes) were not conducted on the same material (composition, texture, grain size, etc.) as that used in torsion tests. Also, data reduction in torsion is often questionable. Canova et al.[51] developed a rigorous technique for reduction of data for torsion of solid rods.

An additional complication is that most of the materials tested had initial textures and were not randomly oriented polycrystals as assumed in the Taylor

---

*Gil-Sevillano et al. use a nonstandard orientation convention for their pole figures. Usually, the direction of the shear-plane normal is vertical (12 o'clock) and the shear direction is to the right (3 o'clock). Gil-Sevillano et al. define the shear-plane normal to the right (3 o'clock) and the shear direction to the bottom (6 o'clock). Their pole figure is simply rotated 90° clockwise from the standard convention.

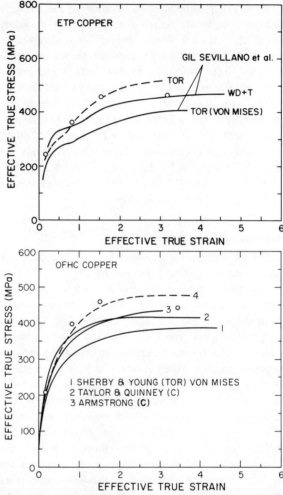

Fig. 18. (Top) Crystallographic effective stress-strain comparison for data on ETP copper from *Gil-Sevillano*.[16] Solid curves labeled WD + T and TOR are von Mises flow curves from the upper graph in Fig. 4. Dashed curve labeled TOR represents the crystallographic effective stress-strain curve, using the evolving Taylor factors of Fig. 17. Data points (circles) represent the Taylor-type effective stress-strain analysis using the method of relaxed constraints for torsion. The Taylor factors were reported by *Jonas et al.*[35] for this method. (Bottom) Crystallographic effective stress-strain comparison for OFHC copper. Torsional flow curve based on von Mises criterion is shown as curve 1. Curve 4 represents the same data based on the crystallographic Taylor-type criterion based in turn on the Taylor factors of Fig. 17. Data points (circles) again represent the predictions of the method of relaxed constraints. Curves 2 and 3 represent the compression curves in the lower graph in Fig. 4.

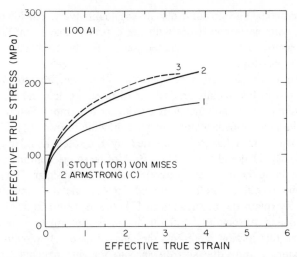

Fig. 19. Crystallographic effective stress-strain comparison for 1100 aluminum. Curve 1 represents the von Mises torsion curve, and curve 3 the crystallographic torsion curve (based on the Taylor factors of Fig. 17). Curve 2 represents uniaxial compression for comparison.

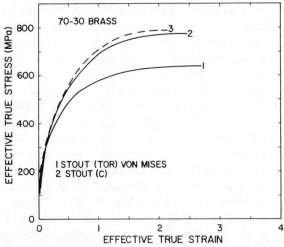

Fig. 20. Crystallographic effective stress-strain comparison for 70-30 brass. Curve 1 represents the von Mises torsion curve, and curve 3 the crystallographic torsion curve (based on the Taylor factors of Fig. 17). Curve 2 represents uniaxial compression for comparison.

analysis. The influence of initial texture is demonstrated in Fig. 7, where the flow curves for uniaxial tension and hoop tension can be seen to differ substantially. These curves are in qualitative agreement with the initial texture.[20] The plane-strain and torsion curves are in general agreement with evolving Taylor factors.

The balanced biaxial tension curve also agrees qualitatively. The lower hardening rate at larger strains is consistent with the expected change in Taylor factors (Fig. 17), because balanced biaxial tension should be equivalent to through-thickness compression.

Figure 6 shows the flow-curve comparison for bcc iron. Texture predictions for {110} ⟨111⟩ slip in bcc metals are similar to those for fcc metals, except that axisymmetric tension in bcc metals is equivalent to compression in fcc metals. Therefore, the predominant texture for wire drawing in iron is [110]. The shear texture in Armco iron has been reported by Backofen and Hundy[50] to be principally (112) [1$\bar{1}$0] and (110) [1$\bar{1}$2]. Both of these orientations place a ⟨111⟩ slip direction along the transverse direction of maximum shear. The evolution of Taylor factors is not as well developed for bcc metals. However, several authors[27,52,53] explain the bcc results in Fig. 6 and 9 by texture development. Gil-Sevillano and Aernoudt[52,53] claim that most deformation modes lead to hardening and that torsion represents the unusual case. They maintain that the torsional texture permits the slip distance to remain unchanged at moderate strains and actually to increase at large strains because of dynamic recovery. Razavi and Langford[27] relate the continued hardening during wire drawing to redundant strain (curling of grains). In strip drawing and torsion, deformation may be accommodated by cooperative rearrangements such as shear banding, leading to lower hardening rates. Young et al.[19,54] and Weertman and Hecker[55] have proposed explanations based on substructure, as explained below.

The path-change experiments (Fig. 13 to 16) provide another comparison for texture-based hardening predictions. We determined the partial pole figures for the brass tube in Fig. 14 prestrained in torsion at $\gamma = 2.4$ ($\varepsilon_{vm} = 1.39$). Figure 21 shows the partial (111) pole figure for the same brass tube along with the results of Backofen and Hundy for brass rod at $\gamma = 3.2$. These pole figures show that shear deformation aligns (111) plane normals in (or close to) the axial direction. The Taylor factor for axial tension for an ideal [111] orientation is at its highest possible value, and hence one would expect the stress required for plastic flow in axial tension following torsion to be high. Qualitatively, this is what is observed at large prestrains. As mentioned below, tensile data subsequent to smaller torsional prestrains is questionable. Williams[47] showed that the torsional textures are well developed at strains greater than 0.5. Therefore, we would expect a "stiff" response following all torsional prestrains greater than 0.5. It is also of interest to note that Williams found the same torsional textures to develop irrespective of original texture. The manifestation of this in terms of flow behavior has not been investigated.

Data for compression following torsion in 1100 aluminum are presented in Fig. 15. Witzel[49] found that the end texture in torsion for aluminum could be described completely by two ideal orientations: (112) [1$\bar{1}$0] and (112) [$\bar{1}$10]. Such a texture again makes the axial direction "stiff," and one should expect the flow

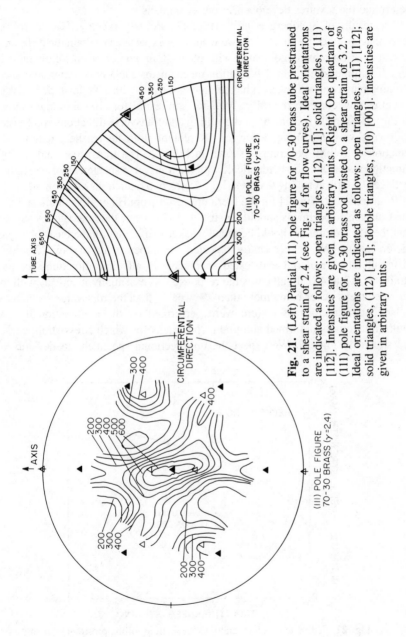

Fig. 21. (Left) Partial (111) pole figure for 70-30 brass tube prestrained to a shear strain of 2.4 (see Fig. 14 for flow curves). Ideal orientations are indicated as follows: open triangles, (112) [11$\bar{1}$]; solid triangles, (111) [11$\bar{2}$]. Intensities are given in arbitrary units. (Right) One quadrant of (111) pole figure for 70-30 brass rod twisted to a shear strain of 3.2.[50] Ideal orientations are indicated as follows: open triangles, (11$\bar{1}$) [112]; solid triangles, (112) [11$\bar{1}$]; double triangles, (110) [001]. Intensities are given in arbitrary units.

stress in compression to be greater than that in continued torsion. We have yet to determine the textures developed in our specimens.

The results of Sundberg et al.[29] (Fig. 13) are very sketchy. For the rolling texture determined by Hu et al.[56] we would not expect such a dramatic difference in response for tension compared with plane-strain compression following rolling. It is quite likely that deformation mechanisms such as twinning and shear banding may control flow and, hence, may not agree with a Taylor-type analysis.

Let us now consider the rolling-plus-tension results for 1100 aluminum shown in Fig. 11. Tensile specimens in the rolling and transverse directions were tested following various rolling prestrains. A plot of ultimate tensile strength as a function of rolling prestrain is shown in Fig. 22. The transverse- and rolling-direction results are nearly identical, with the transverse specimens being slightly stronger. A look at rolling textures (Fig. 23 for two reductions on our material by O'Rourke,[57] and Fig. 24 for the literature result from Hu et al.[56]) indicates that the transverse specimens should be stronger. However, the pole figures suggest that the 45° specimens should be much weaker than either on the basis of Taylor factors alone. We have recently repeated the experiments of Fig. 22 with a different lot of material and tested specimens in all three directions for pre-strains of 1.2 to 4.0. In all cases the response was essentially identical, with the 45° specimens never being more than 2% weaker than the transverse specimens.

No determinations of texture were performed on the unidirectionally and multidirectionally deformed aluminum specimens for which stress-strain curves are shown in Fig. 16. We expect the unidirectional specimens to develop the

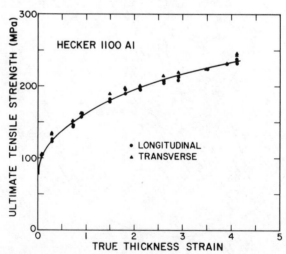

Fig. 22. Ultimate tensile strength subsequent to rolling prestrain (shown here as true thickness strain) for 1100 aluminum. The flow curve represents the best fit through all data points.

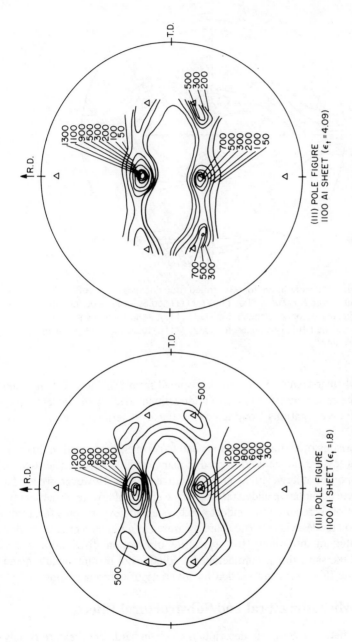

Fig. 23. 1100 aluminum rolled at room temperature to a true thickness strain of (left) 1.8 and (right) 4.09. Both figures represent partial (111) pole figures. An ideal orientation is designated by the open triangles, (113) [332]. Intensities are given in arbitrary units.[57]

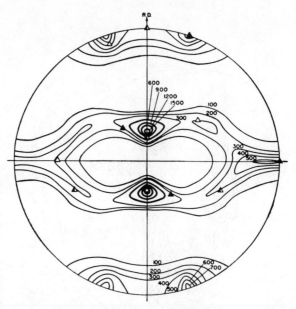

Fig. 24. (111) pole figure for 95% rolled (true thickness strain of 3) 1100 aluminum from *Hu et al.*[56] Positions for (111) poles for ideal orientations are indicated by open triangles for (123) [1$\bar{2}$1], solid triangles for (110) [1$\bar{1}$2], and half-filled triangles for (112) [11$\bar{1}$]. Intensities are given in arbitrary units.

standard [110] fiber texture with a weaker spread from [110] to [311]. It is not clear what kind of texture (if any) the multidirectional loading will develop. The substructural investigations conducted for these experiments will be discussed below.

The comparisons of texture predictions and experimens presented in this section show that texture can be used to predict many of the observed macroscopic hardening responses at large strains in a qualitative manner. However, we are far from a complete quantitative understanding of the effects of texture on hardening. Much more effort is required to conduct good systematic experiments for different deformation modes where textures are measured as a function of deformation and represented quantitatively by crystallite orientation distribution functions.[58-61] At the same time, more detailed predictions of texture evolution and its effect on the flow curve are needed for a variety of deformation modes.

Microstructural and Substructural Effects

We expect that, in a complete description of strain hardening, texture is only one of several important factors, and other microstructural and substructural features often play important roles. These features may include solute atoms, precipitates, second phases or particles, twins, and dislocation structures such as cells, subgrains and microbands. The relationships between flow stress and these microstructural features have been studied extensively at small strains. At large strains there have been few systematic studies. The recent literature is reviewed by Gil-Sevillano et al.[26] In this section we will concentrate on the effects of solutes and dislocation substructures in fcc metals and alloys.

Effect of Purity (Solutes) on Hardening

A careful review of the large-strain literature shows the important effect of purity on solute hardening.[8] Figure 25 presents flow curves for aluminums of different purities and for several dilute aluminum alloys. These curves clearly show that high-purity aluminum tends to saturate at low stress levels (note that the torsion curve of Luthy et al.[65] is for a temperature of −20°C). Low-purity or alloyed aluminum exhibits continued hardening to large strains. The results for copper shown in Fig. 26 are similar. Most of the literature data for fcc metals support the observation that high purity leads to saturation regardless of deformation mode. Furthermore, the effects of small amounts of solutes on flow stress and hardening rate are substantial, as shown in Fig. 25 and 26. The flow curves for various commercial-purity fcc metals in Fig. 27 show that continued hardening at large strains is very common. Few data are available on purity effects

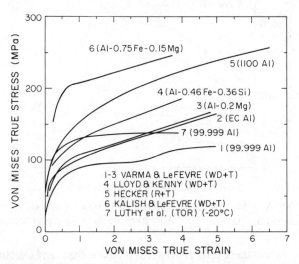

Fig. 25. Von Mises effective stress-strain curves for aluminums of different purities and for several dilute aluminum alloys.[62–65] WD + T denotes wire drawing followed by tension; R + T, rolling plus tension; and TOR, torsion.

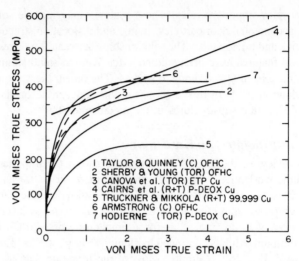

Fig. 26. Von Mises effective stress-strain curves for coppers of different purities.[66-69] C denotes compression.

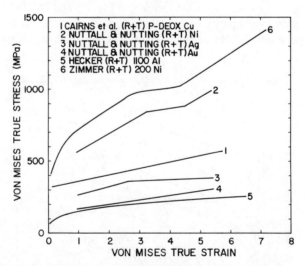

Fig. 27. Von Mises effective stress-strain curves for various fcc metals of commercial purity.[66,69,70] All curves are for rolling plus tension.

in alloys. We expect these effects to be less dramatic than, and overshadowed by, the solute effects from the intentional alloying elements.

Solutes affect hardening at large strains principally by retarding dynamic recovery processes, thereby offsetting the balance between dislocation generation and annihilation required for steady state and saturation. No quantitative description of the solute-dislocation interactions at large strains has yet been developed. In a qualitative sense it is easily recognized that solutes will hinder the cross slip or climb processes required for dynamic recovery. The magnitudes of the large solute effects at large strains had not been previously recognized. Kocks[71] presents some new aspects of solute hardening in this volume.

Alloying to form precipitates, dispersed particles or large inclusion particles can have dramatic effects on the flow curve, ranging from flow softening in some Al-Cu alloys[72] to a fiber-reinforcing effect in high-sulfur steels.[73]

Evolution of Substructure

At small strains the evolution of dislocation substructure depends on crystal structure and SFE in addition to the imposed parameters of temperature and strain rate. At low homologous temperatures $(T/T_m < 0.4)$, fcc metals with high SFE and bcc metals form well-developed dislocation cell structures, whereas fcc metals and alloys with low SFE tend to form planar dislocation arrangements that are typically associated with higher strain-hardening rates. A complete description of how substructure evolves at large strains is not yet available. Only in recent years have investigators performed multisurface transmission electron microscopy. This is particularly important and difficult for rolling and other deformation modes where one specimen dimension becomes very small at large strains. The most complete studies of deformation substructure have been carried out on pearlitic steels[74-77] and on copper and its alloys.[66,69,78-82] Many of these results are reviewed by Gil-Sevillano et al.[26] Fewer studies have been conducted on high-SFE metals. Chandra-Holm and Embury[83] have recently reported studies on aluminum. We will use our work on commercial-purity aluminum and nickel to illustrate some of the important substructural developments.

Our experiments on 1100 aluminum involved prestraining by rolling at room temperature $(T/T_m = 0.32)$ and subsequently testing in tension. (Rolled sheets were immersed in cold tap water immediately after rolling to keep the temperature rise to a minimum.) The flow curves shown in Fig. 3 and 25 show continued hardening. Specimens for transmission electron microscopy (TEM) were prepared from rolled sheets of many different prestrain levels. Most of the TEM observations to date were made through the sheet surface. Figure 28 shows a series of TEM micrographs at different strain levels. Unfortunately, to date we have examined specimens from only two prestrain levels in the other two directions (edge-on in the rolling direction and edge-on in the transverse direction). The thinning techniques for such specimens have been described by Rohr and Hecker.[84]

We will use the through-surface observations first to explain the general substructural evolution. Figure 28(a) shows that by a strain of 0.11 a definite dislocation cell network has developed. With increasing deformation this cell network decreases in size and the cell boundaries sharpen up. By a strain level of 0.28 we find distinct evidence of dynamic recovery (Fig. 28b). Many of the

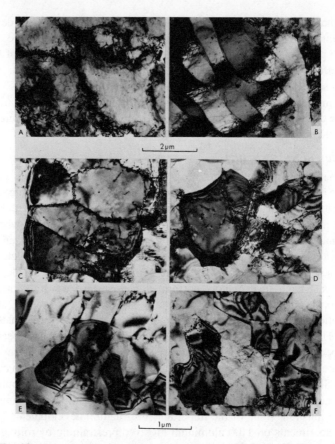

Fig. 28. Transmission electron micrographs of cold rolled 1100 aluminum taken through the sheet surface for different strain levels: (a) 0.11, (b) 0.28, (c) 0.71, (d) 1.49, (e) 4.07 and (f) 6.2.

cell boundaries have recovered into higher misorientation subgrain boundaries that exhibit a distinct boundary contrast instead of the dislocation character of cell walls. With continued deformation we find that more and more of the structure takes on a recovered appearance. However, new cells appear to be created within the newly recovered subgrains. Both cells and subgrains continue to become smaller up to strain levels of ~2. Beyond this strain level, we find little change in the dimensions of through-surface cells and subgrains. The substructural dimensions were measured statistically (many hundreds per strain level) and are shown in Fig. 29. At very large strains the substructure was predominantly of a subgrain character. We also measured the average misorientation of cells and subgrains in the TEM field of view (~2 μm in diameter) and found that the

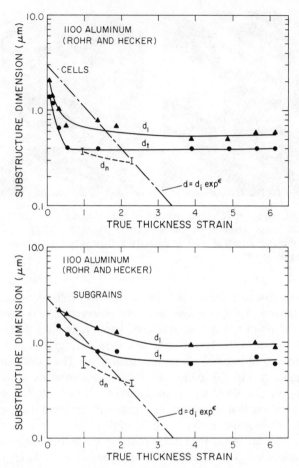

Fig. 29. TEM measurements of substructural dimensions on cold rolled 1100 aluminum. (Top) Measurements for dislocation cells. (Bottom) Measurements for subgrains, showing distinct boundary contrast. d_l represents the dimension in the rolling (longitudinal) direction, d_t in the transverse direction, and d_n in the through-thickness (normal) direction. $d = d_i \exp^\varepsilon$ represents the imposed shape change on some initial diameter, d_i.

misorientation increased continuously with strain, averaging ~10 degrees at a strain of 6.

As expected, the through-surface observations are only of limited value because they do not reflect the most important structural features. TEM micrographs of the two edge-on observations are shown in Fig. 30 and 31. At a strain of 1.0 we found a cell/subgrain structure of elongated ribbon shape with distinct

Fig. 30. Edge-on TEM micrographs of 1100 aluminum cold rolled to a true thickness strain of 1. Micron ˋmarkers are in the rolling direction. Note the microbands at ~35° to the ribbon-shape subgrains aligned in the rolling direction.

evidence of microscopic shear bands (which we will call microbands). Microbands were limited to single grains and were observed only occasionally. The structure at a strain of 2.3 is well developed into a ribbonlike subgrain and cell structure, as shown in the bright field/dark field micrographs of Fig. 31. We found no evidence of microbands at this strain level. The substructural dimensions are shown in Fig. 29 along with those taken from the through-surface observations. We also show an extrapolation of cell and subgrain sizes to zero strain and draw a line that represents the imposed geometrical shape change. The experimental results indicate that at small strains both cells and subgrains are reduced in size more rapidly than the imposed shape change. This demonstrates that new cells and subgrains are being formed continuously. At large strains we have no edge-on measurements.

Similar results were obtained previously by Schuh and von Heimendahl[85] on higher-purity aluminum (99.98% Al). Again, no edge-on measurements are available beyond a strain level of 1.2. However, their in-plane substructural dimensions appeared to saturate at $d_l \cong 1.1$ μm and $d_t \cong 0.7$ μm (see Ref 26). They did not try to distinguish between cells and subgrains, and hence a direct comparison with our data is not conclusive.

We conducted similar experiments on commercially pure nickel (Nickel 200) at room temperature.[70] This material has a much lower homologous temperature (0.16 for Ni versus 0.32 for Al) while still representing high-SFE metals. The flow curve resulting from rolling-plus-tension experiments is shown in Fig. 27. Transmission microscopy observations were reported by Zimmer et al.[70] The general evolution of substructure is similar to that found for aluminum. However, the dislocation cell structures dominate to much larger strains. Dynamic recovery

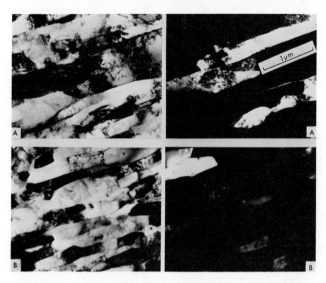

Fig. 31. Edge-on TEM micrographs of 1100 aluminum cold rolled to a true thickness strain of 2.3. (a) Bright field/dark field pair edge-on along rolling direction. (b) Bright field/dark field pair edge-on transverse to rolling direction.

does occur, with subgrains being quite distinct at a strain of 1.6. We also found evidence of microbands at intermediate strain levels (~1.6) and no microbands at very large strains. The results of Zimmer et al.[70] are shown in Fig. 32 along with one large-strain edge-on measurement made since then.[86] The TEM micrographs for this strain level are shown in Fig. 33. The evidence for substantial dynamic recovery and subgrains is clear.

In nickel the substructural dimensions continue to decrease with strain. We also find a continued increase in flow stress (Fig. 27). We are still puzzled by the rapid increase in hardening at strains greater than 4. In our previous analysis[70] we suspected the possible influence of grain size at large strains proposed by Gil-Sevillano et al.[26] Their ideas are represented in Fig. 34. They observed that the decrease in grain size dictated by the geometric shape change is more rapid than the decrease in substructural dimensions observed in the literature. Hence, at large strains it is possible that grain boundaries may again play an important role in hardening. The most recent results for nickel (Fig. 32) are not consistent with the hardening transition at a strain of 4. One must now question the generally accepted observation of a saturation in substructural dimensions at large strains. To our knowledge, the nickel results in Fig. 32 represent the largest strains at which edge-on measurements were made, and they do not show saturation.

The substructures of heavily cold rolled copper and low-SFE copper alloys have been studied extensively.[66,69,78–82] Hatherly and Malin[81] found that micro-

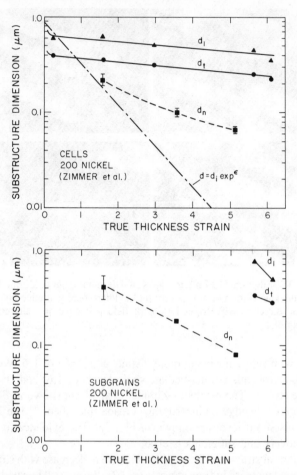

Fig. 32. TEM measurements of substructural dimensions on cold rolled 200 nickel. (Top) Measurements for dislocation cells. (Bottom) Measurements for subgrains. d_l represents the dimension in the rolling (longitudinal) direction d_t in the transverse direction, and d_n in the through-thickness (normal) direction.[70,86]

bands form in copper at strains as low as 0.2. Microbands are long, thin, sheetlike features that form on {111} planes amidst a well-developed cell structure (see Fig. 35). With continued deformation they cluster together and become parallel to the rolling plane. Shear bands develop at larger strains, and dynamic recrystallization has been observed at very large strains.[66] In lower-SFE copper alloys, stacking faults and fine twins develop at low strain levels. In 70-30 brass, twins cluster together to form banded regions, but by a strain of 0.4 most grains are uniformly twinned and shear bands begin to develop.[81] The first shear bands

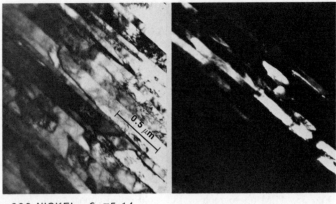

200 NICKEL $\varepsilon_t = 5.14$

Fig. 33. Bright field/dark field pair of TEM micrographs of cold rolled 200 nickel (true thickness strain, 5.14) taken edge-on. Micron marker is in the rolling direction.

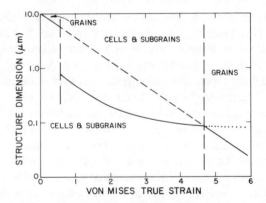

Fig. 34. Potential role of grain and substructure sizes in controlling flow stress. The dashed diagonal line represents the decrease in transverse grain size imposed by the external shape changes due to rolling. The solid curve for cells and subgrains is schematic.

develop in the grains that are most profusely twinned. Macroscopic shear bands that cut across the full sheet thickness and that are oriented at $\sim 35°$ to the rolling plane eventually develop. The exact roles of microbands, twins and shear bands in accommodating the imposed shape change and in controlling the flow stress have not yet been determined. From our work on aluminum and nickel, and from the work of Hatherly, Malin and co-workers,[79–82] we have drawn up a summary of microstructural observations, which is presented in Table 3.

Fig. 35. TEM micrograph of 18% cold rolled copper from *Malin and Hatherly*.[80] Beam direction, approximately [$\bar{1}\bar{1}0$]; diffraction vector (g), [$1\bar{1}\bar{1}$]. (The 1-μm marker is parallel to trace of rolling plane.)

Relationship Between Substructure and Flow Stress

The relationship between substructural features and flow stress is not well understood. A Hall-Petch type of relation ($\sigma = \sigma_0 + kd^{-m}$, where σ_0 is a frictional stress and k is a strength constant) is generally acknowledged.[87-89] There is still much controversy over the exponent m. It appears that m = 1 is well accepted where d represents dislocation cells. However, for recovered subgrains a value of m = 1/2 is often found.[88] In the case of cold worked aluminum, both cells and subgrains are present. A detailed correlation between flow stress and the substructural features shown in Fig. 29 is not yet possible because insufficient through-thickness measurements exist. However, preliminary attempts indicate that at large strains the misorientation across subgrain boundaries needs to be incorporated. Armstrong et al.[15] also found that they were able to correlate flow stresses in unidirectional versus multidirectional compression by separately including the density of dislocation tangles, cell size and subgrain size multiplied by the square root of the misorientation. The complex behavior of nickel (Fig. 27) has not yet been explained.

The even more complicated behavior resulting from changes in deformation mode has not been studied extensively by electron microscopy. Detailed investigations such as those performed in fatigue studies[30,90] are in order. Weertman and Hecker[55] have recently proposed a possible explanation of the difference in strain hardening at very large plastic strains for torsion compared with axisymmetric deformation. It is shown that, in torsion, deformation is accommodated principally by dislocations of the same Burgers vector whereas axisymmetric deformation requires dislocations from several different Burgers vectors. In torsion this results from the evolution of a texture in which the predominant slip planes

Table 3. Summary of Substructural Evolution for Several fcc Metals and Alloys

Material	T/T_m	SFE[96] (erg/cm^2)	Low strains (<0.3)	Moderate strains (<2.3)	Large strains (>2.3)
1100 aluminum	0.32	166	Dislocation tangles; cell network by $\varepsilon = 0.1$; dynamic recovery and subgrains above 0.2.	Cells and subgrains decrease in size, new ones form, some evidence of microbands in edge-on section.	Structure looks very recovered; more subgrains than cells; misorientation between subgrains continues to increase. No edge-on substructural dimensions available. Only occasional microband at $\varepsilon = 2.3$.
200 nickel	0.17	128	Dislocation tangles; cell network by $\varepsilon = 0.1$.	Cells decrease in size, definite recovery at $\varepsilon > 1$ with distinct subgrain boundaries. Cells and subgrains continue to decrease. Structure looks ribbon-like. Some evidence of microbands.	Continued recovery; cells and subgrains continue to decrease in size (edge-on). No evidence of microbands or shear bands.
Copper[82]	0.22	78	Dislocation tangles; cell network by $\varepsilon = 0.1$; microbands evident by 0.2.	Microbands along {111}; rotate toward rolling plane; new microbands form. Much of deformation appears by microband mechanism. Macro shear bands form at $\varepsilon > 1$.	Shear bands assume dominant role in deformation. Dynamic recovery occurs. Nutting and co-workers[66,69] found evidence of dynamic recrystallization at very large strains (>4).
70-30 brass[82]	0.2	14	Slip by partials; develop planar dislocation array, followed by microbands.	At strains of 0.5 to 1.3, twinning is major deformation mode observed; twins form and rotate to align with rolling plane; by $\varepsilon = 1.6$ most of volume is twinned.	Extensive macro shear banding, especially in twinned areas. Subgrains form in shear-band areas, and again deform by slip.

share the same slip direction (tangential to the specimen surface). Because the dislocations tend to clump into cell walls, the torsion case leads to easier annihilation and hence an earlier approach to the steady state (saturation). This behavior is particularly dramatic for iron (Fig. 6), but also appears to hold true for most fcc metals and alloys.

We are convinced that a true correlation of flow stress with microstructural features will need to include both textural and substructural features. As mentioned above, more systematic experiments are required to sort out the contribution of texture to hardening. Modeling of texture evolution by the method of relaxed constraints[40-42] holds great promise. However, careful experiments are required in which textures are measured as a function of deformation, converted to crystallite orientation distribution functions, and related to flow stress through a Taylor-type analysis. In any case, in comparing flow curves for different deformation modes, the texture correction must be made first. To assess the proper role of substructure, more through-thickness measurements of substructural dimensions are required. One must also examine the objections raised by Truckner and Mikkola[67] about TEM measurements and evaluate the benefits of their x-ray line-broadening measurements.

Some Properties of Heavily Cold Worked Metals

Elongation After Rolling

In our work on commercially pure aluminum and nickel we found that the flow stress as determined by subsequent tensile tests increased with increasing prestrain. One would expect the elongation in tension following rolling to decrease with increasing prestrain. The rolling experiments were conducted to provide specimens of identical thickness after different amounts of prestrain to avoid potential complications due to thickness effects. This was accomplished by reducing the original aluminum bar to different starting thicknesses, annealing at 343°C for one hour, and then rolling to final thickness. (The starting grain sizes of all specimens prepared in this way were similar.) In the case of aluminum we produced specimens of two different final thicknesses, 0.127 mm and 0.635 mm. The results for aluminum and nickel are shown in Fig. 36 and 37, respectively.

As expected, both uniform and total elongations dropped rapidly with increasing prestrain. In aluminum, the total elongation increased for both thicknesses at large strains. The uniform elongation definitely increased for the 0.127-mm sheet but appeared to remain constant for the 0.635-mm sheet. However, the prestrain levels for the thicker sheets were lower. In nickel, both uniform and total elongations decreased rapidly, remained constant, and then increased at strain levels greater than 4. This is most intriguing, because the flow stress also increased dramatically at these strain levels (Fig. 27). In fact, the combinations of strength and total elongation for the very large prestrains (250 MPa and 4% for aluminum; 1400 MPa and 5% for nickel) are remarkable.

The rapid decrease in tensile elongation after rolling is easily understood by referring to Fig. 38. For rate-insensitive materials, tensile instability occurs when $d\sigma/d\varepsilon = \sigma$. Figure 38 shows the curves for σ and $\theta = d\sigma/d\varepsilon$ as functions of strain. They cross at the point where tensile instability is observed in an annealed

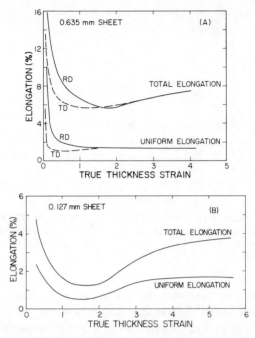

Fig. 36. Elongation in tension following rolling prestrain for 1100 aluminum. (a) Sheets rolled to a final thickness of 0.635 mm. RD and TD refer to specimens in rolling and transverse directions. (b) Sheets rolled to a final thickness of 0.127 mm. RD and TD results are averaged.

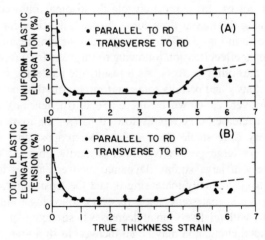

Fig. 37. Elongation in tension following rolling prestrain for 200 nickel.

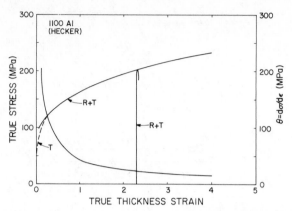

Fig. 38. Stress-strain curve and hardening rate (θ) for 1100 aluminum. Dashed curve labeled T represents uniaxial tension on annealed material. The curve at a true thickness strain of 2.3 represents tension following a prestrain of 2.3.

specimen ($\varepsilon \cong 0.25$). As shown, the flow stress can be increased much beyond the ultimate tensile strength by rolling prestrain. However, the intrinsic hardening rate ($d\sigma/d\varepsilon$) decreases. For the prestrain of 2.3 (90% rolling reduction) chosen in Fig. 38, we demonstrate how the flow stress has been increased much above the level that could be supported by the intrinsic hardening rate. Hence, a tensile specimen should go unstable immediately upon yielding, and little tensile elongation is to be expected.

The actual tensile loading response is shown in Fig. 39. The prestrained tensile specimen does not exhibit completely elastic loading followed by an abrupt plastic transition to the master flow curve. Instead, the flow curve is somewhat rounded, as shown in Fig. 39. This more gradual yielding results from a generalized Bauschinger effect (tension following rolling) and from recovery occurring during or after the rolling process. As a result, the hardening rate θ is initially increased enormously and permits some uniform plastic flow before the onset of instability. However, as shown by the curve θ_t (for tension) in Fig. 39, the hardening rate decreases very rapidly toward θ_R (for the master flow curve in rolling) and very little stable flow occurs. We propose that the increase in elongation at very large prestrains results primarily from the more complete dynamic recovery at large strains. Dynamic recovery allows the tensile flow curve to bend over gradually increasing θ_t and the tensile uniform elongation. These results suggest that some static recovery following rolling may greatly increase ductility with relatively small decreases in strength. An additional factor increasing the total elongation may be an increase in strain-rate sensitivity with prestraining. A few preliminary tests showed that prestrained aluminum was more rate-sensitive than annealed aluminum.

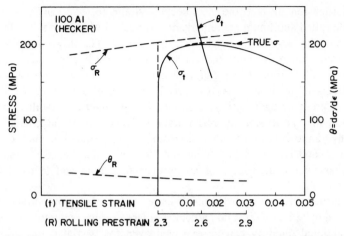

Fig. 39. Enlargement of Fig. 38 near the prestrain level of 2.3. Subscript R refers to rolling curve, and subscript t to subsequent tension. The solid σ_t curve represents the engineering stress-strain curve. True stress-strain behavior is labeled TRUE σ.

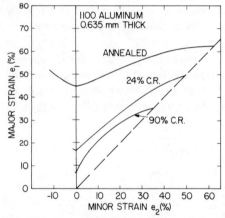

Fig. 40. Failure-limit curves for 1100 aluminum under annealed and cold rolled conditions.[92] Failure is defined by local necking. Sheet thickness is 0.635 mm.

Biaxial Ductility of Cold Worked Metals

Biaxial ductility or resistance to local necking can be characterized by failure-limit (or forming-limit) curves (FLC's).[91] Hecker[92] previously reported the influence of prior cold work (by rolling) on FLC's for 1100 aluminum. The results are shown in Fig. 40 for several prestrain levels. It is quite apparent that pre-straining lowers the ductility much more in tension and plane strain ($e_2 = 0$) than in biaxial tension. In fact, the local necking strains ($e_1 \times e_2$) of $32 \times 32\%$ for 90% cold rolled aluminum are remarkable. We can now explain this behavior on the basis of the large-strain results discussed above. The flow curve of Fig. 39 demonstrates unequivocally that 1100 aluminum still possesses substantial intrinsic ductility after 90% prestrain because of continued work hardening to strain levels of at least 7. As explained above, the lack of tensile ductility following prestraining is strictly a problem of geometric instability, which is greatest in uniaxial tension and plane strain. However, in biaxial tension of sheet (either hydraulic bulging or punch stretching) there is added geometric stability (see Ghosh and Hecker[93] and Stout and Hecker[94]) and hence the material is able to take better advantage of its intrinsic hardening and to exhibit much greater ductility.

The intrinsic hardening behavior at large strains is very important. For instance, if a material exhibits no hardening at large strains (saturation), then very little additional ductility can be expected even for biaxial deformation. The importance of the intrinsic hardening curve at large strains was demonstrated convincingly by Bird and Duncan[95] in comparing the behaviors of 1008 steel, 1100 Al and 2036-T4 Al. The FLC's for these three materials, as determined by Hecker,[91] are shown in Fig. 41. The strain-hardening exponents (n-value from $\sigma = k\varepsilon^n$) in uniaxial tension were between 0.21 and 0.26 for all three materials. The low FLC for 2036-T4 Al was explained on the basis of the negative strain-rate sensitivity (m-value of -0.005 from $\sigma = k'\dot{\varepsilon}^m$) of this alloy. However, steel has a positive rate sensitivity of m = 0.012 whereas 1100 aluminum is rate-insensitive, and hence the FLC for steel should be higher. Bird and Duncan[95] showed that strain hardening at large strains provides the answer. Figure 42 shows the hardening rate normalized with respect to stress during

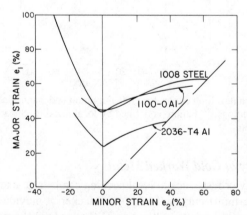

Fig. 41. Failure-limit curves for three materials from Ref 91 and 92. Failure is defined by local necking.

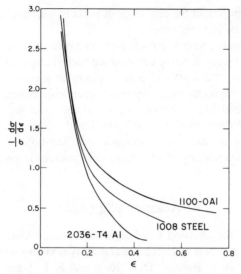

Fig. 42. Hardening rate normalized by stress as a function of strain for hydraulic bulge tests from *Bird and Duncan*.[95]

biaxial tension[95] (hydraulic bulging). At strains greater than the level where $(1/\sigma)(d\sigma/d\varepsilon) = 1$ (beyond uniform strain in tension), the hardening rates for the three materials differ substantially. The higher rate in 1100 Al apparently offsets the zero strain-rate sensitivity in the failure-limit-curve behavior.

Concluding Remarks

We have shown sufficient evidence to demonstrate that strain hardening in torsion cannot be correlated with axisymmetric deformation by the von Mises effective stress-strain criterion. In fcc materials, the flow-stress levels and strain-hardening rates are typically lower in torsion, and saturation, if it occurs at all, is observed at lower stress levels. In bcc iron, a low saturation stress is observed for torsion, whereas linear hardening is observed for axisymmetric extension. Experiments also suggest that flow-stress levels and strain-hardening rates are also low for other plane-strain deformation modes when compared with those for axisymmetric deformation.

Much of the discrepancy in flow curves can be explained by texture. We demonstrated that a crystallographic effective stress-strain criterion based on evolving average Taylor factors provides the proper magnitude correction for torsional flow curves in fcc materials. However, the details of the hardening behavior have not been fully explained. This is, in part, a result of inadequate systematic experimental work at large strains. The crystallographic analysis must also be extended to predict the responses of initially anisotropic materials and to account for changes in grain shape. The simple crystallographic analysis presented here also does not fully explain the hardening response following deformation-path changes and multidirectional loading. Moreover, the strong effects of purity and grain size (discussed in Ref 8) suggest an important microstructural dependence of hardening in addition to the texture effect. Much remains to be learned about the evolution of substructure and the influences of cells, subgrains, and substructural features such as microbands, twins and shear bands on hardening. Additional three-surface transmission electron microscopy at very large deformations is needed.

Acknowledgments

The support of the Division of Materials Sciences, Office of Basic Energy Sciences, U.S Department of Energy, is gratefully acknowledged. We also thank K. A. Johnson, J. A. O'Rourke, D. L. Rohr and K. P. Staudhammer for their experimental assistance, and G. R. Canova, U. F. Kocks and J. J. Jonas for helpful discussions.

References

1. R. McMeeking and J. R. Rice, "Finite-Element Formulations for Problems of Large Elastic-Plastic Deformation," *Int. J. Solids Struct.*, Vol 11, 1975, p 601-616.

2. E. H. Lee, R. L. Mallett and T. B. Wertheimer, "Stress Analysis for Kinematic Hardening in Finite-Deformation Plasticity," SUDAM Report No. 81-11, Stanford University, Dec 1981.

3. V. A. Lubarda and E. H. Lee, "A Correct Definition of Elastic and Plastic Deformation and Its Computational Significance," SUDAM Report No. 80-1, Stanford University, Jan 1980; also to appear in *J. Appl. Mech.*

4. J. C. Nagtegaal and J. E. deJong, "Some Aspects of Nonisotropic Workhardening in Finite Deformation Plasticity," in *Proceedings of Workshop on Plasticity of Metals at Finite Strain: Theory, Experiment and Computation*, edited by E. H. Lee and R. L. Mallett, Quicksilver Printing, 1982, p 65-106.

5. E. H. Lee, "Some Comments on Elastic-Plastic Analysis," SUDAM Report No. 80-5, Stanford University, Oct 1980.

6. U. F. Kocks, "Laws for Work-Hardening and Low Temperature Creep," *J. Eng. Mater. Tech., Trans. ASME*, Vol 98, 1976, p 76-85.

7. H. Mecking and A. Grinberg, "Discussion on the Development of a Stage of Steady-State Flow at Large Strains," in *Proceedings of 5th International Conference on Strength of Metals and Alloys*, edited by P. Haasen, V. Gerold and G. Kostorz, Pergamon Press, 1980, p 289-294.

8. S. S. Hecker, M. G. Stout and D. T. Eash, "Experiments on Plastic Deformation at Finite Strains," in *Proceedings of Workshop on Plasticity of Metals at Finite Strain: Theory, Experiment and Computation*, edited by E. H. Lee and R. L. Mallett, Quicksilver Printing, 1982, p 162-205.

9. J. H. Hollomon, "Tensile Deformation," *Trans. AIME*, Vol 162, 1945, p 268-290.
10. P. Ludwik, "Elemente der Technologischen Mechanik," Springer Verlag, Berlin, 1909.
11. H. Swift, "Plastic Strain in an Isotropic Strain-Hardening Material," *Engineering*, Vol 162, p 381.
12. E. Voce, "The Relationship Between Stress and Strain for Homogeneous Deformation," *J. Inst. Metals*, Vol 74, 1948, p 537-562.
13. E. Voce, "A Practical Strain-Hardening Function," *Metallurgia*, Vol 51, 1955, p 219-226.
14. J. E. Hockett and O. D. Sherby, "Large Strain Deformation of Polycrystalline Metals at Low Homologous Temperatures," *J. Mech. Phys. Solids*, Vol 23, 1975, p 87-98.
15. P. E. Armstrong, J. E. Hockett and O. D. Sherby, "Large Strain Multidirectional Deformation of 1100 Aluminum at 300 K," *J. Mech. Phys. Solids*, Vol 30, 1982, p 37-58.
16. J. Gil-Sevillano, Ph.D. Thesis, Katholieke Universiteit Leuven, 1974.
17. O. D. Sherby and C. M. Young, "Some Factors Influencing the Strain Rate-Temperature Dependence of the Flow Stress in Polycrystalline Solids," *Rate Processes in Plastic Deformation of Materials*, edited by J. C. M. Li and A. K. Mukherjee, American Society for Metals, 1975, p 497-541.
18. G. I. Taylor and H. Quinney, *Proc. Royal Soc. London*, A, Vol 143, 1934, p 307.
19. C. M. Young, L. J. Anderson and O. D. Sherby, "On the Steady State Flow Stress of Iron at Low Temperatures and Large Strains," *Met. Trans.*, Vol 5, 1974, p 519-520.
20. M. G. Stout and S. S. Hecker, 29th Sagamore Army Materials Research Conference, Lake Placid, NY, July 19-23, 1982.
21. A. K. Ghosh, "Plastic Flow Properties in Relation to Localized Necking in Sheets," in *Mechanics of Sheet Metal Forming–Material Behavior and Deformation Analysis*, edited by D. P. Koistinen and N.-M. Wang, Plenum Press, 1978, p 287-312.
22. R. H. Wagoner, "Plastic Behavior of 70-30 Brass Sheet," *Met. Trans. A*, Vol 13A, 1982, p 1491-1500.
23. DGM, Wire Drawing Commission, "Fliesskurven verschiedener Werkstoffe," 1975.
24. P. Blanchard, D. Whitwham and J. Herenguel, *Ecrouissage, Restauration, Recristallisation*, Presses Universitet de France, Paris, 1963, p 41.
25. L. E. Murr and F. I. Grace, "Residual Structure and Mechanical Properties of Alpha Brass and Stainless Steel Following Deformation by Cold Rolling and Explosive Shock Loading," *Trans. AIME*, Vol 245, 1969, p 2225-2235.
26. J. Gil-Sevillano, P. van Houtte and E. Aernoudt, "Large Strain Work Hardening and Textures," *Prog. Mater. Sci.*, Vol 25, 1980, p 69-412.
27. A. Razavi and G. Langford, "Strain Hardening of Iron: Axisymmetric vs. Plane Strain Elongation," in *Proceedings of 5th International Conference on Strength of Metals and Alloys*, edited by P. Haasen, V. Gerold and G. Kostorz, Pergamon Press, 1980, p 831-836.
28. H. Ford, "The Working Properties of Metals and Their Relationship to Mechanical-Test Behavior," *J. Inst. Metals*, Vol 96, 1968, p 65-74.
29. R. Sundberg, L.-E. Olsson and B. Thundal, "Microstructure of Heavily Deformed Brass and Nickel Silver," *Jernkont. Ann.*, Vol 155, 1971, p 434-436.
30. H. Mughrabi, "The Cyclic Hardening and Saturation Behavior of Copper Single Crystals," *Mater. Sci. Eng.*, Vol 33, 1978, p 207-223.
31. G. Sachs, *Z. Verein Deut. Ing.*, Vol 72, 1924, p 734.
32. G. I. Taylor, "Plastic Strain in Metals," *J. Inst. Metals*, Vol 62, 1938, p 307-324.
33. U. F. Kocks, "The Relation Between Polycrystal Deformation and Single Crystal Deformation," *Met. Trans.*, Vol 1, 1970, p 1121-1143.

34. G. Y. Chin, "The Role of Preferred Orientation in Plastic Deformation," in *Inhomogeneity of Plastic Deformation*, edited by R. E. Reed-Hill, American Society for Metals, 1973, p 83-112.
35. J. J. Jonas, G. R. Canova, S. C. Shrivastava and N. Christodoulou, "Sources of the Discrepancy Between the Flow Curve Determined in Torsion and in Axisymmetric Tension and Compression Testing," in *Proceedings of Workshop on Plasticity of Metals at Finite Strain: Theory, Experiment and Computation*, edited by E. H. Lee and R. L. Mallett, Quicksilver Printing, 1982, p 206-229.
36. J. F. W. Bishop and R. Hill, "A Theory of the Plastic Distortion of a Polycrystalline Aggregate Under Combined Stresses," *Phil. Mag.*, Vol 42, 1951, p 414-427.
37. G. Y. Chin and W. L. Mammel, "Computer Solutions of the Taylor Analysis for Axisymmetric Flow," *Trans. AIME*, Vol 239, 1967, p 1400-1405.
38. G. Y. Chin, W. L. Mammel and M. T. Dolan, "Taylor Analysis for {111} ⟨110⟩ Slip Under Conditions of Axisymmetric Flow," *Trans. AIME*, Vol 245, 1969, p 383-388.
39. J. Gil-Sevillano, P. van Houtte and E. Aernoudt, "Calculation of Shear Texture with the Taylor-Analysis," *Z. Metallkunde*, Vol 66, 1975, p 367-373.
40. H. Honneff and H. Mecking, in *Textures of Materials*, edited by G. Gottstein and K. Lücke, 1978, p 265.
41. U. F. Kocks and H. Chandra, "Slip Geometry in Partially Constrained Deformation," *Acta Met.*, Vol 30, 1982, p 695-709.
42. U. F. Kocks and G. R. Canova, "How Many Slip Systems, and Which?" *Proceedings of 2nd Riso International Symposium on Metallurgy and Materials Science*, Riso, Denmark, 14-18 Sept 1981.
43. G. Y. Chin, W. F. Hosford and D. R. Mendorf, "Accommodation of Constrained Deformation in f.c.c. Metals by Slip and Twinning," *Proc. Roy. Soc.*, Vol A309, 1969, p 433-456.
44. P. van Houtte, "Simulation of the Rolling and Shear Texture of Brass by the Taylor Theory Adapted for Mechanical Twinning," *Acta Met.*, Vol 26, 1978, p 591-604.
45. C. S. Barrett, *Structure of Metals*, McGraw-Hill, New York and London, 1952.
46. W. A. Backofen, "The Torsion Texture of Copper," *Trans. AIME*, Vol 188, 1950, p 1454-1459.
47. R. O. Williams, "Shear Textures in Copper, Brass, Aluminum, Iron and Zirconium," *Trans. AIME*, Vol 224, 1962, p 129-140.
48. P. Regenet and H.-P. Stüwe, "Zur Entstehung von Oberflachentexturen beim Walzen kubisch flachen zentrierter Metalle," *Z. Metallkunde*, Vol 54, 1963, p 273-278.
49. W. Witzel, "Die Endtextur bei der Scherverformung des Aluminums," *Z. Metallkunde*, Vol 64, 1973, p 813-817.
50. W. A. Backofen and B. B. Hundy, "Torsion Texture of 70-30 Brass and Armco Iron," *Trans. AIME*, Vol 197, 1953, p 61-62.
51. S. C. Shrivastava, J. J. Jonas and G. R. Canova, "Equivalent Strain in Large Deformation Torsion Testing: Theoretical and Practical Considerations," *J. Mech. Phys. Solids*, Vol 30, 1982, p 75-90.
52. E. Aernoudt and J. Gil-Sevillano, "Influence of the Mode of Deformation on the Hardening of Ferritic and Pearlitic Carbon Steels at Large Strains," *J. Iron Steel Inst.*, Vol 211, 1973, p 718-725.
53. J. Gil-Sevillano and E. Aernoudt, "On the Influence of the Mode of Deformation on the Hardening of Iron at Low Temperature and Large Strains," *Met. Trans.*, Vol 6A, 1975, p 2163-2164.
54. C. M. Young, L. J. Anderson and O. D. Sherby, Reply to "On the Influence of the Mode of Deformation on the Hardening of Iron at Low Temperature and Large Strains," *Met. Trans. A*, Vol 6A, 1975, p 2164-2165.

55. J. Weertman and S. S. Hecker, "Theory for Saturation Stress Difference in Torsion vs. Other Types of Deformation at Low Temperatures," *J. Mech. Mater.*, Vol 2, 1983, p 89-101.

56. H. Hu, P. R. Sperry and P. A. Beck, "Rolling Textures in Face-Centered Cubic Metals," *Trans. AIME*, Vol 194, 1952, p 76-81.

57. J. O'Rourke, Los Alamos National Laboratory (unpublished work).

58. J. S. Kallend and G. J. Davies, "The Prediction of Plastic Anisotropy in Annealed Sheets of Copper and α-Brass," *J. Inst. Metals*, Vol 98, 1970, p 242-244.

59. J. S. Kallend and G. J. Davies, "The Elastic and Plastic Anisotropy of Cold-Rolled Sheets of Copper, Gilding Metal, and α-Brass," *J. Inst. Metals*, Vol 99, 1971, p 257-260.

60. H. J. Bunge and W. T. Roberts, "Orientation Distributions, Elastic and Plastic Anisotropy in Stabilized Steel Sheet," *J. Appl. Cryst.*, 1969, Vol 2, 1969, p 116-128.

61. I. L. Dillamore and R. E. Smallman, "The Status of Research on Textures in Metals," *Met. Science*, Vol 6, 1972, p 184-188.

62. S. K. Varma and B. G. LeFevre, "Large Wire Drawing Plastic Deformation in Aluminum and Its Dilute Alloys," *Met. Trans. A*, Vol 11A, 1980, p 935-942.

63. D. J. Lloyd and D. Kenny, "The Structure and Properties of Some Heavily Cold Worked Aluminum Alloys," Prepublication paper from Aluminum Co. of Canada, Ltd., Research Center, Kingston, Ontario, Canada.

64. D. Kalish and B. G. LeFevre, "Subgrain Strengthening of Aluminum Conductor Wires," *Met. Trans. A*, Vol 6A, 1975, p 1319-1324.

65. H. Luthy, A. K. Miller and O. D. Sherby, "The Stress and Temperature Dependence of Steady-State Flow at Intermediate Temperatures for Pure Polycrystalline Aluminum," *Acta. Met.*, Vol 28, 1980, p 169-178.

66. J. H. Cairns, J. Clough, M. A. P. Dewey and J. Nutting,"The Structue and Mechanical Properties of Heavily Deformed Copper," *J. Inst. Metals*, Vol 99, 1971, p 93-97.

67. W. G. Truckner and D. E. Mikkola, "Strengthening of Copper by Dislocation Substructures," *Met. Trans. A*, Vol 8A, 1977, p 45-49.

68. F. A. Hodierne, "A Torsion Test for Use in Metalworking Studies," *J. Inst. Metals*, Vol 91, 1963, p 267-273.

69. J. Nuttall and J. Nutting, "Structure and Properties of Heavily Cold-Worked fcc Metals and Alloys," *Met. Science*, Vol 12, 1978, p 430-437.

70. W. H. Zimmer, S. S. Hecker, L. E. Murr and D. L. Rohr, "Large-Strain Plastic Deformation of Commercially-Pure Nickel," *Met. Science*, Vol 17, 1983, p 198-206.

71. U. F. Kocks, "Solution Hardening and Strain Hardening at Elevated Temperatures," Chap. 3, this volume.

72. S. Nourbakhsh and J. Nutting, "The High Strain Deformation of an Aluminum-4% Copper Alloy in the Supersaturated and Aged Conditions," *Acta Met.*, Vol 28, 1980, p 257-365.

73. R. L. Aghan and J. Nutting, "Structure and Properties of Free-Cutting Steels After Deformation to High Strains, *Met. Tech.*, Vol 8, 1981, p 41-45.

74. J. D. Embury and R. M. Fisher, "The Structure and Properties of Drawn Pearlite," *Acta Met.*, Vol 14, 1966, p 147-159.

75. G. Langford and M. Cohen, "Strain Hardening of Iron by Severe Plastic Deformation," *Trans. Quart.*, ASM, Vol 62, 1969, p 623-638.

76. H. J. Rack and M. Cohen, "Strain Hardening of Iron-Titanium Alloys at Very Large Strains," *Mater. Sci. Eng.*, Vol 6, 1970, p 320-326.

77. G. Langford, "A Study of the Deformation of Patented Steel Wire," *Met. Trans.*, Vol 1, 1970, p 465-477.

78. J. Nutting, *Proceedings of 8th International Congress on Electron Microscopy*, J. V. Sanders and D. J. Goodchild, Vol 1, Australian Academy of Science, Canberra, 1974, p 580.
79. P. T. Wakefield, A. S. Malin and M. Hatherly, "The Structure and Texture of a Rolled Low Stacking Fault Energy Alloy," *J. Aust. Inst. Met.*, Vol 22, 1977, p 143-151.
80. A. S. Malin and M. Hatherly, "Microstructure of Cold-Rolled Copper," *Met. Science*, Vol 13, 1979, p 463-472.
81. M. Hatherly and A. S. Malin, "Deformation of Copper and Low Stacking-Fault Energy, Copper-Base Alloys," *Met. Tech.*, Vol 6, 1979, p 308-319.
82. P. T. Wakefield and M. Hatherly, "Microstructure and Texture of Cold-Rolled Cu-10Zn brass," *Met. Science*, Vol 15, 1981, p 109-115.
83. H. Chandra-Holm and J. D. Embury, "The Development of Substructure in Aluminum and Aluminum Alloys," to appear in *Met. Science*, 1983.
84. D. L. Rohr and S. S. Hecker, "Three Surface TEM Observations of Cold Rolled 1100 Aluminum," *39th Am. Proc. Electron Microscopy Soc. Amer.*, edited by G. W. Bailey, Atlanta, GA, 1981, p 50-51.
85. F. Schuh and M. von Heimendahl, "Die Ausbildung der Versetzungsstruktur in Aluminum und der Beziehungen zum Verformungsverhalten," *Z. Metallkunde*, Vol 68, 1974, p 260-265.
86. K. P. Staudhammer, D. L. Rohr, K. A. Johnson and S. S. Hecker, Los Alamos National Laboratory (unpublished work).
87. J. J. Jonas, C. M. Sellars and W. J. McG. Tegart, "Strength and Structure under Hot-Working Conditions," *Met. Rev.*, Vol 14, 1969, p 1-24.
88. R. J. McElroy and Z. C. Szkopiak, "Dislocation-Substructure-Strengthening and Mechanical-Thermal Treatment of Metals," *Int. Met. Rev.*, Vol. 17, 1972, p 175-202.
89. A. W. Thompson, "Substructure Strengthening Mechanisms," *Met. Trans. A*, Vol 8A, 1977, p 833-842.
90. Z. S. Basinski, A. S. Korbel and S. J. Basinski, "The Temperature Dependence of the Saturation Stress and Dislocation Substructure in Fatigued Copper Single Crystals," *Acta Met.*, Vol 28, 1980, p 191-207.
91. S. S. Hecker, "Experimental Studies of Sheet Stretchability," in *Formability: Analysis, Modeling, and Experimentation*, edited by S. S. Hecker, A. K. Ghosh and H. L. Gegel, The Metallurgical Society of AIME, 1978, p 150-182.
92. S. S. Hecker, "Sheet Stretching Experiments," in *Application of Numerical Methods to Forming Processes*, edited by H. Armen, American Soc. Mech. Engng., AMD – Vol 28, 1978, p 85-94.
93. A. K. Ghosh and S. S. Hecker, "Failure in Thin Sheets Stretched Over Rigid Punches," *Met. Trans. A*, Vol 6A, 1975, p 1065-1074.
94. M. G. Stout and S. S. Hecker, "Role of Geometry in Plastic Instability and Fracture of Tubes and Sheet," *Mech. Mater.*, Vol 2, 1983, p 23-31.
95. J. E. Bird and J. L. Duncan, "Strain Hardening at High Strain in Aluminum Alloys and its Effect on Strain Localization," *Met. Trans. A*, Vol 12A, 1981, p 235-241.
96. L. E. Murr, *Interfacial Phenomena in Metals and Alloys*, Addison-Wesley Publishing Co., Reading, MA, 1975, p 145-148.

Structure, Properties, and Strain Hardening of Dual-Phase Steels

D. K. Matlock, F. Zia-Ebrahimi, and G. Krauss
Department of Metallurgical Engineering
Colorado School of Mines
Golden, Colorado 80401

Abstract

This paper describes the effects of processing and microstructure on the properties and strain-hardening behavior of dual-phase steels. The importance of low yield strength and the early stages of strain hardening on the attainment of the good ductility and high tensile strengths characteristic of dual-phase steels is developed. Also, it is shown that a number of processing and microstructural factors, including intercritical annealing temperature, cooling rate after intercritical annealing, martensite volume fraction and ferrite grain size, can be varied to provide the balance of yield strength and strain-hardening behavior required for good dual-phase steel properties.

Introduction

Two approaches to the thermomechanical processing of low- and medium-carbon steels have traditionally been used. Both approaches incorporate cooling from above the upper critical temperature (A_3) where only austenite is stable. Alloy composition and cooling rate then establish various desirable combinations of microstructure and properties. For example, low-carbon sheet steels, used in applications where good ductility is required, after being hot rolled as austenite, are cooled at rates which produce microstructures consisting primarily of ferrite and small amounts of pearlite or carbides.[1,2] On the other hand, medium-carbon bar steels, used in applications where high strength and wear resistance are required, are fully austenitized and are cooled rapidly enough to produce martensitic microstructures.[3]

In the last decade, microalloying and/or controlled hot rolling have dramatically increased the strength of low-carbon sheet steels compared with traditional processing.[4] The steels produced by these new approaches are referred to as high-strength low-alloy (HSLA) steels and develop high strength through grain-size refinement and precipitation of alloy carbonitrides. However, the microstructures of HSLA steels still largely consist of ferrite with some pearlite and are generally formed by cooling of a fully austenitic structure from above the A_3 temperature. Although HSLA steels are high in strength, their ductilities are low compared with those of traditionally processed low-carbon ferrite-pearlite steels—a major drawback in sheet steel applications which require high formability.

Recently, processing approaches other than those which produce exclusively either ferrite-pearlite or martensitic microstructures have been developed. These approaches result in microstructures consisting of ferrite and martensite. Any steel with such a microstructure is now referred to as a dual-phase steel even though phases other than ferrite and martensite may be present. The first papers to demonstrate effectively the advantages of dual-phase sheet steels were published in 1977 by Rashid[5] and by Hayami and Furukawa,[6] but earlier discussions of the processing and properties of steels with ferrite-martensite microstructures may also be found in the literature.[7,8] The 1977 papers generated intense practical and theoretical study of dual-phase steels, much of which was recorded in the proceedings of three symposia devoted to dual-phase steels in 1977, 1979 and 1981.[9-11]

Figure 1, after Rashid and Rao,[12] compares the stress-strain curve for a dual-phase steel with those for a plain carbon steel and an HSLA steel. The

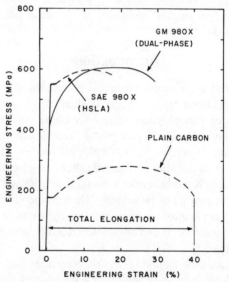

Fig. 1. Comparison of the stress-strain curve for a dual-phase steel with those for a plain carbon steel and an HSLA steel.[12]

compositions of the dual-phase and HSLA steels are identical (0.1 C, 1.5 Mn, 0.5 Si, 0.1 V). The dual-phase and HSLA steels have the same ultimate strength, but the dual-phase steel has significantly higher ductility. The conventionally processed plain carbon steel has much lower strength, but higher ductility, than the other two steels. Unique features of the stress-strain curve for the dual-phase steel are the absence of discontinuous yielding and a very high rate of strain hardening at low strains. Many combinations of strength and ductility may be developed in dual-phase steels, depending on processing conditions. Nevertheless, Fig. 2 shows that dual-phase steels have higher ductilities at all strength levels than do HSLA steels or cold rolled steels subjected to recovery annealing (RA) treatments.[13]

The ferrite-martensite microstructures of dual-phase steels are generally produced by intercritical annealing, i.e., heating between A_1 and A_3 in the ferrite-austenite phase field. Hot rolled or cold rolled low-carbon steels may be intercritically annealed. In each type of steel the pearlite is replaced by austenite on heating and much of the ferrite of the initial microstructure is retained. Some of the austenite then transforms to martensite on cooling. The microstructures which are produced by intercritical annealing are quite complex and strongly affect mechanical properties and strain hardening of dual-phase steels, as will be discussed in the remainder of this chapter.

Dual-phase microstructures may also be formed directly after hot rolling in alloys containing chromium and molybdenum.[14] Polygonal ferrite first forms directly after rolling on the hot strip mill. Then the remaining austenite transforms to martensite during slow cooling in the low-temperature sheet-coiling operation because of the high hardenability induced in the steel by the Cr and Mo additions.

The brief introduction above shows that dual-phase steels represent a significantly different approach to development of structure and good combinations of

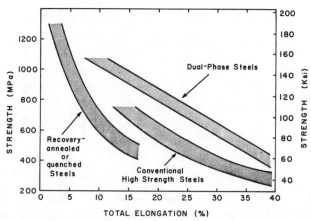

Fig. 2. Tensile strength as a function of total elongation for three types of high-strength steels.[13]

strength and ductility in low-carbon steels. The remainder of this chapter describes in more detail the unique microstructures and properties of dual-phase steels, the multistage strain-hardening behavior of the ferrite-martensite microstructures, and some of the processing parameters which affect the microstructures and properties of dual-phase steels.

Phase Transformations and Microstructure in Dual-Phase Steels

Austenite Formation in Dual-Phase Steels

The first step in the development of dual-phase microstructures by intercritical annealing consists of austenite formation from a hot or cold rolled microstructure of ferrite and pearlite. The formation and transformation of this austenite within a matrix of ferrite result in several features unique to low-carbon dual-phase steels — namely, the chemical partitioning of alloying elements between the austenite and ferrite, epitaxial ferrite formation, retained austenite, differences in precipitation within epitaxial ferrite and the ferrite retained at the intercritical annealing temperature, and quench aging. In dual-phase steels produced directly after hot rolling, more conventional continuous cooling transformation behavior and hardenability considerations apply because transformation occurs from fully austenitic microstructures rather than from ferrite-austenite mixtures.[14,15] The unique microstructural and transformational considerations associated with intercritical annealing will be emphasized in this section.

Intercritical annealing times for development of dual-phase properties are quite short (often on the order of minutes), and, therefore, nonequilibrium mixtures of austenite and ferrite result. Figure 3 shows a section through the Fe-C-Mn phase

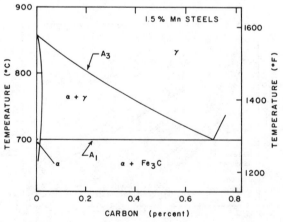

Fig. 3. A section through the Fe-C-Mn phase diagram, calculated on the basis of paraequilibrium, at a Mn content of 1.5%.[16]

diagram at a manganese content of 1.5%, an amount typically used for dual-phase steels. The phase diagram shown has been calculated on the basis of para-equilibrium, i.e., assuming that austenite formation is dependent only on carbon diffusion.[16] The latter assumption is reasonable in view of the short times desired for commercial intercritical annealing treatments. Thus, Fig. 3 may be used to estimate the amounts and composition of austenite formed by intercritical annealing treatments of plain carbon steel with a manganese content of 1.5%. Other alloying elements will, of course, displace somewhat the boundaries of the ferrite-austenite field, and the tie lines in the $\alpha + \gamma$ field may not lie in the section shown in Fig. 3. Nevertheless, Fig. 3 shows that high intercritical annealing temperatures produce large amounts of low-carbon austenite whereas intercritical annealing at low temperatures produces small amounts of high-carbon austenite.

As shown by Speich et al.,[16,17] the formation of austenite during intercritical annealing occurs by the following steps:

1. Very rapid nucleation and growth of austenite at carbide particles in pearlite or at ferrite grain boundaries until the carbides are dissolved.
2. Slower growth of austenite into ferrite. At high temperatures — around 850°C — this growth is controlled by diffusion of carbon in austenite, while at lower temperatures the growth is controlled by diffusion of manganese in ferrite.
3. Very slow growth controlled by diffusion of manganese in austenite as final equilibrium is attained.

Rarely does step 3 progress to a significant extent in dual-phase steels. However, partitioning of manganese does begin after several minutes at an intercritical annealing temperature. Evidence obtained by scanning transmission electron microscopy (STEM) shows that manganese is enriched in the austenite and depleted in the ferrite immediately adjacent to the austenite-ferrite interface.[17,18] Silicon, another element present in dual-phase steels, concentrates in the ferrite rather than in the austenite.[19,20] The partitioning of manganese and silicon during intercritical annealing affects quench aging of the ferrite in dual-phase steels: precipitation is retarded in the Mn-rich epitaxial ferrite and stimulated in the Si-rich retained ferrite.[21]

Austenite Transformation and Microstructure

The essential feature of dual-phase behavior and microstructure is transformation of some of the austenite formed during intercritical annealing to martensite. Not only does the martensitic transformation introduce a hard, deformation-resistant phase into a dual-phase microstructure, but the shear mechanism of martensite formation also introduces a high density of mobile or unpinned dislocations into the surrounding microstructure. Figure 4 shows a transmission electron micrograph of a martensite patch formed in

Fig. 4. Martensite patch and surrounding ferrite in 0.08C-Mn-Si-Mo-V dual-phase steel. TEM micrograph.

0.08C-1.47Mn-0.57Si-0.10Mo-0.04V steel air cooled from 790°C. The martensite is best characterized as a lath martensite with a fine structure consisting of a very high density of dislocations.[22] This morphology is consistent with the relatively high concentration of carbon in the austenite (Fig. 3) produced by low-temperature intercritical annealing or by the concentration of carbon in the austenite by rejection from ferrite during cooling from high intercritical annealing temperatures. Sometimes fine transformation twins, more typical of a high-carbon plate martensite, make up a portion of the fine structure of the martensite in dual-phase steels.

Light microscopy shows none of the substructural detail present in dual-phase steels. However, light microscopy is extremely important in characterizing the distribution of the martensite and other austenite decomposition products relative to the ferritic matrix grain structure. Figure 5 shows a light micrograph of a specimen etched in nital. The nital etch defines the martensite patches but develops very little contrast between the martensite and the ferrite. Also, microstructural features and phases other than martensite and ferrite might be present but are not clearly revealed in metallographic specimens etched in nital.

A major development in the characterization of dual-phase microstructures was the application of a two-step etching technique which first reveals the ferrite grain structure by picral etching and then the austenite decomposition products by immersion in boiling alkaline chromate.[23] Huppi[24,25] first showed that the picral–boiling alkaline chromate etch not only enhanced the contrast between martensite and ferrite, but also differentiated between two types of ferrite and delineated the size of the austenite pool present at the intercritical annealing temperature prior to cooling.

Figure 6 shows microstructures of a niobium-containing steel which has been oil quenched after intercritical annealing at 760 and 810°C.[26] The gray areas

Fig. 5. Martensite and ferrite in 0.08C-Mn-Si-Nb dual-phase steel. Light micrograph. Nital etch. (*Courtesy of G. Huppi.*)

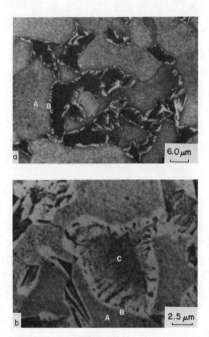

Fig. 6. Microstructures of 0.08C-Mn-Si-Nb steel intercritically annealed (a) for 8 min at 760°C and (b) for 16 min at 810°C.[26] Boiling alkaline chromate etch.

marked with the letter A are ferrite that was retained during intercritical annealing and will henceforth be referred to as retained ferrite. The white areas marked with the letter B are ferrite which has grown epitaxially on the retained ferrite during cooling after intercritical annealing; this ferrite will henceforth be referred to as epitaxial ferrite. Transmission electron microscopy (Fig. 7b) shows that there

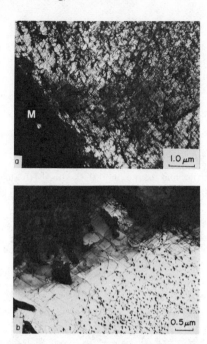

Fig. 7. Ferrite substructure adjacent to transformed austenite in 0.08C-Mn-Si-Nb steel intercritically annealed (a) for 8 min at 760°C and (b) for 16 min at 810°C.[26] TEM micrographs.

is no structural interface or boundary between the retained and epitaxial ferrite and provides evidence for the epitaxial growth of ferrite on the retained ferrite acting as a substrate. A fine precipitate dispersion, most probably a niobium carbonitride, is present in the retained ferrite but not in the epitaxial ferrite. The precipitates have formed in the retained ferrite during intercritical annealing and reflect differences in thermal history and perhaps composition between the two types of ferrite.

The dark-etching structures within the epitaxial ferrite are either martensite (Fig. 6a) or mixtures of ferrite and carbide and some martensite (area C in Fig. 6b). The micrographs of Fig. 6 illustrate an important effect of intercritical annealing temperature on the hardenability of the austenite in dual-phase steels. The higher carbon content of the austenite produced at lower intercritical annealing temperatures promotes hardenability and results in more martensite formation than occurs in specimens cooled at the same rate from higher annealing temperatures.[26-28] Figure 7(a) shows that a consequence of large amounts of martensite formation is a very high dislocation density in the ferrite surrounding the martensite.

A wide variety of austenite decomposition products other than epitaxial ferrite and martensite have been observed in dual-phase steels. In the niobium steel,

TEM analysis of areas, such as that marked C in Fig. 6(b), shows a sequence of transformation from epitaxial ferrite to pearlite with a coarse interlamellar spacing to degenerate pearlite with discontinuous carbide lamellae and/or carbide-ferrite arrays produced by interphase or row precipitation.[26]

In vanadium-containing dual-phase steels, rows of fine vanadium carbonitride precipitates formed by the decomposition of the austenite by an interphase mechanism are frequently observed.[12,29] Also, under certain alloying and cooling conditions, acicular or Widmanstätten ferrite is a common ferrite morphology which develops between epitaxial ferrite and martensite.[25]

Another feature unique to dual-phase steels is the presence of small islands of retained austenite dispersed throughout the ferrite matrix. Figure 8 shows a retained austenite island in a C-Mn-Si dual-phase steel.[30] A corner of the retained austenite had started to transform to pearlite, but the remainder of the austenite was resistant to transformation of any kind. The remarkable stability of some of the austenite in dual-phase steels appears to be a complex function of the partitioning of carbon and such elements as manganese to the austenite, the small size of the austenite crystals, and the constraints of the partially coherent interface between the ferrite and austenite.[19] Although the retained austenite is quite stable at room temperature, it transforms under stress to martensite,[19,31,32] and during tempering transforms to a coarse bainitic structure.[12]

Systematic Development of Dual-Phase Microstructure

The previous section described various dual-phase microstructural components which may be produced when austenite transforms after intercritical annealing. For a given starting condition, three controllable parameters essentially determine the dual-phase microstructure: intercritical annealing temperature, cooling rate from the intercritical annealing temperature, and alloy content. Time at intercritical annealing temperature will be short, on the order of minutes, for applica-

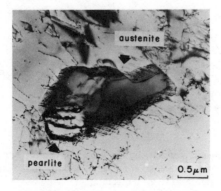

Fig. 8. Retained austenite island, partially transformed to pearlite, in 0.06C-Mn-Si dual-phase steel.[30] TEM micrograph.

tions of commercial interest, and therefore the effect of time at temperature will be relatively constant. The microstructure present prior to intercritical annealing is extremely important, particularly in hot rolled steels, because it fixes the ferrite grain size and the sites at which austenite will form on intercritical annealing.

Figure 9 is a schematic diagram of the effects of extremes in cooling rate on the microstructures of dual-phase steels. High cooling rates cause almost all of the austenite to transform to martensite; intermediate cooling rates result in significant amounts of epitaxial ferrite and other decomposition products in addition to martensite; and low cooling rates produce only epitaxial ferrite and ferrite-carbide mixtures. Figure 10 shows microstructures in an Fe-Mn-Si-C steel which correspond directly to those illustrated schematically in Fig. 9. Quantitative metallographic analyses of the amounts of the various microstructural components as functions of cooling rate result in microstructural maps, such as that shown in Fig. 11 for the Fe-Mn-Si-C steel intercritically annealed at 810°C. Retained ferrite, of course, dominates the microstructure. The amount of retained ferrite is constant for a given intercritical annealing treatment, as indicated by the horizontal line. The regions below the horizontal line show how the austenite has decomposed as a function of cooling rate.

Increased alloy content and low intercritical annealing temperatures increase the hardenability of the austenite. Therefore, the martensite phase boundary in Fig. 11 would be shifted to lower cooling rates. Alloying and variations in intercritical annealing temperatures would, of course, also vary the amounts of retained ferrite and austenite formed on intercritical annealing.

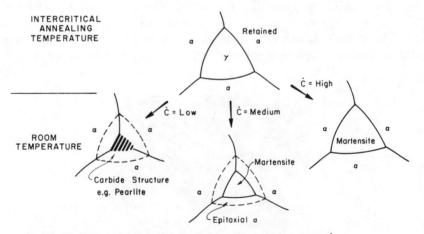

Fig. 9. Schematic diagram showing effect of cooling rate ($\dot{C}$) on transformation of an austenite grain during cooling from the intercritical annealing temperature to room temperature.

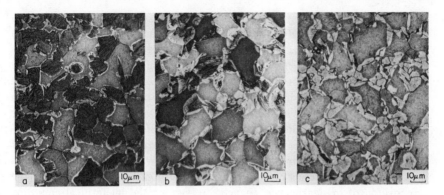

Fig. 10. Microstructure of 0.06C-Mn-Si steel intercritically annealed at 810°C and cooled at (a) 1000°C/s, (b) 300°C/s and (c) 60°C/s.[27] Light micrographs. Nital–picral–alkaline chromate etch.

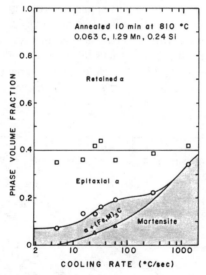

Fig. 11. Quantitative microstructural map showing the effects of cooling rate on the amounts of various microstructural constituents.[27]

Microstructure — Tensile Property Relations in Dual-Phase Steels

The unique tensile properties of dual-phase steels, i.e., low yield strength, high tensile strength, and large uniform and total elongations, result from complex interactions of the yielding and strain-hardening behavior of the microstructural components of these steels. The microstructural parameters, such as volume fraction, composition, size and distribution of phases (mainly ferrite and martensite), can be varied by changes in alloy composition, initial microstructure and processing, as discussed in the previous section.

The strengths of dual-phase steels are expected to be raised by an increase in the volume fraction of the harder martensite phase. Davies[33] has shown that the strengths of Fe-1.5Mn-C dual-phase steels increase linearly with martensite volume fraction as shown in Fig. 12, independent of the composition or strength of the martensite. The volume fraction of martensite was varied by quenching a series of alloys of different carbon contents from a constant temperature, and the composition of martensite was changed by varying the intercritical annealing temperature. Similar results have been reported by other investigators.[34-36]

As can be seen in Fig. 12, a range of experimental scatter larger than 200 MPa is associated with these curves. More detailed analyses of the strengths of dual-phase steels have shown that the composition, strength, size and distribution of the martensite phase, or the amount of epitaxial ferrite, may cause variations in strength which are within the scatter band of the curves presented by Davies.[33] For example, Speich and Miller[37] have shown that an increase in annealing temperature, which results in a martensite of lower carbon content, reduces strength in dual-phase steels (Fig. 13). Note that the variations in both yield and tensile strengths are within the scatter band of the curves in Fig. 12.

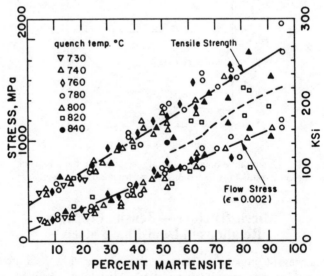

Fig. 12. Flow stress (0.2%) and tensile strength as functions of martensite volume fraction for Fe-Mn-C alloys.[33]

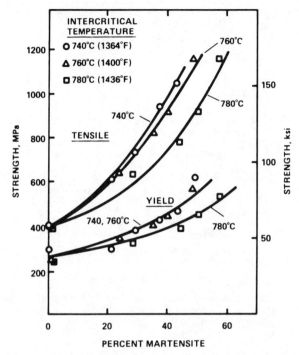

Fig. 13. Flow stress (0.2%) and tensile strength as functions of martensite volume fraction for Fe-C-1.5Mn-0.5Si-0.15Mo alloys.[34]

Data presented by Lawson et al.[27] show that an increase in annealing temperature from 760 to 810°C results in higher values of yield and tensile strengths as functions of martensite volume fraction, as shown in Fig. 14. These results may seem contradictory to data presented in Fig. 13. However, it should be noted that in the Speich and Miller[37] investigation various volume fractions of martensite were produced by use of steels with different carbon contents and all specimens were water quenched from the annealing temperature. Lawson et al.[27] used a constant composition and varied the cooling rate to achieve various martensite volume fractions. Therefore, the inconsistency in evaluation of the effect of annealing temperature may be related to changes in the amounts of other phases, mainly epitaxial ferrite, with cooling rate. Furthermore, the carbon content of the martensite varies with a change in cooling rate. While Fig. 13 reflects the effect of carbon content in martensite, the results shown in Fig. 14 include the effects of a number of parameters.

Figure 14 shows a minimum in 0.2% offset yield strength at low volume fractions of martensite. This minimum in yield strength has also been reported by other investigators, who changed the volume fraction of martensite by applying

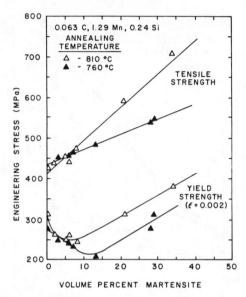

Fig. 14. Flow stress (0.2%) and tensile strength as functions of martensite volume fraction for an Fe-0.063C-1.29Mn-0.24Si steel annealed for 10 min at 810 and 760°C. [27]

various annealing temperatures [38] or cooling rates. [39] Thus, minima in yield strength can also be found in curves of yield strength versus cooling rate [27,39–41] or annealing temperature. [38] The volume fraction of martensite at which the minimum yield strength is achieved appears to be related to the disappearance of discontinuous yielding or to the lack of sufficient numbers of mobile dislocations. [39]

Hansen and Pradhan [41] have suggested that not only is continuous yielding controlled by production of a high dislocation density in ferrite due to austenite-to-martensite transformation, but also a sufficient cooling rate below the M_s temperature is required to maintain the mobility of these dislocations.

The other parameter, which contributes to an increase in hardening of dual-phase steels, is the strength of the ferrite matrix, which can be increased by refining grain size, precipitation, and solid-solution hardening. Smaller ferrite grain size can be achieved by employing a finer initial microstructure or a lower annealing temperature in the case of cold rolled steels. Figure 15 shows the effect of grain size on 0.2% flow stress in dual-phase steels with a constant volume fraction of martensite and a constant silicon content. Such a normal linear dependency of strength of dual-phase steels on the inverse square root of grain diameter (Hall-Petch relation) has been reported by several investigators. [44,45]

Alloying elements in dual-phase steels, besides their effects on hardenability, can increase the strength of ferrite by solid-solution hardening or precipitation.

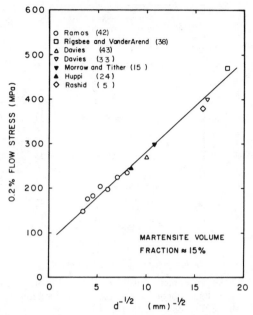

Fig. 15. Flow stress (0.2%) as a function of $d^{-1/2}$ for dual-phase steels with a martensite volume fraction of 15%.[42]

For example, Davies[44] has shown that significant hardening can be obtained by adding silicon to an Fe-C-Mn dual-phase steel without a loss in uniform elongation. The beneficial effect of silicon on the strength/ductility combination has been attributed to a reduction in ferrite carbon content, which will lead to a "cleaner" and more ductile ferrite.[44]

Microalloying elements (V, Nb, etc.) are usually added for the purpose of refining the initial microstructure through controlled rolling, and they may precipitate in retained ferrite as carbonitrides during cooling from rolling and intercritical annealing temperatures.[26,29] However, the epitaxial ferrite formed on cooling from the intercritical annealing temperature has been shown to be precipitate-free.[26] The importance of this precipitate-free epitaxial ferrite is shown in Fig. 16. This figure shows an increase in elongation and a moderate decline in strength as the amount of epitaxial ferrite is increased by means of an increase in the intercritical annealing temperature. In general, it is often desirable to produce a very ductile ferrite matrix with a high density of mobile dislocations. Therefore, any pinning of dislocations by interstitial elements or fine precipitates is not desirable.

Another microstructural constituent which is believed to improve ductility in dual-phase steels is retained austenite.[31,38,47] A mild rate of cooling after intercritical annealing increases the amount of retained austenite, and volume frac-

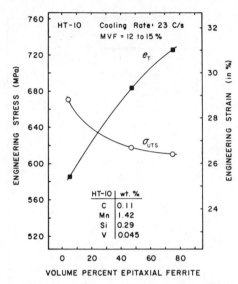

Fig. 16. Variation in ultimate tensile strength and total elongation with epitaxial ferrite volume fraction for a vanadium-microalloyed steel with a constant volume fraction of martensite.[46]

tions as high as 10% may be retained in some dual-phase steels.[47] During tensile deformation, most of the retained austenite transforms to martensite with the first few percent of plastic strain,[38,48] thereby increasing the effective volume fraction of martensite. Rigsbee and VanderArend[38] have suggested that the high work-hardening rate in the regime of 1 to 7% plastic strain is due to the continuous transformation of retained austenite to martensite, which generates more dislocations in the ferrite matrix. As will be shown in the following sections, for constant yield strength, an increase in the work-hardening rate in the early stages of plastic deformation reduces the extent of uniform strain. Therefore, the mechanism proposed by Rigsbee and VanderArend[38] cannot explain the improvement in ductility that accompanies an increase in the volume fraction of retained austenite. Similar to TRIP materials,[48] the improvement in uniform ductility may be related to a *softening mechanism* due to the *transformation-induced strain* in the early stages of plastic deformation, and to a *static-hardening effect* of the transformation product, i.e., an increase in the volume fraction of martensite, which increases work hardening in the later stages of plastic deformation.

An improvement in strength, regardless of the means by which it is achieved, is usually gained at the expense of a reduction in ductility. Figure 17 shows tensile properties of an Fe-Mn-C steel as functions of cooling rate. An increase in cooling rate initially decreases yield strength and slightly increases tensile strength, whereas uniform strain is not changed considerably. At high rates of

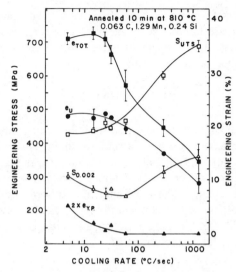

Fig. 17. Summary of tensile data as a function of cooling rate for Fe-0.063C-1.29Mn-0.24Si steel annealed 10 minutes at 810°C.[27]

cooling, tensile and yield strengths increase rapidly and, correspondingly, uniform and total elongations decrease drastically. The best combination of strength and ductility is achieved at an intermediate cooling rate. The relationship between strength and ductility in dual-phase steels will be discussed in terms of work-hardening behavior in the following sections.

Strain Hardening in Dual-Phase Steels

As discussed above, the mechanical properties of dual-phase steels are significantly different from those of similar steels with ferrite-pearlite microstructures. In this section, the unique properties of dual-phase steels, i.e., good combinations of strength and ductility, will be shown to result from the effects of dual-phase processing on yield-strength modification and multistage strain-hardening behavior. In the following section, multistage strain hardening will be shown to be predictably dependent on the processing history–dependent microstructure. Prior to a consideration of strain hardening in dual-phase steels, a general review of the effects of strain hardening and instability conditions will be presented.

A Correlation of Strain Hardening and Instability

The strain-dependent strain-hardening behavior in a tensile test controls both the strength level for a given strain and the limiting magnitude of the uniform strain. In particular, the maximum uniform true strain is the strain at which plastic instability occurs, i.e., the increase in load-bearing capacity due to strain hardening is insufficient to counteract the decrease in load-bearing capacity that results from a decrease in cross-sectional area. This condition correlates to the point in a tensile test where the change in load with strain is zero and results in the well-known criterion for instability

$$\frac{d\sigma}{d\varepsilon} = \sigma \text{ when } \varepsilon = \varepsilon_u \tag{1}$$

where $d\sigma/d\varepsilon$ is the strain-dependent true strain-hardening rate, σ is the strain-dependent true stress, and ε_u is the true uniform strain. This condition is graphically illustrated for a typical parabolic stress-strain curve in Fig. 18, in which the strain-dependent functions for $d\sigma/d\varepsilon$ and σ are plotted. The point of intersection defines, according to Eq 1, the maximum true uniform strain, ε_u. It is also apparent in Fig. 18 that the true stress at instability can be viewed as the sum of two terms: the yield stress, σ_y; and the change in stress due to strain hardening, $\Delta\sigma(\varepsilon)$. Accordingly, Eq 1 is rewritten as

$$\frac{d\sigma}{d\varepsilon}(\varepsilon_u) = \sigma_y + \Delta\sigma(\varepsilon_u) \tag{2}$$

For the most general case, each term in Eq 2 will vary with microstructure, and $(d\sigma/d\varepsilon)(\varepsilon)$ and $\Delta\sigma(\varepsilon)$ will be independent functions of strain.

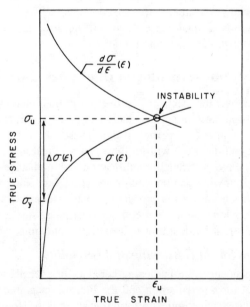

Fig. 18. Schematic diagram showing the instability condition, $\sigma = d\sigma/d\varepsilon$.

To illustrate the significance of independent variations in each term of Eq 2, we will consider the schematic stress-strain curves presented in Fig. 19. These two stress-strain curves involve different yield strengths but equivalent strain-dependent strain-hardening-rate functions. In this case an increase in yield strength corresponds to a decrease in uniform strain due to the decrease in the magnitude of $\Delta\sigma$ required to satisfy instability. The behavior presented in Fig. 19 closely approximates the effect of grain size on ductility in alpha brass.[49]

Figure 20 illustrates the importance of the magnitude of the true strain-hardening rate at high strains. In this figure, samples A and C have the same yield strength whereas samples A and B have the same dependence of strain-hardening rate on strain. Clearly, for constant σ_y, ε_u increases with $d\sigma/d\varepsilon$ because the strain required to achieve the appropriate $\Delta\sigma$ in Eq 2 also increases. However, as shown by a comparison of samples B and C, if both σ_y and $(d\sigma/d\varepsilon)(\varepsilon)$ are different then it is possible for two samples that differ widely in deformation behavior, and thus in microstructure, to exhibit the same ductility.

In contrast to the stress-strain curves presented in Fig. 18 to 20, several materials, such as mild steels, exhibit more complex strain-hardening behavior in which multiple strain-dependent stages in strain hardening are observed.[50,51] To illustrate the effects of multiple stages in strain hardening, consider the stress-

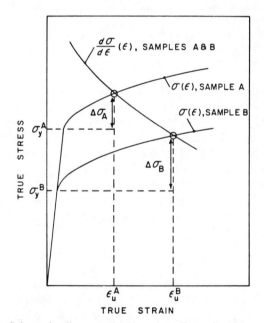

Fig. 19. Schematic diagram showing the effect of yield strength on maximum uniform strain for equivalent strain-dependent strain-hardening-rate function.

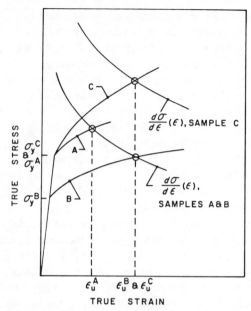

Fig. 20. Schematic diagram showing the effect of strain dependency of strain-hardening rate on maximum uniform strain for equivalent yield strength (samples A and C), and the possibility of achieving the same maximum uniform strain in materials differing widely in deformation behavior (samples B and C).

strain curves presented in Fig. 21. This figure comprises a series of stress-strain curves with equivalent yield strengths but with systematic variations in strain-hardening behavior at low strains. Also, the strain dependence of the strain-hardening rate for strains greater than ε^* is assumed to be the same for all three samples. For this series of samples, the effect of an increase in the average strain-hardening rate at low strains (i.e., less than ε^*) is a decrease in the maximum true uniform strain. As shown in Fig. 21, the flow stress at ε^*, σ_{ε^*}, directly mirrors the strain-hardening behavior at low strains, and with σ_{ε^*} the instability condition in Fig. 21 can be rewritten as

$$\frac{d\sigma}{d\varepsilon} = \sigma_{\varepsilon^*} + \Delta\sigma(\varepsilon_u - \varepsilon^*) \tag{3}$$

where $\Delta\sigma(\varepsilon_u - \varepsilon^*)$ represents the additional stress above σ_{ε^*} required to satisfy instability. It should be noted that the effects on ε_u of variations in the multistage strain-hardening-dependent σ_{ε^*} are analogous to those presented in Fig. 19 for variations in σ_y. Thus the primary effect of altering the strain-hardening behavior at low strains is a significant increase in flow stress.

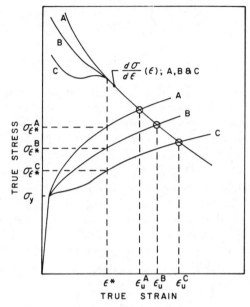

Fig. 21. Schematic diagram showing the effect of strain hardening behavior at low strains on the maximum uniform strain for equivalent yield strength, and strain hardening at high strains.

An example illustrating the concepts shown in Fig. 21 was presented by Pickering.[52] He showed that in mild steels an increase in pearlite volume fraction, which does not alter strain hardening at high strains, causes a decrease in yield-point elongation, an increase in the rate of strength accumulation at low strains, and a corresponding decrease in the uniform strain at instability.

The preceding discussion has shown that the true uniform strain at instability depends, in a complex way, on both the yielding and strain-dependent strain-hardening behavior. In attempts to simplify discussions of strain hardening, deformation behavior is commonly analyzed with the aid of idealized mathematical stress-strain equations and modified plotting techniques. The most common stress-strain equation is the Holloman equation:[53]

$$\sigma = K\varepsilon_p^{\,n} \tag{4}$$

where σ is true stress, ε_p is true plastic strain, and K and n are empirical parameters. For materials which obey Eq 4, a plot of log σ versus log ε_p is linear with a slope of n. Furthermore, n is equal to ε_u. Equation 4, with n constant, has been used to describe the stress-strain behaviors of several alloy systems.[53] However, for materials which exhibit stages in strain hardening, n varies with strain, and Eq 1 and 4 can be combined to describe an instantaneous value of n_i:

$$\left(\frac{\varepsilon}{\sigma}\frac{d\sigma}{d\varepsilon}\right)_i = n_i \tag{5}$$

This expression is often used to describe an "incremental work-hardening parameter".[54]

The effects of variations in stress-strain behavior on common plots used to evaluate and compare strain-hardening behavior are illustrated schematically in Fig. 22. In this figure, sample A is assumed to follow Eq 4 with n constant, and sample B exhibits stages in strain hardening. As shown in Fig. 22(a), sample A exhibits a smooth parabolic stress-strain curve, which translates to a linear function of log σ versus log ε_p in Fig. 22(b). Correspondingly, the value of n_i in Eq 5 is constant with strain, as shown in Fig. 22(c). For sample B, which exhibits stages in strain hardening, the plot of log σ versus log ε_p in Fig. 22(b) is not described by a single slope. Correspondingly, the strain-hardening parameter in Fig. 22(c) varies in a complex way with strain.

The distinct regions in strain hardening are further amplified by the following analysis. Similar to the Holloman equation (Eq 4), the constitutive equation due to Ludwik[55] is given by

$$\sigma = \sigma_0 + B\varepsilon_p^m \tag{6}$$

where σ_0 is the yield strength and B and m are empirical parameters. Following Monteiro et al.,[56] the stages in strain hardening were delineated by application

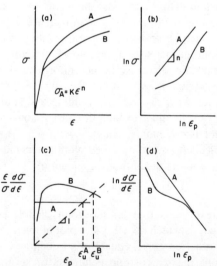

Fig. 22. Schematic diagrams showing the common plots used to evaluate strain-hardening behavior: (a) stress-strain curves; (b) ln σ vs ln ε_p; (c) n = $(\varepsilon/\sigma)(d\sigma/d\varepsilon)$ vs ε_p; and (d) ln dσ/dε vs ln ε_p (*Jaoul-Crussard plot*).

of the Jaoul[57]-Crussard[58] analysis, in which Eq 6 is differentiated and rewritten as

$$\ln \frac{d\sigma}{d\varepsilon} = \ln (Bm) + (m - 1) \ln \varepsilon_p \qquad (7)$$

Following Eq 7, the stress-strain data of Fig. 22(a) are plotted in Fig. 22(d). For sample B this figure clearly delineates stages in strain hardening. It should be noted that this procedure is particularly sensitive to variations in strain hardening at low strains but tends to obscure possible strain-hardening variations at high strains. Note that, as in Fig. 22(b) and (c), sample A in Fig. 22(d) is described by a single straight line.

In addition to Eq 4 and 5, constitutive stress-strain relations based on the analyses of Voce[59] and Swift[60] have also been utilized. However, as discussed elsewhere, the resulting analytical plots yield results similar to those shown in Fig. 22 and thus will not be considered further.

This section has clearly identified the importance of multistage strain hardening, which can lead to modified strain-hardening rates at low strains. In particular, stages in strain hardening which result in decreased strain-hardening rates at low strains lead to increased uniform strains. Furthermore, plotting techniques which clearly delineate the stages in strain hardening have been demonstrated. In the following section the strain-hardening behavior which characterizes deformation of dual-phase steels is reviewed.

Strain Hardening in Dual-Phase Steels

The deformation behavior of dual-phase steels with ferrite-martensite microstructures differs significantly from that of similar steels with ferrite-pearlite microstructures. At low strains, dual-phase steels exhibit decreased yield strengths, continuous yielding and increased strain-hardening rates in comparison with ferrite-pearlite steels. At high strains, dual-phase steels exhibit high ultimate tensile strengths without significant losses in ductility. For example, Fig. 23 shows a series of true stress–true strain curves as functions of cooling rate for a 0.063C-1.29Mn-0.24Si steel annealed for 10 minutes at 810°C and cooled at the rates indicated.[27] All curves show characteristic dual-phase deformation behavior, but it is clear that many variations are possible depending on variations in processing history, such as the changes in cooling rate noted here.

The significant differences in the strain-hardening behavior of dual-phase steels relative to that of other steels are further amplified in Fig. 24 and 25. Figure 24 presents the results of Magee and Davies,[61] in which standard tensile data are plotted as $d\sigma/d\varepsilon$ versus ε for three classes of steel: conventional low-carbon steels, conventional HSLA steels, and dual-phase steels. At all levels of strain, the strain-hardening rates of the dual-phase steels are the highest.

Figure 25 shows the effect of strain due to wire drawing on the peak flow stress of 5.59-mm-diam 1008 steel rod processed with either a conventional hot rolled

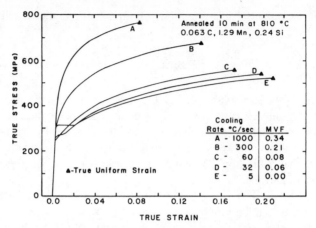

Fig. 23. True stress–true strain curves for Fe-0.063C-1.29Mn-0.24Si steel intercritically annealed at 810°C and cooled at different rates.

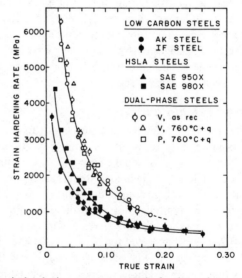

Fig. 24. Strain-hardening rate versus strain for three classes of steel.[61] The strain-hardening rates of the dual-phase steels are the highest at all levels of strain.

ferrite-pearlite microstructure or a dual-phase steel microstructure produced by an iced-brine quench after a 30-min anneal at 760°C. The resulting microstructure contained approximately 25 vol % martensite. This figure shows that the martensite acts to increase both the flow stress in the as-processed condition and the rate of strain hardening even for true strains up to approximately 1.4.

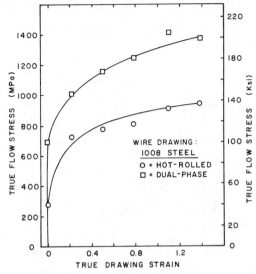

Fig. 25. Comparison of flow curves obtained by wire drawing of 1008 steel with ferrite-pearlite (hot rolled) and ferrite-martensite (dual-phase) microstructures.

Figures 24 and 25 clearly indicate that strain-hardening rates are higher in dual-phase steels. However, it has also been well documented[25,32,62] that dual-phase steels also exhibit distinct stages in strain hardening that are analogous to those discussed in conjunction with Fig. 21 and 22 in the preceding section. For example, consider the results of Magee and Davies[61] in Fig. 26, which presents plots of $d\sigma/d\varepsilon$ versus ε for three dual-phase steels with martensite volume fractions (MVF) of 7, 12 and 20%. Note that at low strains the strain-hardening rate varies with MVF, increasing with an increase in MVF, but that at high strains the strain-hardening rate is essentially the same, independent of MVF. The stress-strain curves for these three steels would then be similar to those shown schematically in Fig. 21.

Differences in strain-hardening behavior at low strains are clarified when tensile data are plotted according to the Jaoul-Crussard analysis discussed in conjunction with Fig. 22. For example, the tensile data shown in Fig. 23 are replotted according to Eq 7 and are presented in the Jaoul-Crussard plots of Fig. 27.[27] This figure shows, in dual-phase steels, the existence of multistage strain hardening, as clearly evidenced by the three linear regions for samples B and C. Furthermore, the average strain-hardening rates at low strains significantly decrease with a decrease in cooling rate. The decrease in cooling rate produces a decrease in MVF, an increase in epitaxial ferrite and a corresponding decrease in average flow stress at low strains. Thus, the variations in strain hardening

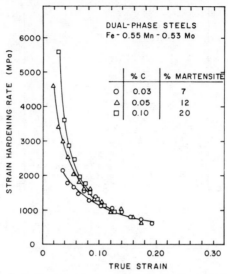

Fig. 26. Strain-hardening rate versus strain for three dual-phase steels with martensite volume fractions of 7, 12 and 20%.[61]

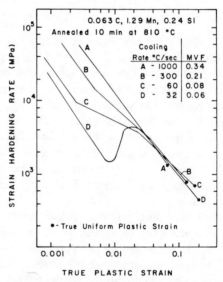

Fig. 27. Jaoul-Crussard analysis of stress-strain curves shown in Fig. 23 according to Eq 7.

at low strains directly reflect variations in the processing-dependent microstructures.

As shown by a comparison of the true stress–true strain curves of Fig. 23 with the Jaoul-Crussard plots of Fig. 27, an increase in the average strain-hardening rate at low strains corresponds directly to both an increase in strength and a decrease in ductility. The importance of the early stages of strain hardening on controlling the over-all deformation behavior of dual-phase steels is further amplified by the correlations among multistage strain hardening, strength, and ductility discussed here. These results,[24,63] which represent tensile data from five different alloys processed with significantly different microstructures, show that the strain-hardening behavior of dual-phase steels can be summarized by the trends indicated in the series of schematic stress-strain curves shown in Fig. 28. This figure shows that all dual-phase steels, regardless of processing history, exhibit similar strain-hardening rates at higher strains. At low strains, however, the strain-hardening rates are significantly different because of the multistage strain-hardening behavior of dual-phase steels. It is this multistage strain-hardening behavior which produces the distinctive properties of dual-phase steels.

The validity of Fig. 28 as an over-all summary of the deformation behavior of dual-phase steels is substantiated by the correlations presented in Fig. 29 to 32. In Fig. 29, a significant increase in the strain-hardening rate at a strain of 0.01 is shown to be directly correlated with an increase in the corresponding flow stress. However, with the microstructure-dependent and processing history–dependent strain-hardening rates at a strain of 0.01 as a reference, Fig. 30 shows that the flow stress at a strain of 0.002 is much less sensitive than the flow stress at a strain of 0.01 to the factors which produce the wide variations in strain-hardening rates at the 0.01 strain level. In Fig. 31 the flow stress at a strain of 0.10 is plotted against the corresponding strain-hardening rate at the same strain and shows that

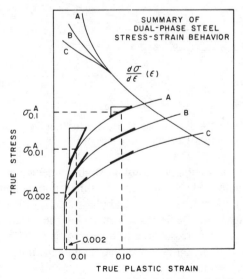

Fig. 28. A summary of the deformation behavior of dual-phase steels.

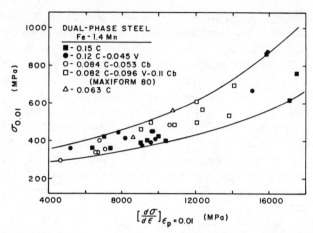

Fig. 29. The 1% flow stress, $\sigma_{0.01}$, as a function of strain-hardening rate for five different steels processed with a variety of dual-phase microstructures.

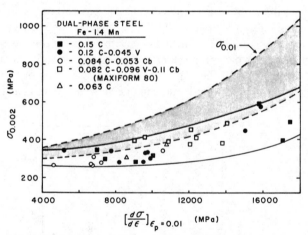

Fig. 30. The 0.2% flow stress, $\sigma_{0.002}$, as a function of strain-hardening rate at a strain of 0.01 for the specimens in Fig. 29.

all samples exhibited essentially a constant strain-hardening rate independent of flow stress. In addition, Fig. 31 indicates that the strain-hardening rates at high strains are similar for a wide variety of dual-phase steels, which is consistent with the results of Magee and Davies[61] shown in Fig. 26. Thus, as shown schematically in Fig. 28, it is reasonable to assume that a single strain-dependent strain-hardening-rate function describes the strain-hardening behavior of dual-phase steels at high strains.

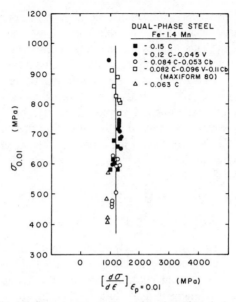

Fig. 31. The 1% flow stress, $\sigma_{0.01}$, as a function of strain-hardening rate at 0.1 strain for the specimens in Fig. 29.

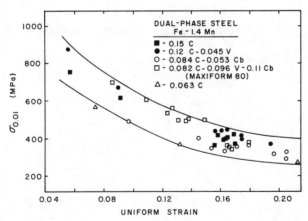

Fig. 32. The dependency of maximum uniform strain on $\sigma_{0.01}$ for the specimens in Fig. 29.

In conjunction with Fig. 29, 30 and 31 it is clear that Fig. 28 correctly summarizes dual-phase steel deformation behavior. Furthermore, the primary parameter which varies with processing history is the strain-hardening behavior at low strains, i.e., at strains less than 0.04, as shown in the Jaoul-Crussard plots of Fig. 27. The consequence of a significant increase in strain-hardening rate at low strain is a large decrease in the strain increments required for significant increases in strength. This implies that the total strain required to increase the stress to the point where the instability condition is satisfied also decreases. Thus, rapid increases in flow stress at low strains will directly result in decreases in ductility. This correlation is verified in the plot of ε_u versus $\sigma_{0.01}$ shown in Fig. 32. Besides the experimental error, the scatter band observed in the ε_u-versus-$\sigma_{0.01}$ curve can be associated with many microstructural parameters. As a first attempt, $\sigma_{0.01}$ was corrected for the variation in ferrite grain size according to the following equations for the dependency of the lower yield strength on grain size for Armco iron:[64]

$$\sigma_y(\text{MPa}) = 25.35 + 34.11 \, d^{-1/2} - 1.079 \, d^{-1} \tag{8}$$

$$\sigma_d(\text{MPa}) = 34.11 \, d^{-1/2} - 1.079 \, d^{-1} \tag{9}$$

where d is grain size (in mm) and σ_d is the grain-size-dependent part of Eq 8. As can be seen in Fig. 33, a tighter scatter band resulted in the plot of ε_u versus $\sigma_{0.01} - \sigma_d$. Thus, it is concluded from the above analysis that the primary processing-dependent variable in dual-phase steels is the strain-hardening behavior at low strains. In the following section the mechanisms responsible for the strain-hardening behavior at low strains will be discussed and correlated with processing parameters.

Processing/Microstructure Correlation

As shown in the previous section, the primary property of dual-phase steels which is controllable by processing is the strain-hardening behavior at low

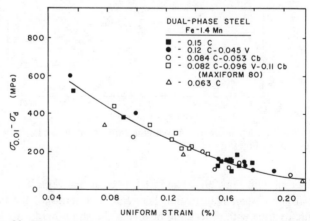

Fig. 33. The dependency of maximum uniform strain on the 1% flow stress normalized for the variation in grain size for the specimens in Fig. 29.

strains. In this section it will be demonstrated that the strain-hardening behavior of dual-phase steels results from strain hardening within the ferrite as modified by the martensite and that, therefore, the initial microstructure of the ferrite controls the processing-dependent initial strain-hardening rate.

Effect of Processing History on Early Stages of Strain Hardening

Dual-phase steel microstructures can be produced in most low-carbon steels by appropriate control of processing history. The variables of interest include alloy content, annealing temperature, annealing time, cooling rate and initial microstructure. Depending on alloy content, which controls the hardenability of the austenite, dual-phase steel microstructures can be produced by continuous annealing and appropriate cooling of cold rolled or hot rolled low-alloy strip,[65] by controlled cooling after hot rolling of steels of high hardenability,[14] or by batch annealing of high-manganese sheet steel coils.[35]

While dual-phase steel microstructures can be produced by a wide variety of processing techniques, previous studies[2,9,11] have shown that systematic variations in a number of processing variables result in similar systematic variations in early-stage strain-hardening behavior. For example, Fig. 23 and 27, considered earlier, showed that multistage strain-hardening behavior of dual-phase steels is controlled by cooling rate after intercritical annealing. Decreases in cooling rate produce decreases in average strain-hardening rate at low strains and corresponding increases in ductility. Variations in multistage strain-hardening behavior similar to those produced by decreases in cooling rate are also produced by increases in intercritical annealing temperature, increases in ferrite grain size, decreases in MVF, and low-temperature quench aging. Figure 34 is a schematic diagram of Jaoul-Crussard plots typical of the systematic changes in strain hardening that are produced by varying the processing and microstructural parameters as itemized above. Generally, the transition in strain-hardening behavior from curve A to curve D is associated with the development of reduced MVF and, therefore, reduced fractions of ferrite with a high density of unpinned dislocations. For example, in specimens processed to develop large MVF, most of the matrix ferrite will be saturated with a high density of dislocations produced by the martensitic transformation, and these specimens will exhibit behavior corresponding to curve A in Fig. 34. As the MVF decreases, less of the ferrite is impacted by the martensitic transformation, more of the ferrite is essentially free of dislocations, and lower strain-hardening rates, as typified by curves B, C and D in Fig. 34, develop. Examples of the heterogeneity of the dislocation substructure and the effects of strain hardening on dislocation substructure in dual-phase steels will be presented later.

Examples of the effects of grain-size changes and of manganese content on strain hardening are shown in Fig. 35 and 36, respectively (in both cases, all other heat treatment factors were held constant). Increasing grain size effectively separates the constant volume fraction of martensite, thus creating a more hetero-

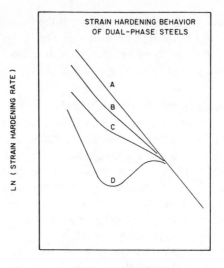

Fig. 34. Schematic diagram of transition in strain-hardening behaviors of dual-phase steels produced by variation of several processing and microstructural parameters.[25]

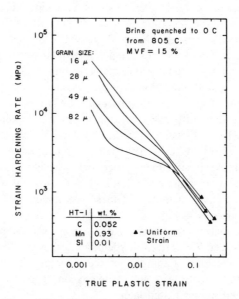

Fig. 35. Transition in strain-hardening behavior produced in 0.05C-Mn-Si dual-phase steel by variations in ferrite grain size.[25]

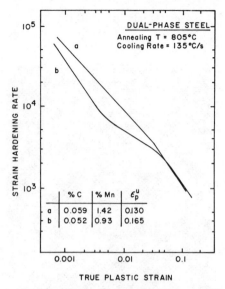

Fig. 36. Transition in strain-hardening behavior due to differences in Mn content in two dual-phase steels treated identically as shown and with approximately the same ferrite grain size.[42]

geneous as-processed dislocation density in the ferrite. Decreasing manganese content decreases hardenability and causes less martensite to be formed at a constant cooling rate. The lower martensite content again results in a lower volume fraction of ferrite with high dislocation density.

The importance of the distribution of dislocations in the ferrite to the strain hardening of dual-phase steels is amplified by the fact that most deformation of dual-phase steels is restricted to the ferrite phase. This aspect of dual-phase deformation behavior is shown in Fig. 37. In this specimen (wire drawn to a strain of 1.39), considerable flow of the ferrite is apparent whereas martensite islands have remained largely undeformed. Although the martensite does not deform significantly, the constraints it places on the ferrite deformation[27,66–68] produce significantly higher flow stresses and work-hardening rates than are attainable during deformation of ferrite-pearlite microstructures (as was pointed out relative to Fig. 25).

Figures 38, 39 and 40 show the development of the dislocation substructure with increasing strain in an Fe-Mn-Si-C dual-phase steel heat treated to produce well-defined early stages of strain hardening. Initially (Fig. 38) the dislocation density is high only around the martensite patches. With increasing strain the dislocation density increases, and does so more rapidly at the martensite-ferrite interfaces than within areas well removed from the martensite. Figure 39 shows

30 min at 760 C

Ċ ≈ 1000 C/s

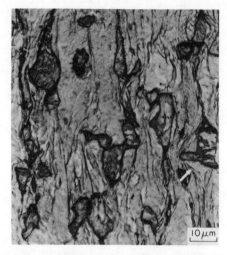

Fig. 37. AISI 1008 dual-phase steel drawn to a total strain of 1.39.[66] Light micrograph. Nital etch.

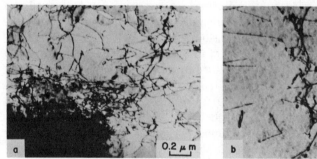

Fig. 38. Dislocation substructures of as-annealed ferrite in 0.06C-Mn-Si dual-phase steel intercritically annealed at 810°C and cooled at 60°C/s.[30] (a) Adjacent to martensite. (b) Away from martensite. TEM micrographs.

areas, in a specimen strained 2%, where concentric semicircular dislocation lines apparently mark dislocation sources at the martensite interfaces. Eventually, a well-developed dislocation cell structure is present throughout the matrix. Figure 41 shows examples of the dislocation cells and the gradients in cell size

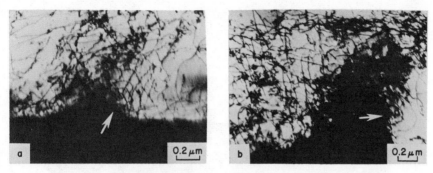

Fig. 39. Dislocation substructures produced at martensite-ferrite interfaces in dual-phase microstructure shown in Fig. 38 after 2% strain.[30] Arrows point to possible dislocation sources. TEM micrographs.

Fig. 40. Dislocation substructures in dual-phase steel of Fig. 38 after 2% strain.[30] (a) Adjacent to martensite. (b) Removed from martensite. TEM micrographs.

as a function of distance from the martensite. Generally, the finer cells are closer to the martensite because of the enhanced dislocation multiplication and interaction at these locations, but sometimes, as shown in Fig. 41, the reverse is true or the cells may be only poorly developed even at large plastic strains.[30]

Summary

This paper has reviewed the microstructures, the mechanical properties and the strain-hardening behavior which are characteristic of dual-phase steels. Although a wide variety of structures and deformation behavior fits into the dual-phase category, the best combinations of ultimate strength and uniform elongation result when structures with low yield strengths and moderate rates of strain hardening

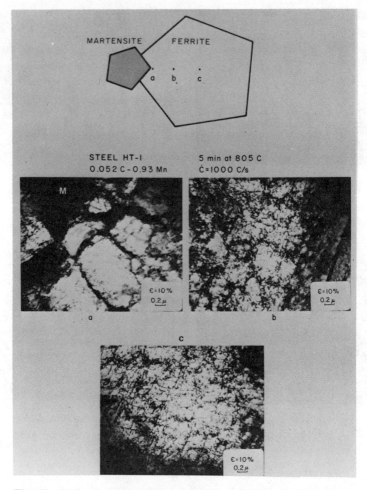

Fig. 41. Dislocation cell structures formed in 0.05C-Mn dual-phase steel strained 10% at locations shown in schematic diagram.[42] TEM micrographs.

in the early stages of deformation are built into a steel. The desirable yielding and strain-hardening behavior result from dislocation gradients introduced by martensite into the matrix ferrite — high densities of dislocations immediately adjacent to martensite-ferrite interfaces and low densities in the ferrite removed from the martensite. On straining, the development of a well-defined dislocation cell structure in the ferrite within the constraints of the nondeforming martensite accounts for the multiple stages of strain hardening, high ultimate strengths and good ductilities of dual-phase steels. Systematic variations of processing and microstructural factors (particularly grain size and MVF) which control multi-stage strain hardening have led to identification of conditions which successfully produce characteristic dual-phase steel behavior. As a result, several processing approaches to the manufacture of dual-phase steels are available. The most effective application of these processing approaches will depend on the development of a deeper understanding of the yielding and strain-hardening behavior of the complex ferrite matrix in dual-phase steels. That matrix may consist of retained and epitaxial ferrite, perhaps of different compositions and susceptibilities to aging, and significant amounts of retained austenite. Each of these components may contribute differently to strain-hardening behavior within the constraints of the nondeforming martensite. Additional work on aging and tempering of dual-phase steels and the interactions between recrystallization and austenite formation during intercritical annealing of cold rolled steels — subjects that are not treated here — also are required for deeper understanding and successful commercial application of dual-phase steels.

Acknowledgments

The authors thank the American Iron and Steel Institute and the Metallurgy Program, Division of Materials Research, National Science Foundation, for their support of the dual-phase research program in the Department of Metallurgical Engineering at the Colorado School of Mines.

References

1. W. C. Leslie, *The Physical Metallurgy of Steels*, McGraw-Hill, New York, 1981, p 142-164.
2. D. J. Blickwede, "Sheet Steel – Micrometallurgy by the Millions," *Trans. ASM*, Vol 61, 1968, p 653-679.
3. G. Krauss, *Principles of Heat Treatment of Steel*, ASM, Metals Park, OH, 1980, p 127-160.
4. *Microalloying '75*, Proceedings of an International Symposium on High-Strength Low-Alloy Steels, Union Carbide Corp., New York, 1977.
5. M. S. Rashid, "GM 980X – Potential Applications and Review," SAE Preprint 770211, 1977.
6. S. Hayami and T. Furukawa, "A Family of High Strength, Cold-Rolled Steels," in *Microalloying '75*, Union Carbide Corp., New York, 1977, p 311-321.
7. R. A. Grange, "Fibrous Microstructures Developed in Steel by Thermomechanical Processing," in *Proceedings of Second International Conference on the Strength of Metals and Alloys*, Vol II, ASM, 1970, p 861-876.
8. P. R. Mould, "Methods for Producing High-Strength Cold-Rolled Steel Sheet," *Met. Eng. Quart.*, Vol 15, No. 3, 1975, p 22-31.
9. *Formable HSLA and Dual-Phase Steels*, edited by A. T. Davenport, TMS-AIME, Warrendale, PA, 1977.
10. *Structure and Properties of Dual-Phase Steels*, edited by R. A. Kot and J. M. Morris, TMS-AIME, Warrendale, PA, 1979.

11. *Fundamentals of Dual-Phase Steels,* edited by R. A. Kot and B. L. Bramfitt, TMS-AIME, Warrendale, PA, 1981.

12. M. S. Rashid and B. V. N. Rao, "Tempering Characteristics of a Vanadium Containing Dual-Phase Steel," in *Fundamentals of Dual-Phase Steels,* edited by R. A. Kot and B. L. Bramfitt, TMS-AIME, Warrendale, PA, 1981, p 249-264.

13. R. G. Davies and C. L. Magee, "Physical Metallurgy of Automotive High Strength Steels," *J. Metals,* Vol 31, No. 11, 1979, p 17-23.

14. A. P. Coldern, G. Tither, A. Cornford and J. R. Hiam, "Development and Mill Trial of As-Rolled Dual-Phase Steel," in *Formable HSLA and Dual-Phase Steels,* edited by A. T. Davenport, TMS-AIME, Warrendale, PA, 1977, p 205-228.

15. J. Morrow and G. Tither, "Molybdenum in Intercritically Annealed Dual-Phase Strips," *J. Metals,* Vol 30, No. 3, 1978, p 16-19.

16. G. R. Speich, "Physical Metallurgy of Dual-Phase Steels," in *Fundamentals of Dual-Phase Steels,* edited by R. A. Kot and B. L. Bramfitt, TMS-AIME, Warrendale, PA, 1981, p 3-45.

17. G. R. Speich, V. A. Demarest and R. L. Miller, "Formation of Austenite During Intercritical Annealing of Dual-Phase Steels", *Met. Trans. A,* Vol 12A, 1981, p 1419-1428.

18. P. A. Wycliffe, G. R. Purdy and J. D. Emburgy, "Austenite Growth in the Intercritical Annealing of Ternary and Quaternary Dual-Phase Steels," in *Fundamentals of Dual-Phase Steels,* edited by R. A. Kot and B. L. Bramfitt, TMS-AIME, Warrendale, PA, 1981, p 59-83.

19. J. M. Rigsbee, "Inhibition of Martensitic Transformation of Small Austenite Particles in Low-Alloy Steels," in *Proceedings of International Conference on Martensitic Transformations,* Cambridge, MA, 1979, p 381-385.

20. A. D. Romig, Jr. and R. Salzbrenner, "Elemental Partitioning as a Function of Heat Treatment in an Fe-Si-V-C Dual-Phase Steel," *Scripta Met.,* Vol 16, 1982, p 33-38.

21. D. A. Korzekwa, D. K. Matlock and G. Krauss, "Aging Susceptibility of Retained and Epitaxial Ferrite in Dual-Phase Steels," *Met. Trans. A,* Vol 13A, 1982, p 2061-2064.

22. G. Krauss and A. R. Marder, "The Morphology of Martensite in Iron Alloys," *Met. Trans.,* Vol 2, 1971, p 2343-2358.

23. R. D. Lawson, D. K. Matlock and G. Krauss, "An Etching Technique for Microalloyed Dual-Phase Steels," *Metallography,* Vol 13, 1980, p 71-87.

24. G. S. Huppi, "Dual-Phase Microalloyed Steels: Temperature and Cooling Rate Effects," M. S. Thesis No. T-2124, Colorado School of Mines, Golden, CO, 1979.

25. D. K. Matlock, G. Krauss, L. F. Ramos and G. S. Huppi, "A Correlation of Processing Variables with Deformation Behavior of Dual-Phase Steels," in *Structure and Properties of Dual-Phase Steels,* edited by R. A. Kot and J. W. Morris, TMS-AIME, Warrendale, PA, 1979, p 91-117.

26. M. D. Geib, D. K. Matlock and G. Krauss, "The Effect of Intercritical Annealing Temperature on the Structure of Niobium Microalloyed Dual-Phase Steel," *Met. Trans. A,* Vol 11A, 1980, p 1683-1689.

27. R. D. Lawson, D. K. Matlock and G. Krauss, "The Effect of Microstructure on the Deformation Behavior and Mechanical Properties of a Dual-Phase Steel," in *Fundamentals of Dual-Phase Steels,* edited by R. A. Kot and B. L. Bramfitt, TMS-AIME, Warrendale, PA, 1981, p 347-381.

28. H. Mathy, J. Gonzov and T. Gréday, "A Study of the Early Stages of Plastic Deformation of Dual-Phase Steels, Including Microplasticity," in *Fundamentals of*

Dual-Phase Steels, edited by R. A. Kot and B. L. Bramfitt, TMS-AIME, Warrendale, PA, 1981, p 413-426.

29. J. S. Gan, J. Y. Koo, A. Nakagawa and G. Thomas, "Microstructure and Properties of Dual-Phase Steels Containing Fine Precipitates," in *Fundamentals of Dual-Phase Steels,* edited by R. A. Kot and B. L. Bramfitt, TMS-AIME, Warrendale, PA, 1981, p 47-58.

30. David A. Korzekwa, "Deformation Substructure Development in a C-Mn-Si Dual-Phase Steel," M.S. Thesis No. T-2517, Colorado School of Mines, Golden, CO, 1981.

31. A. R. Marder, "Factors Affecting the Ductility of Dual-Phase Alloys," in *Formable HSLA and Dual-Phase Steels,* edited by A. T. Davenport, TMS-AIME, Warrendale, PA, 1979, p 87-98.

32. W. R. Cribb and J. M. Rigsbee, "Work-Hardening Behavior and Its Relationship to the Microstructure and Mechanical Properties of Dual-Phase Steels", in *Structure and Properties of Dual-Phase Steels,* edited by R. A. Kot and J. W. Morris, TMS-AIME, Warrendale, PA, 1979, p 91-117.

33. R. G. Davies, "Influence of Martensite Composition and Content on the Properties of Dual-Phase Steels," *Met. Trans. A,* Vol 9A, 1978, p 671-679.

34. G. T. Eldis, "The Influence of Microstructure and Testing Procedure on the Measured Mechanical Properties of Heat Treated Dual-Phase Steels," in *Structure and Properties of Dual-Phase Steels,* edited by R. A. Kot and J. W. Morris, TMS-AIME, Warrendale, PA, 1979, p 202-220.

35. P. R. Mould and C. C. Skena, "Structure and Properties of Cold-Rolled Ferrite-Martensite (Dual-Phase) Steel Sheets," in *Formable HSLA and Dual-Phase Steels,* edited by A. T. Davenport, TMS-AIME, Warrendale, PA, 1977, p 181-204.

36. A. R. Marder, "Deformation Characteristics of Dual-Phase Steels," *Met. Trans. A,* Vol 13A, 1982, p 85-92.

37. G. R. Speich and R. L. Miller, "Mechanical Properties of Ferrite-Martensite Steels," in *Structure and Properties of Dual-Phase Steels,* edited by R. A. Kot and J. W. Morris, TMS-AIME, Warrendale, PA, 1979, p 145-182.

38. J. M. Rigsbee and P. J. VanderArend, "Laboratory Studies of Microstructures and Structure-Property Relationships in Dual-Phase HSLA Steels," in *Formable HSLA and Dual-Phase Steels,* edited by A. T. Davenport, TMS-AIME, Warrendale, PA, 1977, p 56-86.

39. A. R. Marder, "The Structure-Property Relationships in Chromium-Bearing Dual-Phase Steels," in *Fundamentals of Dual-Phase Steels,* edited by R. A. Kot and B. L. Bramfitt, TMS-AIME, Warrendale, PA, 1981, p 145-160.

40. T. Tanaka, M. Nishida, K. Hashiguchi and T. Kato, "Formation and Properties of Ferrite Plus Martensite Dual-Phase Structures," in *Structure and Properties of Dual-Phase Steels,* edited by R. A. Kot and J. W. Morris, TMS-AIME, Warrendale, PA, 1979, p 221-241.

41. S. S. Hansen and R. R. Pradhan, "Structure/Property Relationships and Continuous Yielding Behavior in Dual-Phase Steels," in *Fundamentals of Dual-Phase Steels,* edited by R. A. Kot and B. L. Bramfitt, TMS-AIME, Warrendale, PA, 1981, p 113-144.

42. L. F. V. Ramos, "A Study of Strengthening Mechanisms of the Fe-C-Mn Dual-Phase Steels," M.S. Thesis No. T-2189, Colorado School of Mines, Golden, CO, 1979.

43. R. G. Davies, "The Mechanical Properties of Zero-Carbon Ferrite-Plus-Martensite Structures," *Met. Trans. A,* Vol 9A, 1978, p 451-455.

44. R. G. Davies, "On the Ductility of Dual-Phase Steels," in *Formable HSLA and Dual-Phase Steels*, edited by A. T. Davenport, TMS-AIME, Warrendale, PA, 1979, p 25-39.

45. H. Hu, "Effect of Silicon on Annealing Texture, Plastic Anisotropy, and Mechanical Properties of Low-Carbon Phosphorus-Containing Steels," in *Formable HSLA and Dual-Phase Steels*, edited by A. T. Davenport, TMS-AIME, Warrendale, PA, 1979, p 109-125.

46. G. S. Huppi, D. K. Matlock and G. Krauss, "An Evaluation of the Importance of Epitaxial Ferrite in Dual-Phase Steel Microstructures", *Scripta Met.*, Vol 14, 1980, p 1239-1243.

47. T. Furukawa, H. Morikawa, H. Takechi and K. Koyama, "Process Factors for Highly Ductile Dual-Phase Sheet Steels," in *Structure and Properties of Dual-Phase Steels*, edited by R. A. Kot and J. W. Morris, TMS-AIME, Warrendale, PA, 1979, p 281-303.

48. G. B. Olson, "Transformation Plasticity and the Stability of Plastic Flow," Chap. 9, this volume.

49. A. W. Thompson, M. I. Baskes and W. F. Flangan, "The Dependence of Polycrystal Work Hardening on Grain Size," *Acta Met.*, Vol 21, 1973, p 1017-1028.

50. R. E. Reed-Hill, W. R. Cribb and S. N. Monteiro, "Concerning the Analysis of Tensile Stress-Strain Data Using Log $d\sigma/d\varepsilon$ versus Log σ Diagrams," *Met. Trans. A*, Vol 4A, 1973, p 2665-2667.

51. H. J. Kleemola and M. A. Nieminen, "On the Strain-Hardening Parameters of Metals," *Met. Trans. A*, Vol 5A, 1974, p 1863-1866.

52. F. B. Pickering, "The Effect of Composition and Microstructure on Ductility and Toughness," in *Toward Improved Ductility and Toughness*, Climax Molybdenum Development Co. (Japan), 1971, p 9-31.

53. J. H. Holloman, "Tensile Deformation", *Trans. TMS-AIME*, Vol 62, 1945, p 268.

54. M. S. Rashid, "Relationship Between Steel Microstructure and Formability," in *Formable HSLA and Dual-Phase Steels*, edited by A. T. Davenport, TMS-AIME, Warrendale, PA, 1977, p 1-24.

55. P. Ludwik, *Element der Technologischen Mechanic*, Julius Springer, Berlin, 1909, p 268.

56. S. N. Monteiro and R. E. Reed-Hill, "On the Double-n Behavior of Iron," *Met. Trans.*, Vol 2, 1971, p 2947-2948.

57. B. Jaoul, "Etude de la Forme des Courbes de Deformation Plastique", *J. Mech. Phys. Solids*, Vol 5, 1957, p 95-114.

58. C. Crussard, *Rev. Met. Paris*, Vol 10, 1953, p 697-710.

59. E. Voce, "The Relationship Between Stress and Strain for Homogeneous Deformation," *J. Inst. Metals*, Vol 74, 1948, p 537.

60. H. W. Swift, *J. Mech. Phys. Solids*, Vol 1, 1952, p 1-18.

61. C. L. Magee and R. G. Davies, "Automotive Sheet Steels for the 1980's," in *Alloys for the Eighties*, edited by R. Q. Barr, Climax Molybdenum Company, Greenwich, CT, 1981, p 25-35.

62. L. F. Ramos, D. K. Matlock and G. Krauss, "On the Deformation Behavior of Dual-Phase Steel," *Met. Trans. A*, Vol 10A, 1979, p 259-261.

63. R. D. Lawson, "Effects of Microstructure on the Deformation of Dual-Phase Steels," M.S. Thesis No. T-2489, Colorado School of Mines, Golden, CO, 1981.

64. E. Anderson, D. Law, W. King and J. Spreadborough, "The Relationship Between Lower Yield Stress and Grain Size in Armco Iron," *Trans. AIME*, Vol 242, 1968, p 115-119.

65. P. R. Mould, "An Overview of Continuous-Annealing Technology for Steel Sheet Products", *J. Metals*, Vol 3, No. 5, 1982, p 18-28.
66. D. A. Korzekwa, R. D. Lawson, D. K. Matlock and G. Krauss, "A Consideration of Models Describing the Strength and Ductility of Dual-Phase Steels", *Scripta Met.*, Vol 14, 1980, p 1023-1028.
67. M. F. Ashby, *Strengthening Methods in Crystals*, edited by A. Kelly and R. B. Nicholson, Elsevier Publishing Co., 1971, p 137.
68. Y. Tomota and K. Kuroki, "Tensile Deformation of Two-Ductile-Phase Alloys: Flow Curves of α-γ Fe-Cr-Ni Alloys," *Mater. Sci. Eng.*, Vol 24, 1976, p 85-94.

Solution Hardening and Strain Hardening at Elevated Temperatures

U. F. KOCKS

Center for Materials Science
Los Alamos National Laboratory
Los Alamos, NM 87545

Abstract

Solutes can significantly increase the rate of strain hardening; and, as a consequence, the saturation stress, at which strain hardening tends to cease for a given temperature and strain rate, is increased more than the yield stress. This is the major effect of solutes on strength at elevated temperatures, especially in the regime where dynamic strain-aging occurs. It is shown that local solute mobility can affect both the rate of dynamic recovery and the dislocation/dislocation interaction strength. The latter effect leads to multiplicative solution strengthening. It is explained by a new model based on repeated dislocation unlocking, in a high-temperature limit, which also rationalizes the stress dependence of static and dynamic strain-aging, and may help explain the yield-stress "plateau" at elevated temperatures.

1. Introduction

Neither solution hardening nor strain hardening occupies a prominent place in most discussions of high-temperature strength. This article is an attempt to show that these phenomena are, in fact, important — particularly in the many instances in which they reinforce each other. In the regime of dynamic strain-aging, both the importance of solutes and their effects on strain hardening have been realized for a long time; but we shall demonstrate that this synergism holds true much more generally.

When "strength" at elevated temperatures is discussed, the concern is not so much with yield stress as with flow stress at large strains (for applications to forming processes) or with "creep strength" (for structural applications). Both relate to a regime of the stress-strain curve in which the rate of strain hardening is low, approaching steady-state flow. However, the stress level at which this steady-state limit is reached is itself a property of *strain hardening:* it is controlled by a dynamic balance between hardening and (dynamic) recovery mechanisms (Sec. 2). Thus, any metallurgical change that affects strain hardening also affects elevated-temperature strength.

The principal effect of *solution hardening* that is usually considered is an additive "friction stress"[1-3] (Sec. 3). Such a stress adds to the steady-state stress just as it adds to the yield stress. Inasmuch as this friction stress decreases rapidly with temperature, however, it should not affect high-temperature strength very much: even though the yield stress usually reaches a "plateau" at elevated temperatures, it is small compared with the stress increment due to strain hardening.

When solutes are mobile, they appear not only to decrease the mobility of mobile dislocations (thus increasing the friction stress) but also to slow down the rate of rearrangement of stored dislocations (thus decreasing dynamic recovery and increasing strain hardening)[4-6] (Sec. 4). The latter effect can, in addition, be caused by a decrease in stacking-fault energy due to the solute addition[1] (Sec. 5). For both reasons, the flow stress at large strains may be significantly increased by solution hardening, even when the yield stress is not.

Finally, there are a number of cases in which the effect of solutes on strain hardening appears to be *multiplicative:* the strain-hardening rate and the flow stress, throughout the stress-strain curve, are multiplied by the same, concentration-dependent factor (Sec. 6). In mechanistic terms, this observation may be expressed as an increase in the effectiveness of forest hardening by the solutes.[7]

A unifying mechanism is proposed that relates multiplicative solution hardening to recent observations in static and dynamic strain-aging.[8,9] It is based on a dislocation unlocking model, which has been successful in explaining many aspects of low-temperature solution hardening.[10,11] The new feature is the high-temperature limit. It arises because the length of the activation bulge, which increases with temperature, cannot exceed the forest dislocation spacing: from that point on, the solutes effectively increase the dislocation/forest interaction strength (Sec. 7).

In summary, strain hardening can be severely influenced by solute elements in two distinct ways: one relates to a change in the evolution of the substructure, the other to the effectiveness of a given structure. Both lead to significant effects on the near-steady-state strength at elevated temperatures. Procedures are proposed for separating the different effects experimentally (Sec. 8).

In this chapter, we shall not discuss effects at either very small or very large strains: yield drops and Lüders front propagation were recently dealt with by the author,[12] and effects of solutes on tensile strength after wire drawing or rolling have also been discussed recently in the literature.[2,13,14] In the mechanistic discussions, we will ignore potential effects of small precipitates or ordered regions in nominal "solid solution;" these have also been amply discussed.[5,2]

2. Strain Hardening and High-Temperature Strength

Figure 1 shows a set of stress-strain curves, as a function of temperature, that is typical at least for pure fcc materials, both as polycrystals and as single crystals in multiple-slip orientations.[16-18] A noteworthy feature of these curves is that they coincide for small strains. This observation, together with more detailed considerations, led Mecking and Kocks[18] to identify an initial "athermal hardening" component Θ_h of the strain-hardening rate $\Theta = d\sigma/d\varepsilon$ (much as a "stage II" of hardening had been identified previously in single crystals[19]). The remaining component is then defined as the rate of dynamic recovery, Θ_r:

$$\Theta = \Theta_h - \Theta_r(\sigma, T, \dot{\varepsilon}) \tag{1}$$

As is indicated in Eq 1, all the dependence on stress, temperature and strain rate is associated with the dynamic-recovery term. (A slight stress dependence of Θ_h is not ruled out, but is assumed to be negligible with respect to that of Θ_r.) In fact, the athermal hardening rate is insensitive even to material; in tension it is on the order of $1/50$ of Young's modulus.

The strain-hardening rate decreases continuously; in fact, it decreases rapidly enough that, by one extrapolation method or another, a limit may be

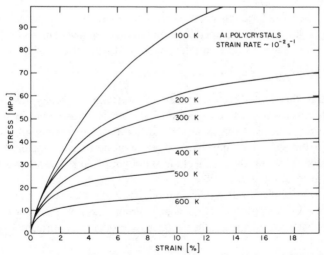

Fig. 1. Nominal tensile stress-strain curves for aluminum (99.99%, grain size d $\simeq$ 0.2 mm). *After Kocks et al.*[15]

defined where it would vanish.[20,17,21] We shall not be concerned here with whether or not this steady-state limit is actually reached:[22] we will use it merely as a semiquantitative indication of flow stress when strain hardening is low. This "saturation stress" and its dependence on temperature and strain rate then follow from a solution of Eq 1 for $\Theta = 0$.

Figure 2 illustrates the temperature dependence of the saturation stress for aluminum polycrystals, as determined from a (short) extrapolation of the curves in Fig. 1.[17] The saturation stress decreases roughly exponentially with temperature (the plot of its logarithm being roughly linear in temperature) and does so much more strongly than does the yield stress.* Despite this rapid decrease, the absolute value of the saturation stress is still significant at two-thirds of the melting temperature (the last point in Fig. 2): it is still about six times the yield stress.

The point that will be made for solution-hardened alloys is that the line for steady-state stress versus temperature is more significantly influenced by solute additions than is the yield-stress line; thus, the "strength," especially as it relates to elevated-temperature applications, can be more efficiently increased through an effect on strain hardening than through an effect on friction stress.

3.   Additive Solution Hardening

The flow stress of solid solutions is usually described by an expression of the form[1-3]

$$\sigma = \sigma_f + \sigma_d \tag{2}$$

Here, σ_f indicates a "friction stress" due to the interaction of solutes with mobile dislocations; it is strongly temperature- and rate-dependent. The second term, σ_d, is due to dislocation/dislocation interactions and generally is of the form

$$\sigma_d = M\alpha\mu b\sqrt{\rho} \tag{3}$$

where M is the Taylor factor converting crystallographically resolved shear stresses (glide resistances) into the relevant stress components in the macroscopic coordinates, μ is the shear modulus, b is the magnitude of the Burgers vector, ρ is the dislocation density, and α is a proportionality factor slightly less than 1 which is also meant to reflect the slight temperature and strain-rate dependence of dislocation cutting.[18]

Equation 2 neglects other contributions to flow stress that may, in certain applications, be significant (e.g., due to grain size). We shall assume here, however, that none of these other processes produces significant *strain hard-*

---

*The yield stresses were evaluated from the same data that were partially reported in Ref 15.

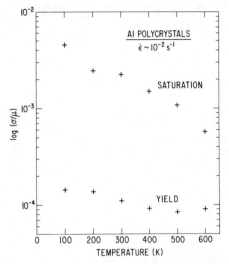

Fig. 2. Yield stress and true saturation stress, divided by shear modulus, as functions of temperature. Data from Fig. 1.

ening. (For this reason, we must specifically exclude second-phase particles.) Then, we can write (from Eq 2 and 3)*

$$\Theta \equiv \left. \frac{\partial \sigma}{\partial \varepsilon} \right|_{T,\dot{\varepsilon}} = M^2 \alpha \mu b \frac{d\sqrt{\rho}}{d\gamma} \tag{4}$$

The important physical process that controls strain hardening is the rate of dislocation accumulation with the glide increment $d\gamma = M \, d\varepsilon$; its two components, statistical storage and dynamic recovery, give rise to the two components of Θ (Eq 1).

The essence of Eq 2 and 4 is that the classical solute friction stress does not contribute to strain hardening: the stress-strain curves should merely be shifted upward. Figures 3 to 5 show the best cases I could find that approximately obey this rule; all relate to deformation at low temperature (77 K) and at rather small strains.[23–25]

In these cases, if the behavior could be properly extrapolated to large strains, the steady-state stress would be augmented by exactly the same amount as the yield stress. At elevated temperatures, this would be proportionately very little. In the following, we will discuss cases in which solutes have been observed to influence strain hardening as well as yield stress. This is possible, according to

---

*Any influence of a strain dependence of the mobile dislocation density (or of the vacancy concentration) on σ_f in Eq 2 has been neglected.

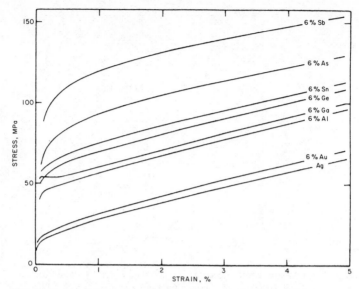

Fig. 3. True tensile stress-strain curves for silver alloys (d ≃ 50 μm), tested at 77 K. *After Hutchison and Honeycombe.*[23]

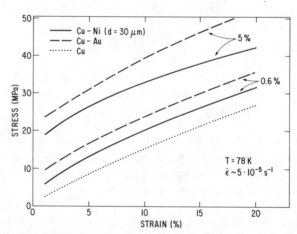

Fig. 4. Tensile stress-strain curves for copper-nickel and copper-gold alloys of two concentrations. *After den Otter and van den Beukel.*[24]

Eq 4, if solutes can affect either the rate of dislocation accumulation $d\sqrt{\rho}/d\gamma$ or the proportionality constant α. Both effects will be shown to exist. (In addition, of course, there are the trivial effects of solutes on the shear modulus and the Burgers vector.)

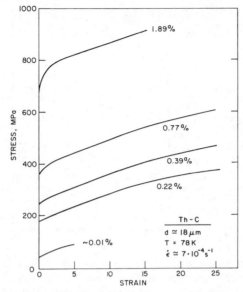

Fig. 5. True tensile stress-strain curves for thorium-carbon alloys. *After Peterson and Skaggs.*[25]

4. Dynamic Strain-Aging and Dynamic Recovery

Figure 6 shows the case of an especially strong influence of solute concentration on stress strain behavior in Ni-C.[9] Note, however, that the curves diverge much more strongly at large strains than at small. In first approximation one can say that the predominant effect of solutes is here to decrease Θ_r.

These stress strain curves were taken in the regime where jerky flow (the Portevin-LeChatelier effect) is observed.* It appears, then, that dynamic strain-aging affects not only the mobility of the mobile dislocations (and thus the character of flow), but also the ease of rearrangement of the previously stored dislocations (and thus dynamic recovery).[4,5] Further evidence for this interpretation is that the rate sensitivity *of strain hardening* is negative in the dynamic-recovery regime, not only the rate sensitivity of the flow stress (Fig. 6).

It has, in fact, long been known that the "humps" in diagrams of flow stress versus temperature in the dynamic strain-aging regime are more pronounced for ultimate tensile strength than for yield strength, e.g., in Fe-N[26] and type 316 stainless steel.[27] Figures 7 and 8 show this feature for Hadfield steel[28] and for Inconel 600.[29] (In the latter, the extrapolated saturation stress is plotted rather than the UTS.)

---

*This phenomenon was not seen (at temperatures up to 550 K) in the other fcc interstitial alloy, Th-C (Fig. 5).[25]

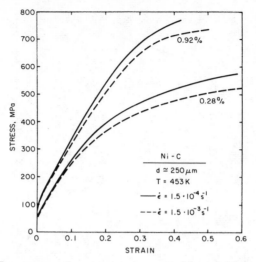

Fig. 6. True compressive stress-strain curves for two nickel-carbon alloys. The higher strain rate (dashed) produces less strain hardening, indicating an influence of dynamic strain-aging on dynamic recovery. New results.[9]

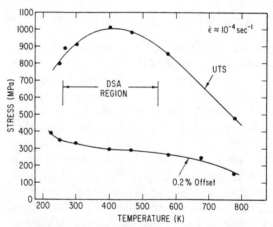

Fig. 7. Yield stress and tensile strength as functions of temperature for Hadfield steel, showing the pronounced influence of dynamic strain-aging on the UTS. *After Dastur and Leslie.*[28]

Figure 9 shows more particularly how strain hardening in this alloy depends on temperature.[29] In the dynamic strain-aging regime, there is a long plateau of constant strain-hardening rate (at a level of about half of the athermal hardening rate, Θ_h, typical of pure materials). It is almost independent of temperature

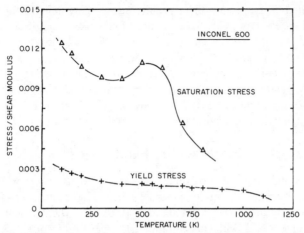

Fig. 8. Yield stress and saturation stress, divided by shear modulus, as functions of temperature, for Inconel 600. *From Mulford and Kocks.*[29]

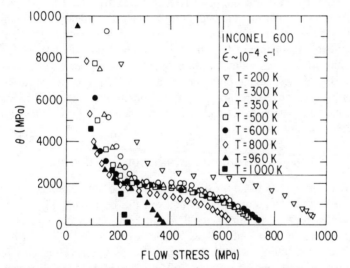

Fig. 9. Strain-hardening rate, $\Theta = d\sigma/d\varepsilon$, as a function of flow stress for Inconel 600. Note the temperature-insensitive plateau between 300 and 600 K. *From Mulford and Kocks.*[29]

in this range. Note, however, that eventually some dynamic recovery process does come in: the hardening rate decreases rather rapidly to zero (much before necking begins).

An interesting observation can be made regarding Fig. 9 that has some generality: the phenomenon of an extended linear hardening stage occurs even at

200 K, which is far below the dynamic strain-aging regime for this alloy. Similarly, Fig. 10 shows how the influence of phosphorus on strain hardening in iron is pronounced (and similar) both in the region of jerky flow and at lower temperatures.[30] We are led to conclude either that phenomena akin to dynamic strain-aging occur far beyond the temperature range where jerky flow is observed[29,11] and exert their influence on dynamic recovery over this wider range, or that an influence of solutes on dynamic recovery is more general than by way of dynamic strain aging (or, most probably, both).

5.   Stacking-Fault Energy

A similar inhibiting effect of solute elements on dynamic recovery has been found in single crystals deforming in single slip, and has been interpreted as being due to a lowering of the stacking-fault energy (SFE).[1] The classical example of a strong change in SFE with concentration is Ni-Co. Results on single crystals of this alloy are shown in Fig. 11,[31] where the effect of solution hardening on easy glide (stage I work hardening) should be ignored for our purposes. It is evident that the stage II work-hardening rate is not significantly affected by the additions of cobalt to nickel, but that the beginning of stage III is delayed — the more so the higher the solute concentration.

It is possible that some of the observations reported in the last section, concerning an influence of solute additions on large-strain behavior, are also due to a lowering of the stacking-fault energy, although this does not seem likely, at least for Ni-C. Conversely, some contribution of strain-aging phenomena to solution hardening in Ni-Co cannot be ruled out and is, in fact, probable in view of results on strain-rate sensitivity.[32]

6.   Multiplicative Solution Hardening

There are some sets of stress-strain curves for solution-hardened alloys that appear to diverge monotonically, right from the beginning. Figure 12 shows such a series for Al-Mg alloys at 78 K; the behavior is qualitatively similar at temperatures up to about 500 K.[4] Alloys of Cu-Zn,[24,33,13] Cu-Al[34] and Al-Mg-Mn[35,36] exhibit similar behavior, as do Fe-C alloys, at least in some observations,[37] and single crystals of Nb-W[38] and Nb-Mo.[39] In some cases, dynamic strain-aging is observed in the same regime, but in many it is not; at least one may say that the phenomenon occurs over a much wider range of temperature than jerky flow (as in Al-Mg, Fig. 12).

Figure 13 shows new results on Ni-Mo alloys.[40] It was found that a replot with a concentration-dependent scale factor on the stress axis brings all of these curves into coincidence, so that these data (as well as similar curves taken at T = 352 K) can be described empirically by the relation

$$\sigma = [1 + k(c)] \cdot \sigma_d \qquad (5)$$

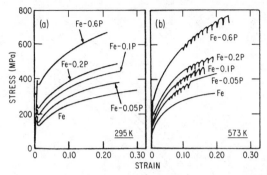

Fig. 10. True tensile stress-strain curves for iron-phosphorus alloys tested (a) below and (b) within the jerky-flow regime. The approximate phosphorus content is given in wt %. *After Spitzig.*[30]

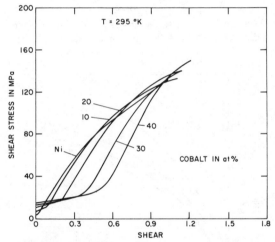

Fig. 11. Resolved shear-stress-vs-shear curves for nickel-cobalt single crystals of a central orientation. Note the continual decrease of dynamic recovery with increasing concentration. *After Meissner.*[31]

This behavior is distinctly different from a divergence of the curves in the dynamic-recovery regime only. On the other hand, Fig. 14 demonstrates, for the 3% alloy, that the dependence of strain hardening on *temperature* still follows the scheme outlined in Sec. 2: coincidence at small strains, and divergence in the dynamic-recovery regime.

The difference between the effects discussed in the previous two sections and those discussed in this section can be characterized with reference to Eq 3. Although in the previous sections we postulated a solute effect on a part ($\propto \Theta_r$) of the net rate of dislocation storage $d\sqrt{\rho}/d\gamma$, it would seem that, in the present

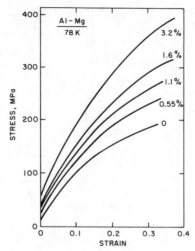

Fig. 12. True tensile stress-strain curves for aluminum-magnesium alloys ($d \simeq 0.3$ mm), tested at 78 K at a strain rate of 2×10^{-3} s^{-1}.

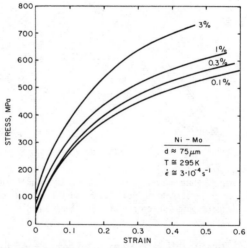

Fig. 13. True compressive stress-strain curves for nickel-molybdenum alloys. Curves diverge monotonically. New results.[40]

alloys, one of the other factors that are constant throughout the strain range must be responsible. The dependence of the shear modulus and the lattice constant on concentration are much too weak for the magnitude of k observed. The only remaining possibility is a concentration dependence of α, which reflects the strength of dislocation/forest interactions. Such an effect has been proposed by Luton and Jonas[41] and by Schmidt and Miller.[7]

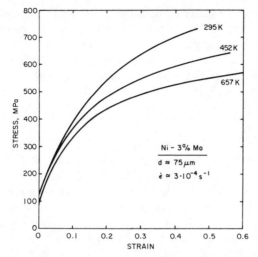

Fig. 14. True compressive stress-strain curves for one Ni-Mo alloy at three temperatures. Curves coincide at low strains, but diverge in the dynamic-recovery regime. New results.[40]

7. A Unifying Mechanism

The empirical equation (Eq 5) has some similarity to relations observed recently in static and dynamic strain-aging experiments. We shall first summarize these results and then relate them to effects outside the dynamic-strain-aging regime and, in fact, outside the regime of solute bulk-mobility. A simple mechanism will then be qualitatively outlined that may underlie all these phenomena.

7.1 Strain-Aging

The term "strain-aging" relates to the appearance, upon aging at (or above) the current deformation temperature, of a yield point the magnitude of which depends on the strain level achieved before aging. It has recently been postulated, and in some cases experimentally verified over a significant regime, that the yield drop $\Delta\sigma_a$ is linearly related to the *stress* reached in the previous straining,[8,9] i.e., is essentially proportional to σ_d:

$$\Delta\sigma_a = k'(c) \cdot \sigma_d \qquad (6)$$

Equation 6 has the same form as the concentration-dependent stress increment in Eq 5.

7.2 Dynamic Strain-Aging

Dynamic strain-aging is currently considered to be due to the same process that brings about static strain-aging, occurring during the normal waiting time of dislocations in their generally jerky progress through the slip plane.[42,29] Because this waiting time is inversely proportional to the imposed strain rate, the aging process makes a *negative* contribution to the strain-rate sensitivity of the flow stress: lower strain rates allow more aging and therefore more hardening.

It was found by Mulford[29,32] that this negative contribution to the strain-rate sensitivity is again proportional to σ_d, at least in some Ni alloys and Al-Mg at strains that are not too high:

$$\Delta \frac{\partial \sigma}{\partial \ln \dot{\epsilon}} \bigg|_T = -k''(c) \cdot \sigma_d \qquad (7)$$

This is in agreement with the observation, on the same alloys,[8,9] of static aging according to Eq 6. The critical strain for the beginning of jerky flow is then interpreted as that strain at which the *total* strain-rate sensitivity becomes negative.[12,42] Note, however, that a negative contribution to strain-rate sensitivity, which is evidence of *dynamic strain-aging*, may not be sufficient to offset other, positive contributions; then, *jerky flow* does not occur. The Ni-Mo alloys quoted above are a case in point, and so is Ni-Co[32] — and, in both cases, the initial negative contribution is proportional to σ_d.

We conclude that the proportionality to the flow stress, which is observed for the stress increments in strain hardening (Eq 5), static strain-aging (Eq 6) and dynamic strain-aging (Eq 7), may be a common feature associated with solute mobility, whether or not this causes jerky flow. It can be observed at temperatures lower than those at which jerky flow occurs.

7.3 Low-Temperature Yield Strength

There have long been difficulties with explanations of the yield strength based on the concept of solute atoms acting as discrete obstacles.[43,44] Principally, the observed rate and temperature dependences are much too low for this mechanism to hold. In terms of dislocation activation, the activation length inferred from measurements of the apparent activation area is large compared with the solute spacing along the dislocation[41,45] and, most importantly, is not proportional to it.[11,41] These problems are pronounced in the regime of the "plateau" in the diagram of yield stress versus temperature (which often occurs at quite moderate temperatures), and have been shown to be substantial at lower temperatures also.

On the other hand, a description of solution hardening in terms of the formulas proposed by Suzuki[10] is much more successful. The essence of the underlying mechanism (whether or not the details of this particular theory apply) is that the solutes behave as if they were continuously, not discretely, distributed along the dislocation. In combination with an assumption of some ability of the solutes to redistribute themselves within the dislocation core (perpendicular to the dislocation[47], this leads to a model in which the dislocation "digs its own trough" while waiting at "hard lines" and must continuously free itself from each new trough.[11]

Thermal activation from such a trough occurs by the formation of a bulge in the dislocation line (Fig. 15), whose length l_b is inversely proportional to the stress[46] (or, more exactly, to the stress in excess of σ_d). This length is generally much larger than the solute spacing, and gets larger as the temperature is raised. This can give an almost plateaulike effect: the unlocking stress becomes inversely proportional to temperature at high temperatures — unless another effect intervenes.

7.4 High-Temperature Limit of the Unlocking Stress

If the activation length in unlocking gets longer and longer as the temperature is raised, there eventually comes a point when the length of the bulge would begin to exceed the forest-dislocation spacing l_d. This cannot happen, and it is at this point that the unlocking stress becomes proportional to σ_d: the strength of the forest junction is augmented by the force to form the bulge. Instead of Eq 2 and 3, we then have

$$\sigma = [\alpha(T, \dot{\epsilon}) + \beta(c, T, \dot{\epsilon})] \cdot M\mu b\sqrt{\rho} = \sigma_d \cdot (1 + \beta/\alpha) \qquad (8)$$

Note that the term proportional to β has *taken the place of* σ_f: no further additive friction stress is needed.

The quantity β in Eq 8 is the angle at which the activation bulge meets the dislocation (Fig. 15). It can be derived when the profile of the line-tension trough is known, and depends on temperature and strain rate.[46]

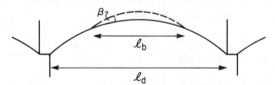

Fig. 15. Locked dislocation segment with activated bulge (dashed). Its curvature is determined by the stress, and the angle β is determined by the line-tension difference.

Equation 8 could form the basis of a unified description of all the phenomena discussed above: it is formally similar, and the mechanism of repeated locking and unlocking seems especially attractive in the elevated-temperature regime. Much quantitative work needs to be done to ascertain the viability of this proposal.

8. Conclusions

We have discussed three distinct effects of solute additions on flow stress, and their relations to the flow-stress contribution from strain hardening, σ_d: an additive friction stress (σ_f); a multiplicative "repeated-unlocking" stress (interpreted as an additive effect on the dislocation/dislocation interaction strength, α); and a decrease in the rate of dynamic recovery (Θ_r). We find that the first effect, which is the most discussed, may be strictly absent in the high-temperature limit, but is certainly negligible at all but the smallest strains. On the other hand, the second effect, a multiplicative solution strengthening (which is the least discussed), probably accounts for the most general contribution over the widest range of elevated temperatures. In the dynamic strain-aging regime, the effect on dynamic recovery becomes dominant.

Although these effects were discussed here for cases v'here one or another of them predominates, they will, in most cases, act in concert. For example, a re-examination of the figures presented in the section on dynamic strain-aging and dynamic recovery will show some evidence of the proportional effect also. However, when this is taken into account, a significant additional effect on dynamic recovery remains.

In general, the flow stress of solution-hardened alloys may then be written, instead of as Eq 2, as

$$\sigma = \sigma_f(c,T,\dot{\varepsilon}) + \mu(T,c) \cdot \alpha_c(c,T,\dot{\varepsilon}) \cdot \int \frac{\Theta}{\mu\alpha_c}\, d\varepsilon \tag{9}$$

where α_c is used for $[\alpha + \beta]$ in Eq 8. It is expected that either the friction stress σ_f or the c-dependence of α_c is negligible in any particular case.

The integrand in Eq 9 represents the (net) rate of dislocation storage (Eq 4) and may be expressed, by way of generalizing Eq 1, as

$$\frac{\Theta}{\mu\alpha_c} = \frac{\Theta_h}{\mu\alpha_c} - \frac{\Theta_r}{\mu\alpha_c}\left(\frac{\sigma - \sigma_f}{\mu\alpha_c}, T, \dot{\varepsilon}, c\right) \tag{10}$$

We cannot envisage any possible influence of solutes on the first term on the right-hand side of this equation, which represents statistical dislocation storage.* The effect on the rate of dynamic recovery may be due in part to a decrease in the stacking-fault energy, but is most significantly influenced by dynamic strain-aging. Thus, it is particularly important in an intermediate, elevated-temperature regime.

The flow stress in the steady-state limit is obtained by setting the right-hand side of Eq 10 equal to zero. The result is of the form

$$\sigma_s = \sigma_f(c,T,\dot{\varepsilon}) + \mu(T,c) \cdot \alpha_c(c,T,\dot{\varepsilon}) \cdot F(T,\dot{\varepsilon},c) \tag{11}$$

where, again, σ_f is expected to be negligible at high temperatures. Equation 11 looks very much like Eq 9, except that the strain-dependent integral has been

---

*Note that Schmidt and Miller[7] combine what we would call "dynamic recovery" with hardening (which they assume to be parabolic in strain) and separate out "recovery" (occurring simultaneously with straining); they expect solutes to influence both of their terms.

replaced by a constant F, depending only on T, $\dot{\varepsilon}$ and c. The latter effect is, again, due to the influence of solutes on dynamic recovery and is potentially the most important at elevated temperatures. It is not possible, in principle, to separate the effects on α_c and F when only deformation at (or near) the steady state is studied.

To distinguish flow-stress effects from evolution effects of solute additions, one must study the strain-hardening behavior. First, stress-strain curves are analyzed according to Eq 9 and 10, manipulating α_c in Eq 10 to produce coincidence for different concentrations at small strains and low temperatures; any remaining difference is interpreted as being due to Θ_r. Second, microscopic investigations of structure can be undertaken on alloys of different concentrations: over the strain range in which no effect of Θ_r is detected, the relevant substructural features should be independent of concentration at the same *strain*, even though the stresses themselves may be significantly different. Finally, the thermomechanical history should be varied: effects on α_c depend only on the current values of the external and structural variables, whereas the integral depends on the entire strain path.[7]

The multiplicative flow-stress effect is new, having been discussed only once, in somewhat different form, by Schmidt and Miller.[7] An explanation is offered that relates it to other solute effects currently under discussion.

The model is based on the concept that "moving" dislocations must continually free themselves from continuous line-tension troughs (rather than from discrete solute atoms). These troughs get deeper as solutes diffuse to the momentarily waiting dislocation (but they exist even at very low temperatures[11]). The stacking-fault ribbon may play a role in the establishment of an effective "trough".[10] The process of unlocking occurs by the formation of a bulge whose length is inversely proportional to the stress. At low stresses, i.e., high temperatures, the bulge length equals the forest-dislocation spacing, which it cannot exceed. From this point on, the solute/dislocation interaction has the effect of merely raising the effective strength of the dislocation/dislocation interaction.[41,7] For this reason, all solute effects are now proportional, as the dislocation density increases, to the strain-hardening contribution to the flow stress, σ_d: the static strain-aging stress peak; the decrease in the strain-rate sensitivity; and the flow-stress increment itself.

The proposed mechanism is as yet of a rather qualitative nature; it suggests, however, a number of experiments for quantitative determinations of the various correlations.

Acknowledgments

It is a pleasure to acknowledge the contribution made by my co-workers T. A. Bloom and R. E. Cook to the unpublished work on Ni-Mo and Ni-C alloys, as well as their valuable comments on the manuscript. This work was supported by the U.S. Department of Energy.

References

1. P. Haasen, in *Fundamental Aspects of Structural Alloy Design,* edited by R. I. Jaffee and B. A. Wilcox, Plenum, 1977, p 3.
2. W. C. Leslie, *Met. Trans.,* Vol 3, 1972, p 5.
3. U. F. Kocks, in *Strength of Metals and Alloys,* edited by P. Haasen, V. Gerold and G. Kostorz, Pergamon, 1980, p 1661.
4. O. D. Sherby, R. A. Anderson and J. E. Dorn, *J. Metals,* Vol 3, 1951, p 643.
5. J. D. Embury, in *Strengthening Methods in Crystals,* edited by A. Kelly and R. B. Nicholson, Elsevier, 1971, p 331.
6. W. G. Truckner and D. E. Mikkola, *Met. Trans.,* Vol 8A, 1977, p 45.
7. C. G. Schmidt and A. K. Miller, *Acta Met.,* Vol 30, 1982, p 615.
8. P. Wycliffe, U. F. Kocks and J. D. Embury, *Scripta Met.,* Vol 14, 1980, p 1349.
9. U. F. Kocks and R. E. Cook, to be published.
10. H. Suzuki, in *Dislocations and Mechanical Properties of Crystals,* edited by J. C. Fisher et al., Wiley, 1957, p 361; and in *Strength of Metals and Alloys,* edited by P. Haasen, V. Gerold and G. Kostorz, Pergamon, 1979, p 1595.
11. U. F. Kocks, to be published.
12. U. F. Kocks, *Prog. Mater. Sci., Chalmers Anniv. Vol.,* 1981, p 185.
13. J. Gil-Sevillano, P. van Houtte and E. Aernoudt, *Prog. Mater. Sci.,* Vol 25, p 90.
14. S. S. Hecker, M. G. Stout and D. T. Eash, in *Proc. Plasticity of Metals at Finite Strain,* edited by E. Lee, Stanford University, 1981.
15. U. F. Kocks, H. S. Chen, D. A. Rigney and R. J. Schaefer, in *Work Hardening,* edited by J. P. Hirth and J. Weertman, AIME, 1968, p. 151.
16. H. Mecking, in *Work Hardening in Tension and Fatigue,* edited by A. W. Thompson, AIME, 1977, p 67.
17. U. F. Kocks, *J. Eng. Mater. Tech.* (ASME series H), Vol 98, 1976, p 76.
18. H. Mecking and U. F. Kocks, *Acta Met.,* Vol 29, 1981, p 1865.
19. J. Diehl, *Z. Metallk.,* Vol 47, 1956, p 331.
20. E. Voce, *J. Inst. Metals,* Vol 74, 1948, p 537.
21. B. Nicklas and H. Mecking, in *Strength of Metals and Alloys,* edited by P. Haasen, V. Gerold and G. Kostorz, Pergamon, 1979, p 351.
22. H. Mecking and A. Grinberg, in *Strength of Metals and Alloys,* edited by P. Haasen, V. Gerold and G. Kostorz, Pergamon, 1979, p 289.
23. M. M. Hutchison and R. W. K. Honeycombe, *Metal Sci.,* Vol 1, 1967, p 70.
24. G. J. den Otter and A. van den Beukel, *phys. stat. sol. (a),* Vol 55, 1979, p 785.
25. D. T. Peterson and R. L. Skaggs, *Trans. AIME,* Vol 242, 1968, p 922.
26. J. D. Baird and C. R. MacKenzie, *J. Iron Steel Inst.,* Vol 202, 1964, p 427.
27. C. F. Jenkins and G. V. Smith, *Trans. AIME,* Vol 245, 1969, p 2149.
28. Y. N. Dastur and W. C. Leslie, *Met. Trans.,* Vol 12A, 1981, p 749.
29. R. A. Mulford and U. F. Kocks, *Acta Met.,* Vol 27, 1979, p 1125.
30. W. A. Spitzig, *Mater. Sci. Eng.,* Vol 16, 1974, p 169.
31. J. Meissner, *Z. Metallk.,* Vol 50, 1959, p 207.
32. R. A. Mulford, *Acta Met.,* Vol 27, 1979, p 1115.
33. A. Korbel, L. Błaz, H. Dybiec, J. Gryziecki and J. Zasadziński, *Metals Technology,* 1979, p 391.
34. L. Gastberger, O. Vöhringer and E. Macherauch, *Z. Metallk.,* Vol 65, 1974, p 17.
35. J. G. Morris, *Mater. Sci. Eng.,* Vol 13, 1974, p 101.
36. H. Herø, Proc. 10th Biennial Congr. Intern. Deep Drawing Res. Group, Warwick, 1979, p 179.

37. D. J. Quesnel, A. Sato and M. Meshii, *Mater. Sci. Eng.*, Vol 18, 1975, p 199.
38. G. Kostorz, *Z. Metallk.*, Vol 59, 1968, p 941.
39. P. Jax, *Z. Metallk.*, Vol 62, 1971, p 284.
40. T. A. Bloom, U. F. Kocks and P. Nash, to be published.
41. M. J. Luton and J. J. Jonas, *Canadian Met. Quart.*, Vol 11, 1972, p 79.
42. A. van den Beukel and U. F. Kocks, *Acta Met.*, Vol 30, 1982, p 1027.
43. R. L. Fleischer, in *The Strengthening of Metals*, Reinhold, 1964, p 93.
44. R. Labusch, *Acta Met.*, Vol 20, 1972, p 917.
45. Z. S. Basinski, R. A. Foxall and R. Pascual, *Scripta Met.*, Vol 6, 1972, p 807.
46. U. F. Kocks, A. S. Argon and M. F. Ashby, *Prog. Mater. Sci.*, Vol 19, 1975, p 1.
47. R. B. Schwarz and L. L. Funk, *Acta Met.*

Dynamic Changes That Occur During Hot Working and Their Significance Regarding Microstructural Development and Hot Workability

WILLIAM ROBERTS

Swedish Institute for Metals Research
114 28 Stockholm, Sweden

Abstract

An overview of the current state of knowledge with regard to dynamic microstructural changes during deformation under conditions of hot working is attempted. Attention is focused on the phenomena of dynamic recrystallization and recovery, plus crack development in association with high-temperature, high-rate deformation. The interrelationships between these processes under hot working conditions are also considered.

The report is polarized toward the importance of dynamic structural transitions in the context of commercial hot work processing. To this end, the survey is presented in three sections:

1. The influence of dynamic recovery and recrystallization on the flow-stress level and on the shape of the flow curve under hot working conditions.
2. The importance of dynamic structural changes in determining the over-all microstructural modification engendered by a hot working operation.
3. The interplay between cracking micromechanisms and dynamic recrystallization in the determination of hot ductility.

LIST OF SYMBOLS

A, A', A'' : constants in $\dot{\varepsilon}(\sigma)$

B, B', B'' : constants in $Z(\sigma)$

C, D : constants in $\sigma(\log_{10}Z)$

C_{MG} : Monkman-Grant coefficient

D* : critical nucleus diameter (m)

D_0 : initial grain size (m)

D_{rex} : recrystallized grain size (m)

F : area fraction cracks

F_I : F at nucleation

F_f : F at fracture

ΔG : activation energy $(J \cdot mol^{-1})$

G : growth velocity $(m \cdot s^{-1})$

H : constant

I_s : specific grain-boundary nucleation rate $(m^{-2}s^{-1})$

L_I : initial wedge-crack length (m)

M : dislocation mobility during recovery $(m^3 \cdot J^{-1}s^{-1})$

$\dot{N}$: frequency of recovery events (s^{-1})

N_A : number of cracks per unit area (m^{-2})

P : constant with dimensions of stress (MPa)

Q : activation energy in Z $(J \cdot mol^{-1})$

Q_{SD} : activation energy for self diffusion $(J \cdot mol^{-1})$

R : gas constant $(J \cdot K^{-1}mol^{-1})$

S : grain-boundary energy, or surface energy $(J \cdot m^{-2})$

S_v : grain-boundary area per unit volume (m^{-1})

T : temperature (K)

X : fraction recrystallized

Y : factor

Z : Zener-Hollomon parameter (s^{-1})

b : Burgers vector (m)

d : thickness of grain-boundary slab (m)

d_{sub} : subgrain diameter (m)

ℓ : dimension in wedge-crack model, or dislocation-link length (m)

ℓ_R : length of dislocation annihilated per recovery event (m)

$k_1 - k_n$: constants

m : grain-boundary mobility $(m^4 \cdot J^{-1}s^{-1})$

n : power-law creep exponent

p : exponent in $\sigma_s(d_{sub})$

q : proportionality constant for strain and stress during recovery, or exponent in $\sigma_s(D_{rex})$

r : recovery rate (MPa)

r_v, r_0 : radius of void, initial void radius (m)

r* : radius of critical nucleus (m)

t : time (s)

t_0 : incubation time for recrystallization (s)

t_I : time at crack initiation (s)

t_f : time to fracture (s)

x : distance coordinate (m)

α : constant in $\dot{\varepsilon}(\sigma)$ and $Z(\sigma)$

α' : dislocation-strengthening parameter

β : constant in $\dot{\varepsilon}(\sigma)$ and $Z(\sigma)$

γ_D : migrating fraction of grain-boundary area

δ : constant in $\psi(X)$

ε : strain

ε_c : critical strain for dynamic recrystallization

ε_f : strain to fracture

ε_g : local strain in grain-boundary regions

ε_h : contribution to ε from work hardening

$\varepsilon(max)$: strain corresponding to stress peak

ε_{sf} : contribution to ε from softening

ε_x : strain required to effect one cycle of dynamic recrystallization

$\dot{\varepsilon}$: strain rate (s^{-1})

$\dot{\varepsilon}_0$: constant in $\dot{\varepsilon}(\sigma)$ (s^{-1})

$\dot{\varepsilon}_g$: strain rate in grain-boundary regions (s^{-1})

$\dot{\varepsilon}_i$: strain rate within grains (s^{-1})

σ : flow stress (MPa)

σ_0 : effective stress, or constant in $\dot{\varepsilon}(\sigma)$ (MPa)

σ_c : critical stress for dynamic recrystallization (MPa)

$\sigma(max)$: maximum stress (MPa)

$\bar{\sigma}$: mean flow stress over a given strain interval (MPa)

σ_h : contribution to σ from hardening (MPa)

σ_{sf} : contribution to σ from softening (MPa)

σ_s : steady-state stress (MPa)

θ : work-hardening rate (MPa)

θ_{II} : athermal, asymptotic work-hardening rate (MPa)

κ : integration variable in Eq 40

ρ : dislocation density (m^{-2})

ρ_0 : dislocation density in unrecrystallized material (m^{-2})

ρ_{0c} : critical dislocation density for dynamic recrystallization (m^{-2})

ρ_i : density of intrasubgrain dislocations (m^{-2})

ρ_{sub} : density of dislocations in subgrain walls (m^{-2})

μ : shear modulus (MPa)

ϕ : dislocation-density distribution function for ρ_i (m^2)

Φ : dislocation-density distribution function for ρ_{sub} (m^2)

ζ : constant related to stacking-fault energy (dimensions of μb^3)

η, ξ : constants in Avrami equation (ξ in s^{-1})

ν_0 : frequency factor (s^{-1})

τ : dislocation line energy $(J \cdot m^{-1})$

ψ : reduction in area in tensile test

ψ_0, ψ_{max} : constants in equation relating ψ to X

ψ_T, ψ_L : reduction in area from specimens with tensile axis perpendicular and parallel to rolling direction

λ_2 : secondary-dendrite-arm spacing

1. Preamble

The awareness that processing via hot working can serve to tailor the service properties of metals, in addition to altering their shape, has promoted intense research interest in this area over the past decade. In fundamental terms, there is a close correspondence between hot deformation at a fixed strain rate and creep under the action of a fixed stress. For both situations, the normal increase of dislocation density, i.e., work hardening, is counteracted by a softening process, i.e., dynamic recovery and/or dynamic recrystallization. Many of the basic characteristics of these restoration phenomena have been established via creep experimentation. For example, in a series of early papers by Gifkins,[1-5] dealing with dynamic recrystallization during creep of polycrystalline lead, the existence of a critical strain for dynamic recrystallization and its dependence on microstructure, purity and stress level were clearly established.

The phenomena of dynamic recovery and recrystallization have important repercussions for the shape of $\sigma(\varepsilon)$ curves under hot working conditions. If recovery proceeds relatively rapidly, the flow stress increases progressively up to a steady-state value (σ_s), which is determined by a balance between the accumulation (due to work hardening) and elimination (via recovery) of dislocations; there is no dynamic recrystallization. Examples of materials exhibiting this behavior are aluminum and its dilute alloys, and ferritic steels (e.g., those containing silicon or chromium as major additions). For fcc metals and alloys characterized by intermediate or low stacking-fault energy (copper, nickel, austenitic steels), recovery proceeds more slowly and the dislocation density can attain a sufficiently high value for dynamic recrystallization to be initiated. In this case, the $\sigma(\varepsilon)$ curve exhibits a characteristic stress maximum followed eventually by a steady state, such that $\sigma_s < \sigma(max)$; at low strain rates and high temperatures, the curves might go through several subsidiary maxima prior to the attainment of a steady state. The transition between the two basic types of $\sigma(\varepsilon)$

behavior can be reconciled with a critical level of dislocation density (stored energy); if the steady state associated with dynamic recovery corresponds to a defect density below this critical value, then no dynamic recrystallization can be expected. A corollary of this argument is that, for a given strain rate and homologous temperature, the normalized flow-stress levels for metals which soften only through dynamic recovery are lower than those associated with dynamic recrystallization. Schematic $\sigma(\varepsilon)$ curves for the separate instances of restoration exclusively by dynamic recovery and restoration by dynamic recovery combined with dynamic recrystallization are presented in Fig. 1.

In general, one can, as a first approximation, apply the same basic concepts of static recovery and recrystallization to the descriptions of the corresponding dynamic processes. However, considerable caution must be exercised in such an endeavor because the effect of concurrent deformation can vastly modify the characteristics of the dynamic phenomena. For example, dynamic re-

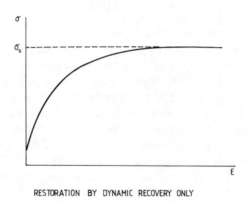

RESTORATION BY DYNAMIC RECOVERY ONLY

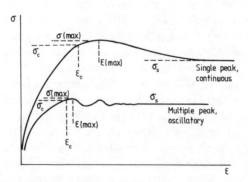

Fig. 1. Shape of $\sigma(\varepsilon)$ curves associated with high-temperature deformation at constant strain rate.

crystallization proceeds in some cases via the continuous nucleation and limited growth of the new grains, whereas in classical (static) recrystallization, grains nucleated early in the process grow continuously until impingement limits further growth. The modified behavior in the dynamic case derives from the reduction in driving force for growth as a result of concurrent deformation, which engenders dislocation accumulation in the developing grains. In a similar vein, it is not feasible to treat dynamic recovery as a purely thermally activated process triggered only by temperature, as is the case with static recovery. In the former, the unit softening events can be activated by the applied stress and/or other moving dislocations (strain-rate effect) with the assistance of thermal activation.

Notwithstanding the fundamental differences of the types quoted above, the phenomena of static and dynamic recrystallization or static and dynamic recovery are characterized by a number of unified features, not the least important being the similar microstructures resulting from the static and dynamic processes.

This report is intended to present an overview of the current standpoint regarding dynamic microstructural changes during hot deformation. The review will be somewhat polarized toward the importance of such structural transitions for commercial hot deformation processing. While some of the more fundamental aspects of dynamic recrystallization and recovery are dealt with where appropriate, the reader requiring fuller insight is referred to the excellent reviews on these subjects which are already available.[6-13] From the point of view of commercial processing, it is convenient to divide consideration of dynamic structural changes during hot deformation into three separate sections:

1. Their significance for the shape of the flow curve under hot working conditions.
2. Their influence in determining the microstructure engendered via hot deformation processing.
3. Their effect on the propensity for cracking during hot working operations, i.e., their effect on hot workability.

An attempt will be made to cover all three of the above aspects, albeit in a cursory manner, in the review which follows.

2. Flow Curves Under Hot Working Conditions

2.1. Practical Significance of Flow-Stress Values

Hot working operations are carried out at temperatures exceeding half the melting point of the metal or alloy in question and at equivalent strain rates ranging from ~ 0.1 s^{-1}, for some forging and extrusion processes, to $>10^3$s^{-1}, for rolling of wire and rod in a high-speed plant. The force (torque, power) required to effect a given step in a hot working procedure depends on geometric restraints imposed by the working tools, on the frictional conditions between the tools and stock, and on the flow curve as it would be measured in, say, a uniaxial test (tension, compression) at an appropriate temperature and strain rate. For a particular hot working operation, it is of importance to be able to evaluate the force/power requirements under the following circumstances:

- In dimensioning of new equipment and plants.
- In estimation of the elastic deflection of the working equipment, knowledge of which is essential in order to attain correct final dimensions for the worked material.
- During development of suitable on-line control software for automated hot working plants.
- In checking that the capacity of the available equipment will not be exceeded in effecting any given step in the operation.

On this basis, metal producers and manufacturers of hot working plants alike are well served by reliable high-temperature flow-stress data. Quite often, the range of materials and processing situations, for which values are required, is appreciable; for example, a manufacturer of specialty steels may process a large number of grades, on one and the same piece of equipment, to a variety of dimensions and shapes. It is thus of considerable interest to derive expressions, either physically based or phenomenological, for the dependence of the flow curve on $T, \dot{\varepsilon}$ in order that as many conceivable situations as possible can be described via a small number of parameters. A further simplification is feasible if different materials can be classified into groups or composition ranges which exhibit rather similar behavior for $\sigma(\varepsilon, \dot{\varepsilon}, T)$; a prerequisite for such a rationalization is detailed information regarding the influences of alloying additions and microstructure on the flow curve.

The above discourse highlights the need for a fundamental understanding of stress-strain behavior at high temperatures and strain rates. The remainder of this chapter attempts to clarify the current state of knowledge in this area.

2.2. Phenomenological Description of Flow Stress at High Temperatures

Independent of whether restoration proceeds via dynamic recovery alone or by a combination of dynamic recovery and recrystallization, the strain-rate and temperature dependence of the steady-state stress during hot deformation can be described by the application of the corresponding relationships for creep[14] at low and high stresses, respectively, i.e.:

$$\dot{\varepsilon} = A \, \sigma_s^n \qquad (\text{low } \sigma_s) \tag{1}$$

and

$$\dot{\varepsilon} = A' \exp (\beta\sigma_s) \qquad (\text{high } \sigma_s) \tag{2}$$

As was first suggested by Garofalo,[14] these expressions can be combined in the unified equation

$$\dot{\epsilon} = A''[\sinh(\alpha\sigma_s)]^n \tag{3}$$

which reduces to Eq 1 and 2 at low ($\alpha\sigma < 0.8$) and high ($\alpha\sigma > 1.2$) stresses, respectively, and where $A = A'' \cdot \alpha^n$, $A' = A'' \cdot 2^n$ and $\beta = \alpha n$. It is generally found that Eq 1 to 3 offer a satisfactory description of $\sigma(\dot{\epsilon})$ as well as $\sigma_s(\dot{\epsilon})$ during hot deformation, i.e., they can account for the strain-rate dependence of the flow stress at any strain and even for that of the maximum stress, in cases where dynamic recrystallization occurs.

In order to correlate $\sigma(\dot{\epsilon})$ over a range of temperature, one makes use of the so-called temperature-compensated strain rate, as originally proposed by Zener and Hollomon:[15]

$$\sigma = f(\dot{\epsilon} \exp(Q/RT) = f(Z)$$

where Q is an activation energy defined by

$$Q = -R\left[\frac{\partial \ln \dot{\epsilon}}{\partial (1/T)}\right]_\sigma$$

Hence, the temperature dependence can be included in Eq 1 to 3 if $\dot{\epsilon}$ is replaced by Z. Although Eq 3 offers a complete description over a wide stress range, it is for practical purposes often more convenient to try to apply one of the modified forms of Eq 1 and 2, because these relationships contain one less parameter, i.e.:

$$Z = B(\sigma^n) \tag{4}$$

or

$$Z = B' \exp(\beta\sigma) \tag{5}$$

However, such simplification is not always possible, and then the unified form must be used:

$$Z = B''[\sinh(\alpha\sigma)]^n \tag{6}$$

For some materials, e.g., pure aluminum,[7] it is possible to apply Eq 6 to the description of $\sigma_s(\dot{\epsilon}, T)$ under both creep and hot working conditions; the activation energy in such cases is very close to that for self diffusion (e.g., for Al, $Q_{SD} = 138$ kJ·mol^{-1}). Normally, however, for materials of commercial interest, such as steels, the Q-values derived from hot working data are appreciably (up to 50%) higher than Q_{SD}. In creep theory, $Q > Q_{SD}$ can sometimes be accounted for via the temperature dependence of the elastic modulus and the stacking-fault energy,[16] which enter into the coefficients B, B' and B'' in Eq 4, 5 and 6, respectively. However, for many materials, and especially for creep-resistant steels,[7] Q(hot working) > Q(creep); e.g., for austenitic 18Cr-8Ni steel, the former is 420 kJ·mol^{-1} [Q(hot working) derived from $\sigma_s(\dot{\epsilon}, T)^{[17]}$] compared with 315 kJ·mol^{-1} for the latter.[18] In such cases, a unified correlation of creep and hot working data is clearly not feasible.

Equations 4 to 6 appear to offer a rationalization of the strain-rate and temperature dependence of flow stress, which is adequate for the evaluation of force and power requirements during hot working. However, an important question is whether or not these relationships are valid at the very high strain rates which characterize certain hot working operations (>1000 s^{-1}). Unfortunately, very few flow-stress data have been published for high-rate forming, but those that are available seem to conform to the above description. This is illustrated by Fig. 2, which demonstrates the application of Eq 5 to M2 high-speed steel (0.85C-4Cr-5Mo-6W-2V) at $\varepsilon = 0.3$. Data from five experimental investigations are presented, including that of Samanta,[19] who determined $\sigma(\varepsilon)$ via instrumented impact-hammer tests; the span of $\dot{\varepsilon}$ is from 0.1 to 430 s^{-1}. The conclusions are that, for the range of conditions considered in Fig. 2, σ is sufficiently "high" for Eq 5 to represent an adequate description, and that this simple correlation between σ and Z is valid even for very high strain rates.

An analytical expression for the strain dependence of σ is likely to be complicated, especially when dynamic recrystallization occurs; this problem is considered later in the present chapter. However, evaluation of load/power requirements for a hot working operation will require a complete description of $\sigma(\varepsilon,\dot{\varepsilon},T)$. One method,[20] which avoids the necessity of an analytical representation for $\sigma(\varepsilon)$ and which is often sufficiently precise, is to determine the coefficients defining $\sigma(\dot{\varepsilon},T)$ at a number of distinct strains; evaluation of an unknown flow stress corresponding to an intermediate strain value can then be effected via simple interpolation. This procedure has been utilized in the establishment of a flow-stress data bank

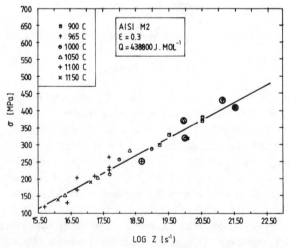

Fig. 2. Illustrating the applicability of the "high-stress" $\sigma(Z)$ formalism to AISI M2 high-speed steel; the points encircled with a heavy ring are from very-high-rate testing ($\dot{\varepsilon} > 100$ s^{-1}).

for high-alloy steels (tool steels, high-speed steels, stainless steels).[20] Two examples of entries in this data bank, i.e., those for austenitic 18Cr-8Ni stainless steel and M2 high-speed steel (both of which undergo restoration via dynamic recovery plus recrystallization), are given in Table 1.

In evaluating the information in Table 1, flow-stress data from a number of sources have been optimized via least-squares fitting. The analysis has been restricted to hot working strain rates and $\sigma > 100$ MPa, and is therefore based on the "high-stress" correlation, Eq 5. $\sigma(\dot{\varepsilon},T)$ at each strain is defined by three parameters — Q, C and D — where C and D are the intercept and slope of a σ-$\log_{10}Z$ plot, i.e., $C = 2.303 \log_{10}B'/\beta$ and $D = 2.303/\beta$. Hence, Table 1 allows evaluation of $\sigma(\varepsilon)$ for any temperature and strain rate combination within the hot working regime. The interval between the strains, for which parameters are defined, is chosen so as to permit an acceptably accurate approximation of the complete flow curve. That such is the case is illustrated by Fig. 3, which presents actual and approximated flow curves for M2 high-speed steel at 1050°C and 5 s^{-1}; for $\varepsilon > 0.05$, the maximum deviation is about 2 MPa, which is considerably less than the scatter in the original data. An acceptable approximation for $\varepsilon < 0.05$ requires information concerning the yield stress at high temperatures and strain rates, which is often not clearly defined on flow curves carried out to large strains (especially in torsion or compression).

Apart from having access to an adequately accurate description of the flow curve, it is often desirable, from the viewpoint of force/power computations, to be able to evaluate the mean flow stress for a given step in a hot working operation, which engenders an effective strain ε. This quantity is defined (T,$\dot{\varepsilon}$ fixed) as

$$\bar{\sigma} = \frac{1}{\varepsilon} \int_0^\varepsilon \sigma \cdot d\varepsilon \qquad (7)$$

Table 1. Description of $\sigma(\varepsilon,\dot{\varepsilon},T)$ for Austenitic 18Cr-8Ni Stainless Steel and M2 High-Speed Steel[20]

	Strain (ε)						
	0.05	0.1	0.2	0.3	0.4	0.6	0.8
AISI 304L: 18Cr-8Ni (C < 0.05); 0.5 < $\dot{\varepsilon}$ < 650 s^{-1}							
Q (J · mol^{-1}) .	311,818	375,174	367,084	398,037	413,296	471,471	460,894
C (MPa)	−85.04	−219.39	−364.67	−489.17	−520.06	−638.11	−652.01
D (MPa)	15.44	22.68	34.03	39.89	40.78	42.46	44.16
AISI M2: 0.85C-4Cr-5Mo-6W-2V; 0.1 < $\dot{\varepsilon}$ < 430 s^{-1}							
Q	266,583	364,791	446,246	428,207	386,410	392,927	268,534
C	−318.26	−461.25	−657.01	−703.66	−699.60	−717.40	−528.61
D	45.13	44.55	48.92	54.04	59.36	58.95	67.03

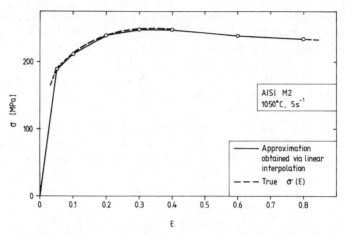

Fig. 3. Stress-strain curve for M2 high-speed steel illustrating the accuracy to which linear interpolation, between the flow stress values specified in Table 2, defines the true $\sigma(\varepsilon)$.

and can be computed approximately as

$$\bar{\sigma} \doteq \frac{1}{2\varepsilon}\Bigg\{\varepsilon_1\sigma_1 + \sum_{n=1}^{n=m-1} (\sigma_n + \sigma_{n+1})(\varepsilon_{n+1} - \varepsilon_n)$$

$$+ [\sigma(\varepsilon) + \sigma_m](\varepsilon - \varepsilon_m)\Bigg\} \tag{8}$$

with σ_1 and ε_1 corresponding to the first column in Table 1 (or a table like it) and where the strain of interest (ε) lies between the mth and (m+1)th column in the table. $\sigma(\varepsilon)$ is evaluated via linear interpolation between σ_m and σ_{m+1}. Some accuracy is sacrificed in Eq 8 because it is assumed that the flow curve starts at the origin; however, this will not be too serious if ε_1 is sufficiently low.

It is interesting to note, from Table 1, how the quantities Q, C and D vary with strain. The activation energy tends to first increase with ε reaching a maximum value at $\varepsilon = 0.6$ to 0.8 for the austenitic steel and at $\varepsilon = 0.2$ to 0.3 for high-speed steel; this would seem to be related to the fact that, for a given Z, the stress peak occurs at considerably lower strains in the latter instance (smaller grain size: see Sec. 2.5). In the case of high-speed steel, Q decreases again and attains a value close to that at small strains when $\varepsilon = 0.8$; for a good portion of the range of T,$\dot{\varepsilon}$ covered by the original data, $\varepsilon = 0.8$ lies comfortably in the steady-state regime for this material. On the basis of our current understanding, it is difficult to rationalize, in physical terms, the strainwise variation of the Zener-Hollomon activation energy. The parameters C and D increase rapidly with ε at small strains and more slowly at larger ones; for materials characterized by low ε(max), e.g., high-speed steel, C can decrease again at high strains. These changes merely reflect the influence of Z on the shape of $\sigma(\varepsilon)$; the principal effect is that the strain to the peak, $\varepsilon(\max)$, increases with Z and so $d\sigma/d(\log_{10}Z)$ ($\equiv D$) increases with ε, at least initially (see Fig. 4).

The rationale for $\sigma(\varepsilon,\dot{\varepsilon},T)$, which has been outlined in this section, is adequate for many requirements in the context of computing the force or power needed to effect a given hot working operation. However, the basic methodology suffers from shortcomings which would need to be obviated if a more accurate correlation were required.

- Since the above description is essentially phenomenological, it will be necessary to perform experimental flow-stress determinations for every new material of interest. Similarly, because $\sigma(\varepsilon)$ is quite sensitive to the initial microstructure, particularly grain size (see Sec. 2.5), one will be forced to determine a parameter list, of the form given in Table 1, for a number of different microstructures in one and the same material. (The possibility that various investigators have studied a material with different initial microstructures has been completely ignored in the evaluation of the data comprising Table 1.) These problems might be avoided if a physically based

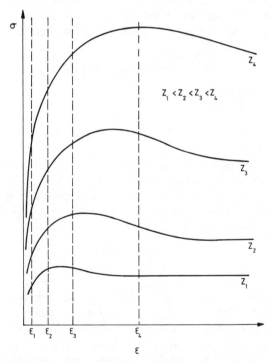

Fig. 4. The effect of increasing Z on the shape of $\sigma(\varepsilon)$ when restoration is by dynamic recovery plus dynamic recrystallization.

description of $\sigma(\varepsilon, \dot{\varepsilon}, T)$ were developed, by means of which the influence of alloy chemistry and microstructure on the flow curve could be evaluated from first principles.

- The approach outlined above is strictly valid only if the hot working operation is performed in a single step, e.g., extrusion. For a multiple-step situation, cognizance must be given to the fact that static restoration processes, following a given deformation step, will modify the form of $\sigma(\varepsilon)$ relevant to the subsequent pass. Furthermore, the temperature and strain rates will normally vary from pass to pass. In the general case, even a phenomenological description is likely to be very complicated.

2.3. Dynamic Recovery

2.3.1. Mechanisms for Dynamic Recovery.
Very little endeavor has been reported pertaining to the operative mechanisms during restoration via dynamic recovery at hot working strain rates. However, it is a reasonable assumption that the basic softening events, characterizing constant-rate testing at low and intermediate temperatures and also creep, are relevant even under the conditions of hot deformation.

Mecking and Lücke[21,22] introduced a formal description of the superposition of hardening and softening during plastic flow, where it is assumed that the individual contributions of these two processes to stress and strain are additive, i.e., that

$$d\varepsilon = d\varepsilon_h + d\varepsilon_{sf} \quad ; \quad d\sigma = d\sigma_h - d\sigma_{sf} \tag{9}$$

In terms of the work-hardening rate, θ,

$$\theta = \theta_{II} - \frac{1}{\dot{\varepsilon}}(\theta_{II}\dot{\varepsilon}_{sf} + \dot{\sigma}_{sf}) \tag{10}$$

where θ_{II} is a constant athermal hardening rate equivalent to that in stage II single-crystal hardening; even at high temperatures, it is presumed that θ approaches θ_{II} asymptotically at small strains. If the softening (dynamic recovery) processes are considered as single dislocation events with frequency $\dot{N}$, then both $\dot{\varepsilon}_{sf}$ and $\dot{\sigma}_{sf}$ are proportional to $\dot{N}$ and Eq 10 is replaced by

$$\theta = \theta_{II} - q\left(\frac{\dot{N}}{\dot{\varepsilon}}\right) \tag{11}$$

where the proportionality constant q is related to the strain increase and stress decrease associated with a single softening event. Equation 11 should be compared with that derivable from the assumptions of the Bailey-Orowan[23,24] theory of creep, in which the unit softening events are considered to be activated by thermal vibrations only, i.e., exactly as for static recovery. On this basis, $\dot{N}$ is constant and Eq 11 can be modified ($\dot{N}$ proportional to r = rate of recovery) to yield

$$\theta = \theta_{II} - \frac{r}{\dot{\varepsilon}} \qquad (12)$$

In the steady-state limit, $\theta = 0$ and $\dot{\varepsilon}_s = r/\theta_{II}$, which is the Bailey-Orowan relation for recovery-controlled creep. By applying Eq 4, we can write for r during creep at "low" stresses

$$r = r_0 \, \sigma^n \exp \, (-Q/RT) \qquad (13)$$

where $r_0 = B\theta_{II}$ (θ_{II} independent of T, $\dot{\varepsilon}$ except via the temperature dependence of the shear modulus). However, since Eq 12 should also apply to the situation away from the steady state, then this relationship together with Eq 13 predicts that the work-hardening rate will exhibit a strong dependence on both stress and strain rate that is much more pronounced than the experimental evidence would indicate.

The physical basis of the Bailey-Orowan approach has been disputed in a number of papers.[25–29] The crux of the problem lies in the assumption that, at a given dislocation structure, the time frequency of recovery events is constant, and that each event only contributes to an elementary reduction in stress but not to a corresponding increase in strain. It would appear that this presumption is questionable. In terms of the formalism given by Mecking and Lücke,[21,22] the proportionality constant q can contain terms due to *both* the elementary stress decrease and the elementary strain increase from a single softening event, i.e.:

$$q = \theta_{II} \cdot d\varepsilon_{el} + d\sigma_{el}$$

Hence, $\dot{N}$ in Eq 11 can depend on the current level of applied stress or strain rate, i.e., the restoration takes place via thermally assisted strain softening rather than the time-dependent, static-type recovery of the Bailey-Orowan approach and, as such, is dynamic recovery in its true sense. The interdependence of strain rate and stress at a given structure makes it difficult to decide, from experimental information, whether the stress level (the driving force) or the strain rate (the motion of dislocations) is the quantity that determines the frequency of softening events. Mecking and Lücke[21,22] considered the $\sigma(\varepsilon)$ behavior in terms of a spectrum of softening centers each with its specific activation stress. The number of active centers then increases with increasing degree of hardening, and the $\sigma(\varepsilon)$ curve flattens continuously. Assistance from thermal vibrations causes the work-hardening rate at a given strain to decrease with increasing temperature and decreasing strain rate. An alternative approach, in which the softening events are considered to be triggered by moving dislocations, has been presented by Kocks;[25] this is given further attention in Sec. 2.4.

In the detailed models for dislocation creep presented by Öström and Lagneborg,[30,31] the dynamic recovery mechanism presumed to operate is one proposed originally by Friedel.[32] He considered the mesh growth of a Frank network (which may exist uniformly throughout the crystal or locally in subgrain walls) via diffusion-controlled climb. The rate of growth of the average link size is

$$\frac{d\bar{l}}{dt} = \frac{M\tau}{\bar{\ell}} \tag{14}$$

where M is a mobility which for diffusion-controlled climb is related to the self-diffusion coefficient. For growth of an individual link of size l, Öström and Lagneborg[30] draw an analogy with grain growth and propose the expression

$$\frac{dl}{dt} = M\tau \left(\frac{1}{\ell_{cr}} - \frac{1}{\ell} \right) \tag{15}$$

where l_{cr} is a critical value above which links increase in size and below which they shrink. In a simple treatment, the dislocation density is functionally related to $\bar{l}$, i.e., $\rho = \bar{\ell}^{-2}$, and the rate of decrease of dislocation density due to recovery is then easily obtained via integration of Eq 14:

$$\left. \frac{d\rho}{dt} \right|_{rec} = -2M\tau\rho^2 \tag{16}$$

Öström and Lagneborg[30,31] have adopted a more advanced formulation based on Eq 15, because in their model they consider a distribution of link sizes which is continuously modified via the accumulation of links due to glide/storage and their disappearance through shrinkage of the smallest meshes and participation in glide (largest links for which $\ell > \alpha' \ \mu b/\sigma$). In its most refined form,[31] the model offers a satisfactory description of the primary and secondary stages of creep in austenitic stainless steels. The steady-state dislocation density and its stress dependence are determined principally by $\ell^* = 2\mu b/\sigma$. In this sense, the simple argument culminating in Eq 16 and the more advanced theory lead to similar results.

Öström and Lagneborg's treatment has been criticized by Kocks and Mecking[28] on a number of points, the principal one being that the Friedel theory, for the growth of meshes in a Frank network, is essentially a model for time-dependent static recovery. However, in a more recent paper,[29] these latter authors have attempted to develop a unified treatment of static (i.e., time-dependent) and dynamic recovery. In this treatment, the problem associated with the simple Bailey-Orowan formalism for recovery creep, i.e., that the wrong dependence of the work-hardening rate on stress and strain rate is predicted (Eq 12 and 13), is avoided because the recovery rate, r, is not controlled by a constant activation energy but rather by one which depends on the local forward internal stress experienced by dislocation segments in tangles. The substance of this model is a distribution of forward internal stresses on dislocation segments. Above a critical forward stress, σ_s, the athermal storage rate (defined by θ_{II}) is balanced by recovery; the rate of recovery under these conditions may then be written as

$$-\frac{df}{dt}\bigg|_{rec} = f(\sigma_s) \, v_0 \, \exp\left\{-\frac{\Delta G(\sigma_s/\sigma_m)}{RT}\right\} \qquad (17)$$

where f is the distribution function for forward stresses and σ_m is the mechanical collapse stress at which dislocations would break free of the tangles even in the absence of thermal activation. On this basis, the net change in dislocation density for *applied* stress levels below σ_s ($\equiv$ steady-state stress) is

$$\frac{d\rho}{d\varepsilon} = h(\sigma, \sigma_s)(\sigma_s - \sigma) \qquad (18)$$

This applies because, above σ_s, $df/d\varepsilon = 0$. The function h depends on the average segment length being stored, $\bar{l}$, and on $df/d\varepsilon$. Kocks and Mecking argue that $h \propto \sigma/\sigma_s$ and so, assuming a proportionality between σ and $\sqrt{\rho}$,

$$\frac{d\sigma}{d\varepsilon} = \text{const.} \, (1 - \sigma/\sigma_s) \qquad (19)$$

which is the formalism for the $\sigma(\varepsilon)$ curve proposed originally by Voce.[33] Equation 19 is characterized by a weak dependence of $d\sigma/d\varepsilon$ on σ and $\dot{\varepsilon}$ (through that of σ_s) and is thus consistent with experimental observations pertaining to the dependence both of the strain-hardening rate prior to σ_s *and* of the steady-state stress on $\dot{\varepsilon}$. However, the authors point out that this model, which involves only time-dependent recovery, will be invalidated if stress- or strain-rate-activated recovery events are rate controlling.

2.3.2. Steady-State Stress. In general, creep and high-temperature deformation at a given rate should, under equivalent conditions, lead to the same value for σ_s; this has been confirmed for pure aluminum by Mecking and Gottstein.[10] However, such a situation is not likely to hold in a comparison of constant-rate tests under hot working conditions and creep, which are often characterized by different activation energies. The temperature and strain-rate dependence of σ_s is defined by Eq 4 to 6. Kocks[25] has proposed that, at low temperatures, n in Eq 4 can be identified with ζ/kT, where ζ is a constant for a given material (fcc) and k is Boltzmann's constant; the former can be derived from a so-called τ_{III} analysis of single-crystal stress-strain curves. This quantity is related to stacking-fault energy in such a way that ζ, and therefore n, decrease with increasing stacking-fault energy. However, as will be made clear in Sec. 2.4, the method of evaluating the σ_s-values, upon which the above correlation is based, is not above criticism.

The $\dot{\varepsilon}$ and T-dependence of σ_s is generally steeper at homologous temperatures greater than 0.6,[13] i.e., n increases. The most plausible explanation for this behavior is a change in the basic softening mechanism from thermally assisted cross-slip at low temperatures to diffusion-controlled dislocation climb at higher ones. Under such circumstances, the stress exponent at high temperatures is predicted theoretically (e.g., see Öström and Lagneborg[30,31]) to be close to 3, whereas for fcc metals and alloys, $4 \leq n \leq 8$ is usually found for both creep and hot working (low stress); the lower the stacking-fault energy, the higher the value of n.[25,27] That the stacking-fault energy has some influence, even at temperatures where diffusive processes are likely to be important, is plausible in view of the fact that constriction is a necessary prerequisite for the climb of extended dislocations. Some investigators (e.g., Ref 34) have interpreted deviations from $n = 3$ in creep as being due to separation of the applied stress into internal and effective components. This controversy is still not completely resolved. Indeed, there are authorities[35] who even question the fundamental basis for diffusional control during power-law creep. Such a reservation would seem to be even more justified in the instance of high-rate hot deformation.

2.3.3. Microstructures Derived From Dynamic Recovery Under Hot Working Conditions. Dynamic recovery is coupled with the generation of dislocation cells, which are roughly equiaxed and which do not change in size much during deformation.[36] This suggests the existence of a dynamic equilibrium involving the continuous depletion (recovery) and storage (hardening) of dislocations in cell walls. The walls of cells formed during low-temperature deformation are rather thick, with a misorientation ranging from a fraction of a degree to a few degrees. However, under hot working conditions, the walls constitute relatively sharp sub-boundaries, while the misorientation remains more or less unchanged. There is considerable evidence for a close dependence of subgrain size on steady-state stress for those materials characterized by restoration via dynamic recovery only,[36-39] e.g., see Fig. 5. The observed relationship is

$$\sigma_s = k_1 + k_2 d_{sub}^{-p} \qquad (20)$$

where values for p usually lie in the range 1 to 1.5 (for a comprehensive review, see Ref 7). A corollary of Eq 20 is that d_{sub} can also be related to the temperature-compensated strain rate (Z) via Eq 4 to 6.

It is an accepted postulate (e.g., Gittus[40] and Kuhlmann-Wilsdorf and Van der Merwe[41]) that dislocations agglomerate into cells or subgrains during deformation because this configuration represents a minimum strain energy for a given dislocation content. An analysis involving minimization of the strain energy of isolated cells leads to the following result:[41]

$$d_{sub} \approx \frac{10\mu b}{\sigma - \sigma_0} \qquad (21)$$

where σ_0 is an effective stress. This is clearly consistent with Eq 20 if $p = 1$, and hence the experimentally based relationship derives some support from theory.

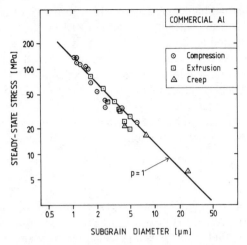

Fig. 5. Relationship between steady-state flow stress and subgrain diameter for commercial aluminum. Data assembled by McQueen and Jonas.[8]

An alternative rationale is that the average size of dislocation links in the cell walls at steady state, which is maintained by a balance between storage (reduction of mesh size) and recovery (enlargement of meshes), is exactly that which can just be activated by the internal stress, $\sigma_s - \sigma_0$, i.e., $\bar{\ell} = $ const. $\mu b/(\sigma_s - \sigma_0)$. According to Kuhlmann-Wilsdorf,[41,42] the subgrain size and mean link size vary in direct proportion to each other, and thus the argument leads to the same form of expression as Eq 21.

Dislocation cells formed at low temperatures during working operations involving large accumulated strains (e.g., wire drawing) are elongated. However, on hot deformation to high strains (as in extrusion), the subgrains remain equiaxed, which is only possible if the subgrain boundaries migrate during deformation. This is, however, unlikely at very high values of $\dot{\varepsilon}$, and the substructure can then only continue to be equiaxed if the individual subgrains are perpetually in a state of annihilation and reformation, such that the spacing of the walls remains constant.[36]

2.4. A Formalism for $\sigma(\varepsilon)$ Under Hot Working Conditions When Restoration Is by Dynamic Recovery Alone

As stated in Sec. 2.3.1, Kocks[25] has presented a treatment for $\sigma(\varepsilon)$ in which recovery events are considered to be controlled by strain rate (moving dislocations). This argument leads to a formula identical to that postulated by Voce,[33] i.e.:

$$\theta = \theta_0 (1 - \sigma/\sigma_s) \tag{22}$$

where θ_0 is an athermal hardening rate to which all $\sigma(\epsilon)$ curves are asymptotic at low strains (related to, but not necessarily equal to, θ_{II}). Apart from minor deviations at small strains, Eq 22 was found to describe accurately the stress-dependence of the work-hardening rate during the tensile testing of aluminum, copper and an austenitic stainless steel (all polycrystalline). Since necking intervened before a steady state could be established, σ_s was evaluated via the extrapolation of $\theta - \sigma$ plots to $\theta = 0$.

Equation 22 is rationalized by Kocks in terms of dislocation storage at a rate, with respect to strain, which is proportional to $\sqrt{\rho}$; in fact,

$$\frac{d\rho}{d\epsilon}\bigg|_{stor} = \frac{k_1\sqrt{\rho}}{b}$$

where k_1 is a proportionality constant between mean-free path and $\sqrt{\rho}$. In evaluating $(d\rho/d\epsilon)|_{rec}$, it is assumed that the probability of a recovery event is proportional to the number of times a potential recovery site is contacted by a moving dislocation. If a length of dislocation ℓ_r is annihilated per recovery event, then for a unit area of slip plane $d\ell_{rec} = \ell_r\rho$, because the number of potential recovery sites is ρ. The shear-strain increment for unit area of slip plane is b, and so

$$\frac{d\rho}{d\epsilon} = \frac{d\rho}{d\epsilon}\bigg|_{stor} - \frac{d\rho}{d\epsilon}\bigg|_{rec} = \frac{1}{b}\left(k_1\sqrt{\rho} - k_2\ell_r\rho\right)$$

which, in terms of hardening rates, becomes

$$\theta = \frac{d\sigma}{d\epsilon} = \frac{\alpha\mu k_1}{2} - \frac{k_2\ell_r}{2b}\cdot\sigma \tag{23}$$

This expression is equivalent to Eq 22. Kocks and Mecking[29] have also derived the Voce law from fundamental arguments using a time-dependent (i.e., static) model for recovery (see Sec. 2.3.1).

In order to check whether Eq 22 applies all the way up to σ_s, and thereby to examine the validity of the extrapolations made by Kocks on the basis of his tensile data, $\sigma(\epsilon)$ was determined for polycrystalline superpure aluminum (200-μm grain size) using compression testing between 450 and 600 K and at $\dot{\epsilon} = 0.01$ and 1 s^{-1}. These experiments were performed by the present author and have not previously been reported. Lubrication with PTFE permitted friction-free compression up to a true strain of 1, which is sufficient to attain steady state for all the conditions examined. Over the relevant strain range, the compressive $\sigma(\epsilon)$ curves are in excellent agreement with the corresponding tensile data reported by Kocks.[25]

Figure 6 shows typical θ-σ ($\theta \equiv d\sigma/d\epsilon$) plots derived from the $\sigma(\epsilon)$ curves for aluminum. These conform accurately to Eq 22 at low stresses, but a systematic deviation from this law is found as σ_s is approached. Extrapolation of the low-

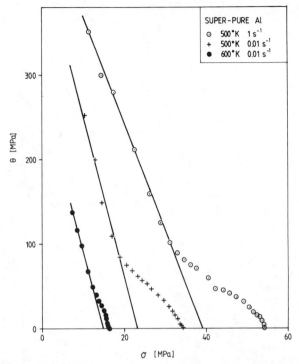

Fig. 6. Plot of work-hardening rate vs. flow stress for superpure aluminum, illustrating the type of extrapolation performed by Kocks[25] on his tensile data.

stress behavior will clearly lead to an underestimation of σ_s. For testing at 500 and 600 K and at 0.01 s^{-1}, an extrapolation after the fashion of that performed by Kocks gives "σ_s" = 23 and 15 MPa, respectively; the corresponding values reported by Kocks for $\dot{\varepsilon} = 1.6 \times 10^{-2}$ s^{-1} are 23 and 14 MPa. However, the correct, experimentally determined levels of σ_s are 34.5 and 16 MPa.

A careful examination of the θ-values suggests that, with the exception of small strains (less than 0.05), θ is linearly related to $1/\sigma$. This proportionality is illustrated in Fig. 7, where all experimental information is included. Some minor deviations from the law are found, especially as σ_s is approached, but the over-all conformity must be regarded as acceptable. The slope of the $1/\sigma$-versus-θ plot increases systematically with increasing temperature and decreasing strain rate, or, more specifically, as $1/\sigma_s$. A plot of $1/\sigma_s$ versus $d(1/\sigma)/d\theta$ is presented in Fig. 8; the data are reasonably well described by a straight line, which passes through the origin. Accordingly, the observed $\sigma(\varepsilon)$ behavior does not conform to Eq 22 but rather follows the law

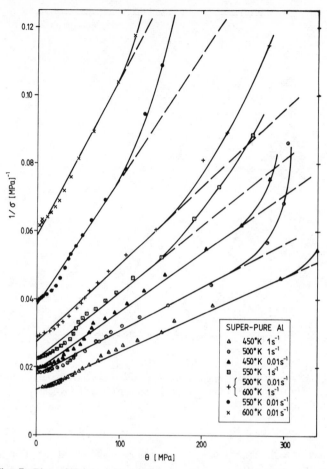

Fig. 7. Plot of the inverse of flow stress vs. work-hardening rate for superpure aluminum.

$$\theta = P\left(\frac{\sigma_s}{\sigma} - 1\right) \qquad (24)$$

where P is the slope of $1/\sigma_s$ versus $d(1/\sigma)/d\theta$ (116 MPa) and is very close to being constant for the range of temperatures and strain rates investigated.

Phenomenologically, Eq 24 is easily shown to be concomitant with the following relationship for the rate of increase of dislocation density with strain:

$$\frac{d\rho}{d\varepsilon} = k_1 - k_2 \sqrt{\rho} \qquad (25)$$

i.e., dislocation accumulation at a constant rate, which is realistic if a cell or subgrain structure is established early in the deformation process, combined

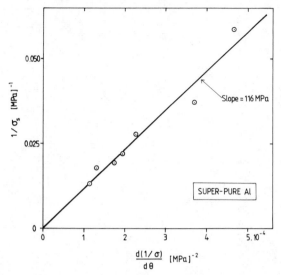

Fig. 8. Relationship between the slope of the lines in Fig. 7 and the inverse of the steady-state stress.

with dynamic recovery at a rate proportional to $\sqrt{\rho}$. The $\sigma(\varepsilon)$ law derived from Eq 24 is

$$\sigma_s \ln \left(\frac{\sigma_s}{\sigma_s - \sigma} \right) - \sigma = P\varepsilon \qquad (26)$$

The constants in Eq 25 are given by

$$\frac{k_1}{k_2} = \frac{\sigma_s}{\alpha' \mu b} \quad ; \quad k_2 = \frac{2P}{\alpha' \mu b} \qquad (27)$$

Examples illustrating the degree of accord between Eq 26 and the experimentally determined $\sigma(\varepsilon)$ behavior are given in Fig. 9. In order to obtain a slightly better fit, the values for P have been determined from the slopes in Fig. 7 ($\equiv 1/P\sigma_s$) rather than taking the unified constant obtained via Fig. 8; the variation between different P-values is quite small. In a more accurate appraisal, it will be necessary to take into account the temperature dependence of elastic modulus in the evaluation of P. The general measure of agreement between the two-parameter formalism (Eq 26) and the experimental $\sigma(\varepsilon)$ is quite good, but hardly perfect, especially at low strains. This reflects the inability of Eq 24 to describe the behavior under these circumstances (Fig. 7); such is not to be expected either, in view of the fact that some initial strain is required before the equilibrium cell size (defining a constant rate of dislocation storage) is established. However, Eq 26 represents the best two-parameter description of $\sigma(\varepsilon)$ under the range of experimental conditions studied.

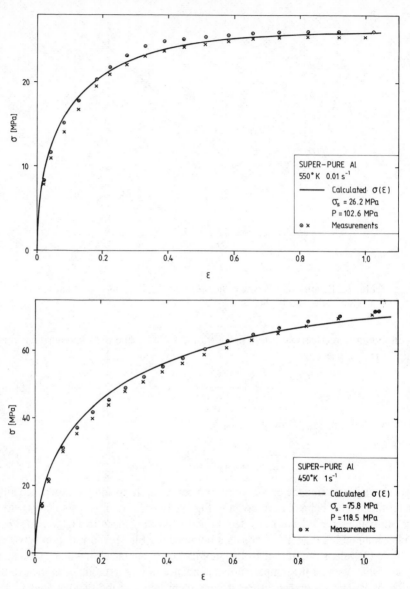

Fig. 9. Correspondence between $\sigma(\varepsilon)$ evaluated on the basis of Eq 26 and the measured curve.

It is instructive to attempt to account for Eq 24 to 27 in mechanistic terms. For a constant rate of dislocation storage, defined by the subgrain size, standard arguments give

$$k_1 = \frac{2P\sigma_s}{(\alpha'\mu b)^2} = \frac{1}{\omega b d_{sub}} \tag{28}$$

where ω is a factor for converting from shear to normal quantities. Since P is virtually independent of T, $\dot{\epsilon}$, then σ_s should be proportional to the inverse of subgrain size, which is approximately the situation found experimentally (Eq 20). In actual fact, the predicted exponent for d_{sub} must deviate somewhat from -1 because of the temperature-dependence of μ (assuming, of course, that σ_s and the subgrain size are varied via changes in test temperature). Turning to the recovery term, it is clear that the proportionality constant, k_2, is approximately independent of T,$\dot{\epsilon}$ (Eq 27); taking $\mu(500\ K) = 2.24 \times 10^4$ MPa, $\alpha' \doteq 1$ and b = 0.286 nm, then $k_2 \doteq 4 \times 10^7$ m^{-1}. Following Kocks,[25] we assume that recovery events are triggered by mobile dislocations; on this basis, the rate of decrease of ρ with respect to strain (see above) is

$$\left.\frac{d\rho}{d\epsilon}\right|_{rec} = const. \frac{\ell_r\rho}{b}$$

where ℓ_r is the length of dislocation lost per recovery event. The constant in the above expression can be identified with $\omega \cdot g$, g being the fraction of encounters between mobile and stationary dislocations which leads to recovery events. Kocks takes ℓ_r to be fixed and independent of total dislocation density. However, since the unit recovery process is likely to involve individual dislocation links, then ℓ_r can be expected to decrease as ρ increases. For both a uniformly distributed network and a subgrain structure, we can thus anticipate that ℓ_r is proportional to $1/\sqrt{\rho}$ and that

$$\left.\frac{d\rho}{d\epsilon}\right|_{rec} = const. \frac{\sqrt{\rho}}{b} \tag{29}$$

in agreement with the formalism derived from experimental observations (Eq 25). Making the rough approximation that $\ell_r = 1/\sqrt{\rho}$, we have $k_2 = \omega g/b$ and so, with the above value for k_2 plus $\omega = 3.1$ (fcc polycrystals), g works out to be 4×10^{-3}, i.e., about one encounter in 200 results in a recovery event.

The two-parameter description of $\sigma(\epsilon)$ embodied in Eq 26 is a very attractive one. Within the range of temperatures and strain rates investigated, the stress-strain behavior up to $\epsilon = 1$ can be described quite accurately in terms of a material constant, P, and the steady-state stress. Furthermore, the experimental data for aluminum indicate that $\sigma_s(\dot{\epsilon},T)$ conforms well to Eq 4 with Q = 130 kJ $\cdot$ mol^{-1}, which is very close to Q_{SD} (138 kJ $\cdot$ mol^{-1}). Hence, the description of $\sigma(\epsilon,\dot{\epsilon},T)$ is simplified further because σ_s can be evaluated for any temperature and strain rate if Q is known. The entire formalism is thus based on just two quantities, which are, at least to a first approximation, independent of temperature or strain rate.

It is of interest to investigate whether or not Eq 26 can be applied to a typical commercial material, characterized by restoration during hot deformation via dynamic recovery alone. Figure 10 shows $\sigma(\varepsilon)$ curves for a ferritic stainless steel (19Cr-0.6Ti) tested in compression at temperatures between 750 and 1150°C and at $\dot{\varepsilon} = 1\ \mathrm{s}^{-1}$. The agreement between the experimental points and the theoretical curves is acceptable. However, in this case, P is not constant but decreases systematically with increasing temperature. The probable explanation lies in the existence of a strain-independent flow-stress component in this commercial steel, i.e., a friction stress, σ_0. This being the case, Eq 24 must be changed to

$$\theta = P \left(\frac{\sigma_s - \sigma}{\sigma - \sigma_0} \right) \tag{30}$$

and Eq 26 must be changed to

$$\sigma_0 - \sigma + (\sigma_s - \sigma_0) \ln \left(\frac{\sigma_s - \sigma_0}{\sigma_s - \sigma} \right) = P\varepsilon \tag{31}$$

It is readily shown that the application of Eq 24 and 26 to $\sigma(\varepsilon)$ behavior characterized by a nonzero σ_0 will result in a P-value which increases with increasing

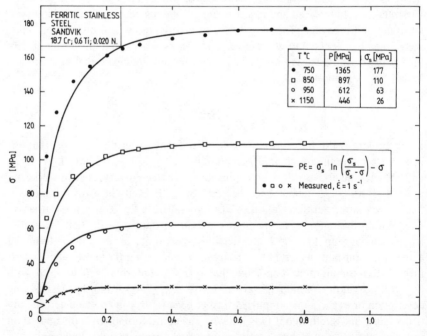

Fig. 10. $\sigma(\varepsilon)$ curves for a commercial ferritic stainless steel at various temperatures; experimental measurements compared with curves evaluated from Eq 26.

σ_0. Hence, the data for P listed in Fig. 10 are consistent with a diminishing σ_0 as the temperature is raised, which seems plausible. For a more accurate correlation with $\sigma(\varepsilon)$ from commercial materials, Eq 30 and 31 should be used; as a first approximation, σ_0 can be set equal to the yield stress under the conditions of interest.

2.5. Dynamic Recrystallization

2.5.1. General Characteristics of Dynamic Recrystallization.

As stated in the introductory chapter, dynamic recrystallization is usually encountered during hot deformation (or creep) of fcc metals and alloys with intermediate or low stacking-fault energies. It has also been observed in bcc metals of very high purity, e.g., α-Fe.[43] The experimental evidence for the occurrence of dynamic recrystallization, which has been furnished via examination of specimens quenched rapidly following hot deformation, is reviewed elsewhere.[6,7]

The following characteristics pertaining to dynamic recrystallization have been gleaned from experiments on a wide range of metals and alloys:

- A critical deformation, ε_c, is necessary in order to initiate dynamic recrystallization; this is somewhat less than $\varepsilon(\max)$. The corresponding stress, σ_c, is accordingly less than $\sigma(\max)$, but the low work-hardening rate in the vicinity of the peak renders these quantities virtually identical (see Fig. 1).
- $\sigma(\max)$ conforms quite well to Eq 4 to 6 in spite of the fact that the stress peak does not occur at a unique strain. In a double-logarithmic plot, $\sigma(\max)$ increases linearly with Z at low and medium Z values, while at high values of Z the slope decreases progressively (this is the normal transition from "low" to "high" stress behavior; see Sec. 2.2). The slope of the linear portion, $d \log_{10}\sigma(\max)/d \log_{10}Z$, lies in the range 0.17 to 0.25 (see Table 2).
- At hot working strain rates, ε_c decreases steadily with decreasing temperature-compensated strain rate, and most experimental data tend toward a minimum ε_c at low Z. However, under creep conditions, the critical strain can actually start to increase again;[9] see Fig. 11.
- The size of dynamically recrystallized grains, D_{rex}, increases monotonically with decreasing $\sigma(\max)$ and σ_s, and the phenomenological relationship

$$\sigma_s = k_1 + k_2 D_{rex}^{-q} \tag{32}$$

has often been reported (e.g., Ref 43, 44, 46 and 52), with the exponent q falling in the range 0.5 to 0.8 independent of temperature and strain rate.
- σ_s and D_{rex} are to all intents and purposes independent of the initial grain size, D_0; however, the kinetics of recrystallization are significantly accelerated in specimens with smaller starting grain size, and both $\sigma(\max)$ and $\varepsilon(\max)$ are lowered, the latter appreciably.[52-54]
- Dynamic recrystallization of polycrystals is usually initiated at pre-existing grain boundaries.[44,52-54] At very low $\dot{\varepsilon}$ and at large D_0, and of course for single crystals,[55] intragranular nucleation becomes relatively more impor-

Table 2. Slopes of $\log_{10}\sigma(\text{max})$ versus $\log_{10}Z$ for Various Materials

Metal or alloy	$\dfrac{d \log_{10}\sigma(\text{max})}{d \log_{10}Z}$	Testing method(a)	Reference(s)
Ni, Ni-5, 10, 20Fe	0.18	T	Luton and Sellars[44]
Zone-refined α-Fe	0.19	T	Glover and Sellars[43]
Cu	0.20	T	McQueen and Bergerson[45]
Cu-8Al	0.24	T	Bromley and Sellars[46]
Low-carbon steel (0.15C-1.3Mn) (γ)	0.18	C, T	Roberts[47]; Le Bon et al.[48]
18Cr-9Ni austenitic steel	0.18	C, T	Ahlblom[49]; Rossard[50]
M2 high-speed steel (0.85C-5Mo-6W-2V)	0.17	C	Carlsson and Roberts[51]

(a) T = torsion testing; C = compression testing.

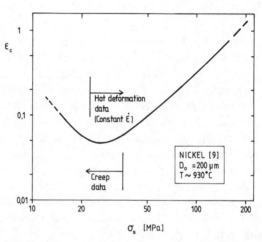

Fig. 11. Relationship between critical strain for dynamic recrystallization of nickel and steady-state stress, illustrating the increase in ε_c at low values of σ_s (*after Sellars[9]*).

tant. Apart from the very initial stages, the recrystallized grain size remains more or less constant during the reaction, with little or no grain growth proceeding simultaneously.[52,54]

In the three sections which follow, the factors governing the initiation of dynamic recrystallization and the influence of this phenomenon on the shape of $\sigma(\varepsilon)$ curves will be considered. The microstructural changes engendered as a result of dynamic recrystallization are given attention in Sec. 3.

2.5.2. Conditions for Initiation of Dynamic Recrystallization. A criterion for the onset of dynamic recrystallization during hot working has been presented by Sandström and Lagneborg[56] and, in a more refined form, by Roberts and Ahlblom.[57] The substance of this treatment is that the reduced driving force, derived from concurrent deformation, modifies the normal energy balance defining the conditions for the nucleation of new grains. The situation envisaged within a potential nucleus for dynamic recrystallization is as depicted in Fig. 12. The dislocation density (ρ_0) generated by the prior strain is reduced to a low value through passage of the high-angle boundary at 0. However, the dislocation density is re-established behind the boundary (due to concurrent deformation) and follows some function $\rho(x)$. This buildup of dislocation density behind the migrating boundary engenders a reduction in the driving force for recrystallization. Roberts and Ahlblom estimate the driving force as a radial average of the dislocation-density difference, i.e., as

$$\frac{\tau}{r} \int_0^r \rho_0 - \rho(x) \cdot dx$$

where r is the radius of a spherical, homogeneous nucleus or, alternatively, the radius of curvature of a bulge engendered by strain-induced grain-boundary migration (the so-called Bailey-Hirsch[58] mechanism for recrystallization). By writing an equation for the net free-energy change associated with the formation of a nucleus with radius r, and combining this with a suitable formalism for $\rho(x)$ based on simple, constant mean-free path dislocation storage (i.e., dynamic recovery ignored), it is possible to obtain an expression for the critical nucleus size, r*. It turns out that a real r* exists only if

$$\rho_0 \geq \text{const.} \left(\frac{S\dot{\varepsilon}}{bd_{sub}m\tau^2} \right)^{1/3} = \rho_{0c} \tag{33}$$

This applies to both homogeneous and bulge nucleation, although the constant is different for these two cases. Hence, a critical dislocation density, which depends

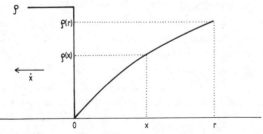

Fig. 12. Situation envisaged within a potential nucleus for dynamic recrystallization;[57] for further discussion, see text.

on the deformation conditions and grain-boundary characteristics (grain-boundary energy and mobility), must be exceeded before a stable nucleus for dynamic recrystallization can be formed. A more realistic approach[57] considers the influence of dynamic recovery on $\rho(x)$; the Friedel relationship (Eq 16) is adopted in the original paper, but in principle any of the treatments of dynamic recovery given in Sec. 2.3 and 2.4 could be applied. A ρ_{0c} exists even in this case, although it is not possible to give an analytical expression for it.

As a rule, the grain-boundary mobility conforms to an Arrhenius law, i.e.:

$$m = m_0 \exp\left(-Q_m/RT\right)$$

Substituting this plus the empirical dependence of subgrain size on Z, $d_{sub} \propto Z^{-p'}$ (cf. Eq 20), into Eq 33 leads to the prediction that the slope of a double-logarithmic plot of σ_c [$\simeq\sigma(max)$] versus Z is $(1 - p')/6$; it is assumed, as usual, that $\sigma_c \simeq \sigma(max) \propto \sqrt{\rho_{0c}}$. Since p' is normally in the range 0.1 to 0.2 (see Ref 7 and Sec. 2.3.2), then the slope d $\log_{10}\sigma(max)/d \log_{10}Z$ is calculated as 0.18 to 0.20, which is in very good agreement with the experimental evidence (see Table 2). It can also be noted that the activation energy derived from the $T,\dot{\varepsilon}$-dependence of σ_c [$\simeq\sigma(max)$] is expected to be the same as that governing the temperature dependence of the grain-boundary mobility, m.

By applying the more refined nucleation criterion, where proper cognizance is given to the influence of dynamic recovery on $\rho(x)$, Roberts and Ahlblom[57] were able to calculate a dependence of σ_c and ε_c on temperature-compensated strain rate that was in acceptable agreement with experimental observations on low-carbon[47] and austenitic[49] stainless steels (see Fig. 13). The predicted deviation from linearity at low values of Z is a consequence of the fact that ρ_{0c} approaches ρ_s, the dislocation density defining the recovery-controlled steady state. The theory therefore forecasts that dynamic recrystallization will not occur once Z has become sufficiently low. However, it is not implausible that dynamic recrystallization nuclei can be developed from a steady-state distribution of subgrains (recrystallization *in situ*); if a sufficiently large subgrain is formed, then it may be capable of functioning as a nucleus even if the model predicts that the *average* dislocation density is too low for dynamic recrystallization to proceed. Gottstein et al.[55] report how, in single crystals of copper, dynamic recrystallization is initiated from a single, large subgrain. A change in the mechanism for dynamic recrystallization when $\rho_{0c} > \rho_s$ might well account for the increase in ε_c observed at low values of Z.[9] However, there will come a stage at which ρ_s is so much less than ρ_{0c} that dynamic recrystallization will not be possible, and restoration during hot deformation then will proceed via recovery only.

The deviation from linearity of $\log_{10}\sigma(max)$ versus $\log_{10}Z$ at high Z, which is predicted from the theory and is observed experimentally, derives from the fact that the critical nucleus size for dynamic recrystallization can never be smaller

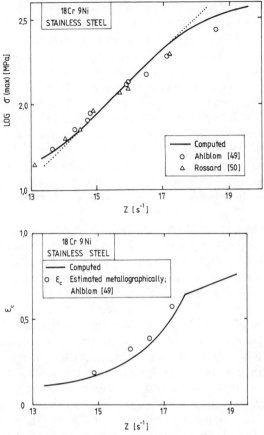

Fig. 13. Comparison of experimentally observed dependence of σ(max) and ε_c on Z for dynamic recrystallization of 18Cr-9Ni austenitic stainless steel, with the behavior predicted on the basis of the nucleation criterion proposed by Roberts and Ahlblom.[57]

than d_{sub}. Hence, at high values of Z, the critical nucleus size is equal to the subgrain size and under these conditions

$$\rho_{0c} = \frac{8S}{\tau d_{sub}} \tag{34}$$

irrespective of whether nucleation proceeds homogeneously or via bulging. This effect is also responsible for the change in slope of the predicted $\varepsilon_c - \log_{10}Z$ behavior at high values of Z (Fig. 13).

In a review, Sellars[11] has stated that the nucleation criterion embodied in Eq 33 does not account for the basic observation that dynamic recrystallization is not generally found in metals and alloys characterized by rapid dynamic recovery. This is not a valid statement! In fact, the predicted value of ρ_{0c}, at least for intermediate and high values of Z, is not much affected even if cognizance is given to dynamic recovery; this is because $\rho_{0c} \ll \rho_s$ under such conditions. If, for a given Z, $\rho_s \ll \rho_{0c}$, then dynamic recrystallization is not possible. The factor in Eq 33 which is likely to vary greatly from metal to metal is m; similarly, the level of ρ_s will be altered depending on the stacking-fault energy, etc. Hence, for very pure metals of different stacking-fault energies, ρ_{0c} at a given strain rate and homologous temperature will be very roughly constant, and whether or not dynamic recrystallization occurs depends on differences in ρ_s. Alternatively, one might expect that the latter is not particularly sensitive to metal purity whereas ρ_{0c} is affected, due to the dependence on m. Hence, dynamic recrystallization becomes more likely as the degree of purity increases; this rationale is confirmed by results on lead[1-5] and α-iron.[43]

One effect, which is not directly obvious from the theoretical considerations outlined above, is the decrease in ε_c engendered by a reduced initial grain size D_0.[52-54] The basic reason for this is the more rapid rate of dislocation accumulation as D_0 is lowered, i.e., ρ_{0c} is attained at progressively smaller strains.[54] This argument is consistent with the fact that σ_c and $\sigma(\max)$ ($\propto \sqrt{\rho_{0c}}$) exhibit very weak dependences on D_0.[54]

2.5.3. Shape of $\sigma(\varepsilon)$ During Restoration via Dynamic Recovery Plus Dynamic Recrystallization. In creep, the occurrence of dynamic recrystallization engenders a discontinuous increase in creep rate. For lead, Gifkins[4] has demonstrated that this change can be accounted for reasonably well if the recrystallized grains are assumed to undergo primary creep once again. Luton and Sellars[44] applied an argument of the same basic substance to hot deformation (torsion) of nickel. Their model is reproduced in Fig. 14. The predicted shape of the flow curves depends on the relative magnitudes of ε_c and the strain, ε_x, over which a cycle of recrystallization can be substantially completed. At low Z (high T, low $\dot{\varepsilon}$), $\varepsilon_c > \varepsilon_x$ and one cycle of recrystallization can proceed to completion before ρ_{0c} is attained once again in the recrystallized grains due to concurrent deformation; the $\sigma(\varepsilon)$ curve is therefore oscillatory due to the separation in time of the alternate cycles of recrystallization and hardening. If, on the other hand, $\varepsilon_x > \varepsilon_c$ (high Z), the various cycles of recrystallization overlap and $\sigma(\varepsilon)$ is, to all intents and purposes, smooth. In this latter case, the drop in flow stress is associated mainly with the first recrystallization cycle. The steady state represents an equilibrium distribution of regions with stored energies between zero and ρ_{0c}, and is associated with a constant average flow stress. If it is assumed that the flow-stress level characterizing fully recrystallized material is invariant, then it is a fairly straightforward matter to evaluate the complete flow curve via the knowledge of ε_c, ε_x and $\sigma(\varepsilon)$ up to ε_c. Luton and Sellars[44] described the strain dependence of the recrystallized fraction in terms of an Avrami-type law, which is realistic (see Sec. 3).

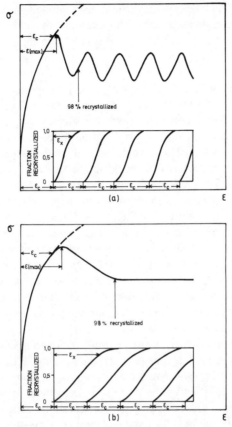

Fig. 14. Model of Luton and Sellars[44] accounting for transition from (a) oscillatory to (b) smooth $\sigma(\varepsilon)$ behavior, when restoration proceeds via dynamic recovery plus dynamic recrystallization.

Sah et al.[59] modified the Luton and Sellars treatment with the aim of correcting a number of deficiencies. The foremost of these is the fact that the oscillatory behavior at low Z is predicted to proceed indefinitely with ε, whereas in practice the oscillations are damped out after a few cycles (see Fig. 1). In addition, the dependence of the dynamic recrystallization kinetics on grain size (Sec. 3) will result in different strain dependences for the recrystallized fraction in the various cycles (unless D_0 and D_{rex} happen to be the same). Hence the main feature of the improved treatment is that ε_c and ε_x are taken to be considerably smaller for the second and subsequent recrystallization cycles than for the first one. Furthermore, in the case of periodic behavior, the specimen is subdivided into successively smaller and smaller volume fractions, each at a different state of recrystallization, in the progression from one cycle to the next. Thus, the first recrystallization cycle takes place synchronously throughout the material, but, as deformation proceeds, the separate cycles overlap more and more; this behavior, which was not especially well justified physically by the authors, engenders a progressive transition from an oscillatory to a smooth $\sigma(\varepsilon)$ curve, i.e., damping of the oscillations.

The basic Luton-Sellars rationale has since been placed on a more formal footing by Stüwe and Ortner[60] and by Sandström and Lagneborg.[61] These authors developed descriptions which permit evaluation of the averge dislocation density when deformation and recrystallization proceed simultaneously. Stüwe and Ortner[60] have adopted a rather idealized model where dynamically recrystallized grains are assumed to nucleate successively at the centers of the original grains. Under such conditions, the average dislocation density, and therefore the flow stress, is determined by a parameter $a = t_n/t_0$, where $t_n = D_0/v$ is the time required for the new grain to grow to size D_0 [v is the average grain-boundary velocity and t_0 is the incubation time for nucleation of new grains; in other words, in the terminology of Luton and Sellars,[44] a is related to the ratio of ε_x to ε_c (constant $\dot{\varepsilon}$)]. If $a \ll 1$ (slow deformation, rapid recrystallization), the $\sigma(\varepsilon)$ curve is a simple repeated sawtooth function; such a situation can be encountered in creep deformation. As a increases, the subsidiary maxima become progressively lower than the first one and the strain corresponding to the first minimum increases.

Although Stüwe and Ortner's model permits the overlapping of several cycles of recrystallization at any one time ($a > 1$), the situation in each individual grain is identical and the process is in exact synchronization. Thus, the theory predicts that oscillations should always be present on $\sigma(\varepsilon)$ even for large values of a, which is, of course, in gross disparity with observations. Another shortcoming is the unrealistic model adopted for the development of dynamically recrystallized grains. In the first place, nucleation is known to occur preferentially at the pre-existing grain boundaries rather than at the grain centers; furthermore, the situation envisaged infers that the dynamically recrystallized grain size, D_{rex}, is always the same as the initial one (D_0).

A physically more plausible approach to describing the effects of dynamic recrystallization on $\sigma(\varepsilon)$ has been taken by Sandström and Lagneborg.[61] In this approach, the fluctuations in ρ characterizing dynamically recrystallizing material are considered in terms of a dislocation-density distribution function, $\Phi(\rho, t)$; the product $\Phi(\rho, t) \cdot d\rho$ is defined as the volume fraction which contains a dislocation density between ρ and $\rho + d\rho$ at some point in time, t. In order to obtain realistic flow-stress values, the authors were forced to assume that this quantity is determined by intrasubgrain dislocations, i.e., that the long-range stress field associated with subgrain boundaries is negligible. Hence, it is necessary to involve two distribution functions, one for the dislocations in sub-boundaries and one for intrasubgrain dislocations, $\Phi(\rho_{sub}, t)$ and $\phi(\rho_i, t)$, respectively. Since $\rho_{sub} \gg \rho_i$, the latter type of dislocations are neglected in evaluating the driving force for recrystallization.

Using the Friedel network-growth model in order to describe dynamic recovery (Eq 16) and the standard expression for grain-boundary velocity,

$$v = m\tau\rho_{sub} \qquad (35)$$

it is quite straightforward to derive differential equations relating ϕ to ρ_i, t and Φ to ρ_{sub}, t. The recrystallization rate can readily be defined in terms of Φ, i.e.,

$$\frac{dX}{dt} = \int_{\rho_{0c}}^{\infty} \frac{3\gamma_D}{D_{av}} \cdot v\Phi(\rho_{sub}, t)d\rho_{sub} \qquad (36)$$

where D_{av} is the mean grain size existing after time t and γ_D is the fraction of the total grain-boundary area which is migrating. The term dX/dt defines the rate at which dislocation-free material is generated and comes in at the beginning of the distribution as $\Phi(0, t)$ or $\phi(0, t)$. At large strains, $\Phi(\rho_{sub}, t)$ approaches a stationary condition, which corresponds to the attainment of a steady-state stress. Once the variation of Φ as a function of time is known, then ϕ can also be determined, because the two distributions are uniquely related. The average value for ρ_i can then be evaluated as

$$\bar{\rho}_i = \int_0^{\rho_s} \rho_i \cdot \phi(\rho_i, t) \, d\rho_i \qquad (37)$$

where ρ_s is here the value of ρ_i corresponding to steady state if no dynamic recrystallization occurred at all. The flow stress is computed as $\alpha'\mu b\sqrt{\bar{\rho}_i}$. Sandström and Lagneborg[61] also give cognizance to the variation of grain size with time between the initiation of dynamic recrystallization at ε_c and the attainment of steady state, i.e., the effects of changes in D_{av} on dX/dt are properly considered (Eq 36).

Curves for $\sigma(\varepsilon)$ computed using the approach outlined above are compared with experimental torsion data for nickel at 934°C[44] in Fig. 15. Corresponding information for $\Phi(\rho_{sub}, t)$ at various strains during deformation at 1.63×10^{-2} s^{-1} is presented in Fig. 16. The positions of the arrows indicate the dislocation densities in material remaining unrecrystallized at the strains for which Φ is given, and their heights represent the volume fractions of material which have not recrystallized at all. Note that, even at the beginning of the third recrystallization cycle, $\Phi(\rho_{sub}, t)$ lies quite close to the steady-state distribution (∞). Considering the rather fundamental nature of the model, the measure of agreement between the computed $\sigma(\varepsilon)$ and the torsional data for nickel is impressive (Fig. 15), although the choices made regarding the magnitudes of some of the parameters in the theory are not completely consistent. In particular, the damping of oscillations at low $\dot{\varepsilon}$ is reproduced satisfactorily. There is no doubt that the Sandström-

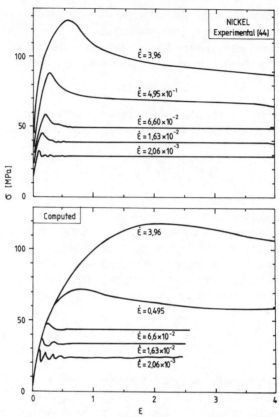

Fig. 15. Comparison of observed $\sigma(\varepsilon)$ curves for hot torsion of nickel[44] with curves computed on the basis of the treatment due to Sandström and Lagneborg.[61]

Lagneborg model represents the most detailed fundamental treatment for the influence of dynamic recrystallization on $\sigma(\varepsilon)$ which is currently available.

2.5.4. Transition From Continuous to Oscillatory $\sigma(\varepsilon)$ Behavior. In terms of the formalism of Luton and Sellars,[44] the transition from a smooth, continuous $\sigma(\varepsilon)$ to an oscillatory $\sigma(\varepsilon)$ occurs when $\varepsilon_c = \varepsilon_x$. In more recent reports, Sakai and co-workers[62–64] have presented convincing evidence to the effect that the nature of $\sigma(\varepsilon)$ depends on the ratio D_{rex}/D_0. This new viewpoint has been compared with the Luton-Sellars approach, and its subsequent modification,[59] by Jonas and Sakai.[65,66] Figure 17 reproduces results from the original work of Sakai and co-workers[62–64] and shows the variation in average grain size with strain during the dynamic recrystallization of low-carbon austenite. The arrows marked P_1, P_2, etc. refer to the positions of flow-stress maxima; the initial grain size is 32 μm.

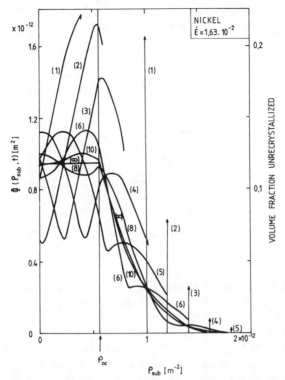

Fig. 16. Theoretical dislocation-density distribution functions at various strains during hot torsion of nickel at 1.63×10^{-2} s^{-1}.[61] (1) $\varepsilon = 0.25$; (2) $\varepsilon = 0.30$ (start of 2nd recrystallization cycle); (3) $\varepsilon = 0.35$; (4) $\varepsilon = 0.40$; (5) $\varepsilon = 0.45$ (3rd cycle started); (6) $\varepsilon = 0.50$; (8) $\varepsilon = 0.60$ (4th cycle started); (10) $\varepsilon = 0.70$; (∞) stationary value.

The general conclusion from this and other experiments is that a multiple-peak $\sigma(\varepsilon)$ is found when $D_0 < 2D_{rex}$; otherwise, the flow curve is smooth, with a single stress maximum. This behavior can be rationalized in terms of a nucleation density argument.[65] For $D_0 < 2D_{rex}$, the pre-existing microstructure can easily furnish the relatively few nuclei that are needed for dynamic recrystallization and there is relatively little spread in ε_c throughout the material. This means that all new grains develop at about the same rate, the recrystallization process is synchronized and multiple-peak behavior is the result. If, on the other hand, $D_0 > 2D_{rex}$, the number of nuclei required during dynamic recrystallization is large and it is not possible for all of them to be formed simultaneously; for example, one might expect that ε_c depends on the orientation difference between grains. The appreciable spread in ε_c means that different regions of the specimen

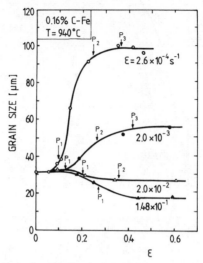

Fig. 17. Variation of grain size during dynamic recrystallization of low-carbon austenite[62] at various strain rates; the arrows indicate the positions of the stress maxima. A single-peak behavior requires a grain refinement ratio of about 2:1.

are at different stages of recrystallization at a given time; hence, the recrystallization, in this case, is nonsynchronized, which is concomitant with a smooth $\sigma(\varepsilon)$ following the single peak. In the multiple-peak case, the increase in grain size during recrystallization (Fig. 17) implies that a nucleus-supply problem gradually develops, the eventual consequence being an increased spread in ε_c and a damping of the oscillations. A corollary of such a rationale is that the equilibrium D_{rex} is not attained in the multiple-peak situation until the oscillations have disappeared (see Fig. 17).

If the above explanation of the transition from multiple to single-peak behavior depending on D_{rex}/D_0 is correct (and the evidence presented in favor of it is convincing), then it is clear that prediction of $\sigma(\varepsilon)$ from first principles, when softening proceeds via dynamic recrystallization, will be difficult. This is because it will be necessary to know the spread in ε_c, i.e., to know how ρ_{0c} in Eq 33 varies with grain-boundary character (affects m). The reason why Sandström and Lagneborg[61] have succeeded in satisfactorily reproducing the damping of oscillations (e.g., see Fig. 15) is because they give proper attention to the variation in grain size during dynamic recrystallization. Hence, the stepwise increase in grain size associated with multiple-peak behavior has, in fact, been given cognizance in their development. That the recrystallization becomes progressively less synchronized with increasing strain is quite clear from the curves for $\Phi(\rho_{sub},t)$ presented in Fig. 16. Sandström and Lagneborg do not consider a spread in ε_c

from one grain boundary to the next, but such effects can readily be included in the dislocation-density distribution function; there is little doubt of the usefulness of this latter concept in the description of $\sigma(\varepsilon)$ during dynamic recrystallization.

Jonas and Sakai[65] draw attention to the difference between torsion testing and testing methods where the deformation is relatively homogeneous (e.g., tension and compression) with regard to damping of the oscillations. The strain gradient during torsion of solid bars means that ε_c is attained successively in different layers from the outside inward. Similarly, the time corresponding to $\varepsilon_c + \varepsilon_x$ varies along the diameter because $\dot{\varepsilon}$ decreases from the outside inward. The net effect is that the desynchronization of recrystallization tends to be more pronounced in torsion, and the oscillations therefore die out faster in this testing mode than in tension or compression. Furthermore, while ε_c is the same for all testing methods (assuming that this is determined metallographically in the outer layer of a torsion specimen), ε_x in torsion will be considerably greater. The chances are that the curves for ε_c and ε_x determined via torsion will intersect, whereas, generally speaking, they do not do so for tension or compression testing; the Luton-Sellars criterion for the multiple- to single-peak transition is thus untenable in these latter instances.

2.6. Summary of Current State of Knowlege Pertaining to $\sigma(\varepsilon)$ Under Hot Working Conditions

Accurate information concerning the flow curve during high-temperature, high-rate deformation is a basic technological need in the context of metalworking computations. A flow-stress data bank based on the phenomenological description of $\sigma(T,\dot{\varepsilon})$ is one way of meeting this requirement. However, the establishment of such a data bank, in its most complete form, will necessitate a testing program of considerable proportions even for one material type (say, high-alloy steels). Development of descriptions of chemical and microstructural effects on $\sigma(\varepsilon)$ from first principles is therefore of interest.

As regards restoration by dynamic recovery alone, the models which have been suggested for the flow curve, including the one presented here for the first time, are at best semiphenomenological. A pre-requisite for the development of a more fundamentally based description is an improved understanding of the unit softening events involved in dynamic recovery under hot working conditions. In particular, the relative importance of time-dependent (i.e., static-type) recovery and true dynamic recovery, as triggered by the applied stress or by mobile dislocations, must be elucidated.

Since a proper treatment of $\sigma(\varepsilon)$ associated with dynamic recrystallization requires an adequate model for dynamic recovery, it will be appreciated that we find ourselves even further from a basic-principles description of the flow behavior in this instance. The models presented for the nucleation of dynamic recrystallization must be extended to take account of the inhomogeneous distribution of dislocations. The next step is then to incorporate such a modified nucleation criterion into a description of the type advocated by Sandström and Lagneborg.[61] Proper attention must be paid to nucleation at pre-existing grain boundaries and the variation of the conditions for nucleation between grain boundaries of varying character. Finally, the way in which the grain size changes in its excursion between D_0 and the steady-state value must be given due cognizance.

In summary, one is forced to the unhappy conclusion that a complete physically based description of the flow curve under hot working conditions is not possible on the basis of current knowledge and is not likely to be realized in the near future. In the meantime, present-day requirements will have to be met via phenomenological descriptions of $\sigma(\varepsilon, \dot{\varepsilon}, T)$ of the type outlined in Sec. 2.2.

3. Evolution of Microstructure in Connection With Dynamic Recrystallization

3.1. Dynamic Recrystallization in Real Hot Working Operations

Hot working processes are normally characterized by high values of the temperature-compensated strain rate; the critical strain for the nucleation of dynamic recrystallization is therefore large (see Fig. 13). On this basis, one expects that the phenomenon will only proceed to any great extent in working operations involving considerable unit deformations, e.g., extrusion, planetary rolling and piercing. For the very important instance of hot rolling, where pass reductions are limited to about 50% ($\varepsilon < 0.7-0.8$), it is less clear whether dynamic recrystallization is important in microstructural development. However, because the grain size is refined progressively by static recrystallization between rolling passes, it is not unfeasible that the dynamic reaction can proceed to a significant extent during the later stages of a rolling schedule; this is because a reduction in D_0 engenders both a lowering of ε_c and an acceleration of the kinetics of dynamic recrystallization (see Sec. 3.2). To some extent, the effect of the diminished D_0 will be counteracted if the temperature falls continuously during rolling (increased ε_c, retardation of kinetics). However, for rolling of rod or wire, the temperature of the stock can actually increase during the process, because of the high $\dot{\varepsilon}$ (high flow stress) and short interpass times; in this case, dynamic recrystallization may well occur during the final passes once the initial (usually coarse) grain size has been reduced via static (interpass) recrystallization.

Even if dynamic recrystallization does proceed in association with a particular hot working operation, one must always remember that the ensuing microstructure will be subject to static changes following the termination of deformation. These structural modifications, undergone by a material after partial or complete dynamic recrystallization during prior deformation, are termed *metadynamic recrystallization*. This is probably that aspect of microstructural development in association with hot working which is the least investigated and understood (see Sec. 3.4).

3.2. Kinetics of Dynamic Recrystallization

As was pointed out when the general characteristics of dynamic recrystallization were dealt with earlier (Sec. 2.5.1), the reaction rate depends strongly on initial grain size. This is a consequence of the way in which the recrystallization reaction proceeds. For normal hot working strain rates, nucleation is effected at pre-existing grain boundaries, usually via local strain-induced grain-boundary migration in the manner first analyzed quantitatively by Bailey and Hirsch.[58] Bulges are discernible quite early during deformation even when $\varepsilon < \varepsilon_c$. Once $\varepsilon_c(\rho_{0c})$ is attained (Sec. 2.5.2), the largest of these can develop as new dynamically recrystallized grains. Nucleation continues until the sites at the initial grain boundaries have been exhausted. From this point on, the reaction proceeds via nucleation at the interface between unrecrystallized and recrystallized material until the regions of the latter, emanating from opposing grain boundaries, impinge at the centers of the pre-existing grains. This sequence of events in 18-9 stainless steel is illustrated in Fig. 18;[54] in this case, nucleation occurs also at twin boundaries. For single-peak $\sigma(\varepsilon)$, different grains may well be at different stages of the above sequence at any given point in time, as a consequence of the variation in ε_c between one grain boundary and the next.

On the basis of the above discussion, one would expect that the rate of dynamic recrystallization decreases strongly with increasing D_0. This is reflected in the grain-size dependence of the additional strain required to attain steady state on smooth $\sigma(\varepsilon)$ curves once the stress peak has been passed. For example, this strain, which is about the same as ε_x in the Luton-Sellars[44] formalism, is much less in high-speed steel, where D_0 is very small because of the presence of primary carbides which limit prior grain growth, than in single-phase austenitic stainless steel with initially large D_0 (see Fig. 19). Furthermore, ε_c decreases noticeably as D_0 is reduced; this effect has been discussed in Sec. 2.5.2. In spite of these considerable differences in the shape of $\sigma(\varepsilon)$ depending on initial grain size, the values of $\sigma(max)$ and σ_s are to all intents and purposes independent of D_0.[52,54]

The strain dependence of the volume fraction of dynamically recrystallized material (X) during single-peak $\sigma(\varepsilon)$ behavior (which is the important case for the context of practical hot working operations) has been determined for nickel by Sah et al.[52] and for stainless steel by Roberts et al.;[54] in particular, the kinetics have been measured for various pre-existing grain sizes. The behavior is well described by the standard Avrami expression

$$X = 1 - \exp\left[-\xi(\varepsilon - \varepsilon_c)^\eta\right] \qquad (38)$$

where ξ and η are constants. For nickel, Sah et al. find $\eta \simeq 1$, whereas the results for stainless steel indicate that $\eta = 1.2$ to 1.4. It would seem that the Avrami

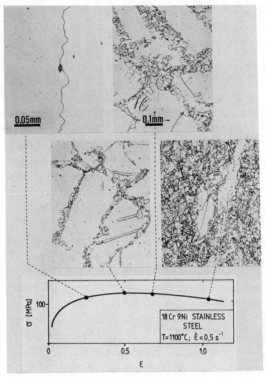

Fig. 18. Micrographs illustrating the progress of dynamic recrystallization in 18Cr-9Ni austenitic stainless steel.[54]

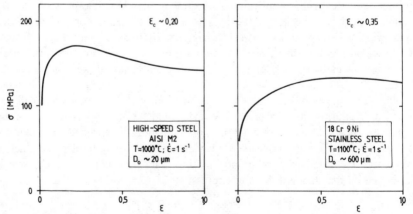

Fig. 19. Comparison of stress-strain curves for M2 high-speed steel and 18Cr-9Ni austenitic stainless steel. The former is characterized by smaller D_0; thus, ε_c is lower and attainment of steady state is faster, in spite of a generally higher level of flow stress.

exponent is approximately independent of T, $\dot{\varepsilon}$ and D_0, at least under conditions where $\sigma(\varepsilon)$ is continuous. The kinetic data of Roberts et al.[54] suggest that the constant ξ varies in inverse proportion to D_0;[54] thus, the grain-size dependence of the rate of dynamic recrystallization resides in this quantity.

Gifkins,[4] who studied dynamic recrystallization in creeping lead, first noted that the size of the recrystallized grains is not completely determined by impingement with other grains. Sah et al.[52] proposed that the dynamic reaction is characterized by the repeated nucleation and limited growth of new grains, which is in sharp contrast to static recrystallization, where the grain size increases progressively during the reaction. This picture is consistent with the sequence of microstructural development depicted in Fig. 18. Nucleation proceeds at a pre-existing grain boundary until the adjacent material on either side of the boundary has completely recrystallized (saturation of boundary sites). Subsequent nucleation occurs in the vicinity of the demarcation between recrystallized and unrecrystallized material. The grains in this second layer grow and impinge sideways, just like the ones nucleated at the original grain boundary did. The process is then repeated until all material is recrystallized. Although it is fairly clear that the initial nucleation at a pre-existing boundary proceeds via local strain-induced grain-boundary migration (bulging), the exact nucleation mechanism operative during formation of the second and subsequent layers of grains is a matter for conjecture. It appears that sideways impingement decelerates growth even in the direction toward the center of the old grain; concurrent deformation then increases the dislocation density behind the boundary of the dynamic grain to such an extent that migration is no longer possible (drastically reduced driving force).

The layerwise progression of the dynamic recrystallization process, along a fairly plane reaction front, has formed the basis of a theoretical description of the kinetics by Roberts et al.[54] This analysis is essentially a modification of Cahn's classical treatment of the kinetics of grain-boundary-nucleated reactions.[67] For $D_{rex} \ll D_0$ [smooth, continuous $\sigma(\varepsilon)$], one can presume that grain surfaces function as the predominant nucleation sites. (The relative significance of nucleation at grain edges or corners is not known in the instance of dynamic recrystallization; these sites may well be important when $D_{rex} > D_0$.) For grain-surface nucleation, Cahn's development leads to the following expression for X(t):

$$X = 1 - \exp[-b_s^{-1/3} \cdot f(a_s)] \tag{39}$$

where the function $f(a_s)$ is

$$f(a_s) = a_s \int_0^1 1 - \exp\left\{-\pi a_s^3\left[\frac{1-\kappa^3}{3} - \kappa^2(1-\kappa)\right]\right\} d\kappa \tag{40}$$

with $a_s = (I_sG^2)^{1/3}t$ and $b_s = I_s/8S_v^3G$. Here, I_s is the specific grain-boundary nucleation rate, G is the radial-growth velocity and S_v is the grain-boundary area per unit volume. Since $f(a_s)$ is easily evaluated via standard numerical methods,

$X(t)$ can be defined. For short times, it turns out that Eq 39 can be approximated as

$$X = 1 - \exp\left(-\frac{\pi}{3}I_s S_v G^3 t^4\right) \tag{41}$$

which is of the form expected for random nucleation on grain surfaces. For $a_s \gg 1$ (long times), nucleation takes place at such a low frequency in relation to the amount of untransformed material that the behavior can be approximated as if all nucleation had occurred at $t = 0$, i.e.:

$$X = 1 - \exp\left(-2S_v Gt\right) \tag{42}$$

Hence, the Avrami exponent, η, changes from 4 to 1 when the original grain-boundary nuclei impinge and the nucleation rate becomes very low. Cahn coined the term *site saturation* for this condition, where the initial grain surfaces are covered by a shell of transformed product. A reasonable estimate of the time to site saturation is given by $a_s = 1$, i.e.:

$$t_{site\ sat.} = (I_s G^2)^{-1/3} \tag{43}$$

which is independent of S_v and, hence, of D_0.

The above relationships cannot be applied directly in a description of dynamic recrystallization kinetics. This is because nucleation has been strictly restricted to pre-existing grain boundaries, the nucleation rate dropping to zero once site saturation sets in; in such a case, the dynamically recrystallized grains would be elongated with a long dimension of $\sim D_0/2$, which is obviously not found in experiments. In the modified treatment,[54] the nucleation of new grains at the reaction front is taken into account, albeit rather simple-mindedly. The basic assumption is that a fresh nucleation event is initiated by each recrystallized grain when the grain has grown over a distance d measured perpendicular to the original grain boundary, i.e., the reaction proceeds blockwise with each elementary slab of material having thickness d; obviously, $d \simeq D_{rex}$. The dependence of fraction recrystallized on t, which is derived in the modified analysis, is rather similar to Eq 39, but $f(a_s)$ has to be summed over all slabs participating in the reaction at any one time. However, even when cognizance is given to repeated nucleation, the predicted Avrami plots, i.e., $\log_{10}[\ln (1/1 - X)]$ versus $\log_{10}t$, exhibit too steep a slope for small values of X, in exactly the same way as the ones derived from the basic Cahn analysis do. The probable reason for this serious discrepancy is that the nucleation efficiency of grain boundaries, and therefore ρ_{0c} and ε_c, vary considerably from one boundary to the next (that such is the case has already been discussed in Sec. 2.5.4). In terms of the Cahn approach, this means that the active grain-boundary area per unit volume (S_v), which is conducive to the nucleation of dynamically recrystallized grains, increases with time. Roberts et al.[54] demonstrate that if S_v is presumed to increase exponentially with time, the resulting Avrami plot approximates a straight line with no gross deviations at small values of X; the Avrami exponent is then 1.3, which is in good agreement with the observations on stainless steel (see Fig. 20). Furthermore, evaluations for different values of D_0 demonstrate that the above approach predicts the correct grain-size dependence of the kinetics of dynamic recrystallization, i.e., $\xi \propto D_0^{-1}$, or $dX/d\varepsilon$. $(\varepsilon - \varepsilon_c)^{n-1}$ at fixed X is proportional to D_0^{-1}.[54] This inverse relationship between reaction rate and the pre-existing grain size is intrinsic to the Cahn approach[67] and will be retained throughout the subsequent modifications taking into account repeated nucleation.

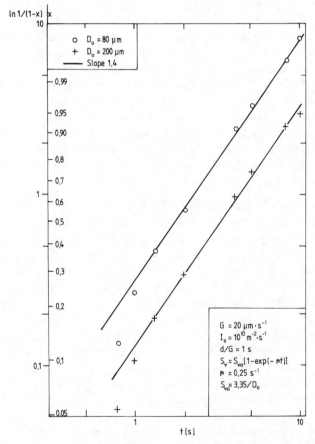

Fig. 20. Avrami plots for kinetics of dynamic recrystallization predicted by repeated-nucleation model with a time dependence of active grain-boundary area.[54] The points are calculated theoretically, whereas the line shown has a slope of 1.4, which corresponds to that found experimentally for 18Cr-9Ni and 16Cr-14Ni-4.5Mo(+N) stainless steels.

As stated earlier, the Avrami η is, during dynamic recrystallization, virtually independent of strain rate and temperature,[49] and the dependence on these variables lies in ξ. For 18Cr-9Ni austenitic stainless steel, Ahlblom[49] reports that temperature exerts a much stronger effect on ξ than does strain rate, suggesting that this quantity is composed of a flow-stress-dependent (driving-force-dependent) component (a component that varies with T,$\dot\varepsilon$) in combination with other contributions, which depend on temperature alone.

An interesting and, from the point of view of commercial hot working, very important question concerns the dynamic recrystallization kinetics characterizing *as-cast* material. Very little data are available, but Ahlblom[49] reports that the critical strain is greater and the rate of dynamic recrystallization appreciably less in specimens taken from a full-scale ingot of 18Cr-9Ni austenitic steel than for the same material previously cold worked and annealed to equivalent grain size ($D_0 = 660$ μm). For example, at 1100°C and 1 s^{-1}, the fraction recrystallized in a wrought specimen at $\varepsilon = 0.95$ is 0.53, whereas the corresponding figure for cast material is 0.17. Similar results have been reported for static recrystallization of austenitic stainless steels.[49] The probable explanation for the sluggish reaction in the cast microstructure is that the original crystal boundaries are characterized by very low mobility (segregation of impurities and/or special orientation relationships); this means that ε_c will be higher than for corresponding wrought material and, further, that the spread in nucleation strain, over all the grain boundaries present, will be greater.

3.3. Dynamically Recrystallized Grain Size

Most investigators have attempted to correlate the grain size produced by dynamic recrystallization with the steady-state stress which can in turn be related to T, $\dot\varepsilon$ via the Zener-Hollomon parameter, Z. The dependence of σ_s on D_{rex} usually takes the form of a power function (Eq 32), where the exponent q is in the range 0.5 to 0.8. For testing under fixed hot working conditions, (T, $\dot\varepsilon$), D_{rex} (and σ_s) are independent of the initial grain size, D_0.[52,54] The characteristic development of dynamic recrystallization when $\sigma(\varepsilon)$ is single-peaked, which involves repeated nucleation and limited growth of the individual grains, has the ramification that the grain size in the recrystallized regions very rapidly attains a constant value and does not change on further straining.[52] In the multiple-peak situation, the steady-state D_{rex} is attained progressively in several steps; however, the grain size does not change much in any one recrystallization cycle.[62-66] Equation 32 suggests that under high-Z conditions, which are often maintained commercially, the dynamically recrystallized grain size should be very small (<10 μm) is not unusual), and much smaller than can be obtained via static recrystallization under equivalent conditions. The reason for this is, of course, that dynamic recrystallization requires a very high driving force, especially if the reaction is to proceed to any great extent during fast straining; consequently, the critical nucleus size is small and the grain refinement powerful. The phenomenon is therefore of some commercial interest for hot working operations where the unit strains are sufficiently large (extrusion, planetary rolling). Unfortunately, the fine microstructure will coarsen or be subjected to metadynamic changes as a result of the residual dislocation density in the grains formed dynamically (see Sec. 3.4), unless the stock is cooled very rapidly following working. Fast cooling is generally applied after hot rolling of steel strip on a continuous mill. The fact that the unit strains in the finishing train can be quite large in combination with the relatively fine grain size, which has been attained via static recrystallization in association with earlier processing stages, implies that dynamic recrystallization might well be an important grain-refinement mechanism in this instance.

The universal applicability of a power-law relationship between σ_s and D_{rex} has recently been questioned by a number of authors.[13,17,68] Ahlblom,[17] working with 18Cr-9Ni austenitic stainless steel, made observations to the effect that D_{rex} at a given Z increases as $\dot{\epsilon}$ is raised; similar results for 316 stainless steel have been reported by Cole and Richardson.[68] For α-brass (70Cu-30Zn), the effect is even more pronounced, as illustrated by the results of Otterberg et al.,[69] which are reproduced in Fig. 21. Roberts and Ahlblom[57] have attempted to rationalize the nonuniqueness of D_{rex} for a fixed Z. They argue that the dynamically recrystallized grain size can be evaluated from the conditions for site saturation, when the original grain boundaries are completely decorated with new grains; this assumes implicitly that $D_{rex} \ll D_0$. On this basis, it is readily shown that D_{rex} is given by

$$D_{rex} = D^* + 2Ym\tau\rho_{0c} \cdot t_{site\ sat.} \qquad (44)$$

where D* is the critical nucleus diameter (which can be evaluated as a function of Z; see Sec. 2.5.2) and Y is a factor which takes into account the increase in dislocation density in unrecrystallized material between the start of the reaction

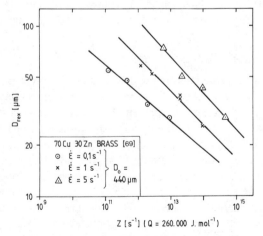

Fig. 21. Dependence of dynamically recrystallized grain size on Zener-Hollomon parameter for 70Cu-30Zn brass.[69]

and site saturation. $t_{site\ sat.}$ can be evaluated from Eq 43,[67] where the growth rate G is equal to $Ym\tau\rho_{0c}$. At a given Z, ρ_{0c} is constant, and it is reasonable, as a first approximation, to assume that the specific areal nucleation rate, I_S, is also uniquely defined by Z (fixed driving force). However, since Eq 44 contains an additional quantity which depends strongly on temperature but not on strain rate, i.e., the grain-boundary mobility, one can expect that, for a given Z, a high-T, high-$\dot\epsilon$ situation will lead to a larger grain size than a low-T, low-$\dot\epsilon$ one, exactly as was found in the investigations quoted above. In fact, making a reasonable estimate of the dependence of I_S on Z, Roberts and Ahlblom[57] were able to apply Eq 44 and reproduce Ahlblom's experimental observations on 18Cr-9Ni steel[17] surprisingly well (these workers evaluated D* and ρ_{0c} as functions of Z from their treatment of the nucleation of dynamic recrystallization; see Sec. 2.5.2).

The arguments outlined in the previous paragraph are all very well, but the fact remains that many investigators *have* found a unique relationship between D_{rex} and Z (σ_s). In this context, it is perhaps significant that anomalous behavior is reported only for materials of very low stacking-fault energy, such as austenitic stainless steels[17,68] and α-brass.[69] Data for nickel,[44,52] α-Fe[43] and low-carbon austenite,[47,48,70] on the other hand, conform to Eq 32. One possible rationalization is that I_S depends on both temperature and driving force; hence, for a given Z, this quantity is greater at high T and high $\dot\epsilon$ than at low T and low $\dot\epsilon$ and compensates for the variation of m in such a way that D_{rex} is approximately constant. This latter argument would apply to materials with relatively high stacking-fault energies. For lower levels of stacking-fault energy, the temperature dependence of I_S is evidently not sufficient to balance the variation in m, and D_{rex} (constant Z) increases with $\dot\epsilon$. The implication is therefore that relatively fast recovery is concomitant with a large temperature dependence of nucleation rate, and vice versa. The above explanation for the different dependences of dynamically recrystallized grain size on Z in materials of relatively high and low stacking-fault energy must be regarded as tentative. The question clearly requires further study.

3.4. Metadynamic Recrystallization

Whenever ϵ_c is exceeded during hot deformation, the presence of pre-existing nuclei implies that any recrystallization after the termination of straining is characterized by zero incubation time and by a driving-force distribution which is heterogeneous. This type of static recrystallization is termed *metadynamic*[71] so as to distinguish it from classical static recrystallization, such as that occurring when $\epsilon < \epsilon_c$. Metadynamic recrystallization will indubitably play an important role in modifying any microstructure which may be developed via dynamic recrystallization in association with industrial hot deformation processing.

The incubation period required for the initiation of static recrystallization decreases with increasing prior strain, if all other factors are equal (T, $\dot\epsilon$, D_0). When $\epsilon = \epsilon_c$, the pre-existence of nuclei created by dynamic recrystallization implies that the incubation time falls to zero. This behavior is well illustrated by results on 18Cr-9Ni austenitic stainless steel reported by Ahlblom[49] and tabulated here in Table 3.

The drastic decrease in the incubation time for static recrystallization, as ε_c is approached, invites the conclusion that the dynamic reaction is, in general, unlikely to be initiated as a result of a number of small accumulated deformation steps, without some measure of static recrystallization proceeding as well. (Of course, it may be possible to engender only dynamic recrystallization via a sequence involving a small strain followed by a much larger one such that $\varepsilon_1 + \varepsilon_2 > \varepsilon_c$.) Furthermore, if the pass strain is greater than ε_c but is insufficient to effect complete dynamic transformation, then the microstructure will be considerably modified by metadynamic recrystallization prior to the next deformation step, even if the interpass time is small, because the metadynamic reaction proceeds very rapidly. The development of microstructure via recrystallization during multipass hot working may thus be very complicated, being controlled to different degrees by static, dynamic and metadynamic processes, the relative contributions from each depending in turn on the sequence of temperature, strain rates, strains and interpass periods.

The progress of the recrystallization reaction following the termination of deformation after previously straining to $\varepsilon > \varepsilon_c$ can be considered in two stages:

1. Recrystallization of regions not previously recrystallized dynamically (if such are present).
2. Reaction of the remaining areas as a result of the residual dislocation density in the dynamically formed grains.

Hence, if one recrystallization cycle has not been completed during previous straining (most likely with single-peak behavior), the metadynamically recrystallized microstructure will be heterogeneous.

Some dispute exists in the literature regarding the grain size resulting from metadynamic recrystallization. Some authors[48,49,72] report that the mean D_{rex} resulting from this reaction is larger than that attainable via complete dynamic recrystallization. Typical data are presented in Fig. 22[49,73] for 18Cr-9Ni stainless

Table 3. Time to X = 0.01 for Static Recrystallization of 18Cr-9Ni Stainless Steel as a Function of Prior Strain

$T = 1000°C$; $\dot{\varepsilon} = 0.96 \ s^{-1}$; $D_0 = 660 \ \mu m$; $\varepsilon_c = 0.58$

Strain (ε)	t(X = 0.01), s
0.17	7 ± 3
0.28	2 ± 0.5
0.50	0.3 ± 0.2

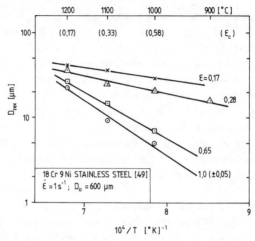

Fig. 22. Temperature dependence of recrystallized grain size in 18Cr-9Ni austenitic stainless steel.[49,73] For the three lower strains, D_{rex} has been measured on specimens held just long enough after straining for completion of recrystallization. For $\varepsilon = 1$, the values refer to specimens quenched directly after deformation.

steel. The figures in brackets refer to ε_c at the various temperatures. Thus, at 1200°C, dynamic recrystallization is virtually complete at a strain of 0.65 and the dynamic and metadynamic points coincide. On the other hand, the dynamic reaction is far from complete at 1000°C for $\varepsilon = 1$; the value given is the grain size in the areas which have recrystallized. According to Ahlblom,[49] full re-crystallization of previously unreacted regions when $\varepsilon = 0.65$ takes 5 s at 1100°C and ~1 s at 1200°C. It is at this point that the specimens have been quenched and the (metadynamic) grain sizes shown in Fig. 22 measured. One can reasonably expect that the grain size developed in areas which have not previously been recrystallized during straining should be larger than that of the dynamically formed grains; in the first place, there is no concurrent de-formation to restrict growth, and secondly, the driving force is progressively diminished as a result of recovery, i.e., the situation is exactly the same as that in static recrystallization.

By way of contrast, Petkovic et al.,[74] studying copper at 500°C, reported observations to the effect that metadynamic recrystallization effects more potent grain refinement than does the dynamic reaction (70 μm as opposed to 170 μm); indeed, in their experiments, even static recrystallization (prestrain $\varepsilon < \varepsilon_c$) resulted in a finer grain size than dynamic recrystallization for the same defor-mation temperature (100 μm for $\varepsilon = 0.15$). The authors' explanation of this behavior rests on the contention that many nuclei become available during meta-dynamic recrystallization, which would not have been viable had straining continued. For the conditions in their tests (500°C, 0.01 s^{-1}), the dynamically recrystallized grain size is, according to them, 170 μm; however, the corresponding value which can be extracted from the results of Bromley and Sellars[46] is 35 μm. One possible rationale for this discrepancy is that the 170-μm grains in the experiments of Petkovic et al. are regions remaining unrecrystallized at the termination of straining. These then disappear at the expense of metadynamic grains with D_{rex} = 70 μm. On this basis, the metadynamic reaction would appear to have resulted in a coarser grain size than the dynamic one, and there is no longer any anomaly. Indeed, the small grains delineating the boundaries of larger ones, in the micrographs corresponding to short holding times shown by Petkovic et al.,[74] undoubtedly give the impression of dynamic recrystallization. Furthermore, it would be most surprising if, as these authors claim, D_{rex} resulting from static recrystallization after ε = 0.15 were actually smaller than that for dynamically formed grains at the same temperature.

Stage 2 of metadynamic recrystallization (see above), i.e., the recrystallization, for a second time, of the areas which have reacted dynamically during prestraining, does not appear to have been studied in any detail. One thing is clear, however, and that is that this part of the reaction is very slow compared with stage 1. For example, Ahlblom[49] has presented evidence to the effect that small grains originating from dynamic recrystallization are present in 18Cr-9Ni stainless steel even after holding for 120 s at 1100°C (ε = 0.65); stage 1 of the metadynamic reaction is completed after 5 s at this temperature.

3.5. Microstructural Changes During Hot Deformation of Coarse Two-Phase Materials

By coarse two-phase materials, we refer to two-constituent alloys characterized by about the same grain size in both phases, e.g., α-β brass and α-γ stainless steels. There have been very few systematic studies aimed at elucidating the microstructural evolution in association with hot deformation of such alloys. In both the instances quoted, one microconstituent undergoes restoration through dynamic recovery plus recrystallization while the other softens via dynamic recovery only. This in itself implies that the two phases are, as separate entities, characterized by quite different flow-stress levels at a given temperature and strain rate.

The hot deformation behavior of α-β brasses has been investigated by Roberts and Otterberg.[75] At normal hot working temperatures (600 to 900°C), standard free-machining 60Cu-40Zn brass either consists of 100% β-phase or contains so much β that the material behaves as though only this constituent were present. However, some special alloys are characterized by large amounts of α even at high temperatures, e.g., dezincification-resistant brasses with a typical analysis of 65Cu-1.7Pb-1.2Al-0.8Sn (rem Zn). Several interesting effects are found in this latter alloy type. Roberts and Otterberg[75] worked on material which had previously been extruded and whose initial microstructure was comprised of bands of α and β parallel to the extrusion direction. Compression cylinders were machined from the extruded bar such that the bands were parallel to or at right angles to the axis of compression.

Some typical σ(ε) curves are shown in Fig. 23 for both types of specimen; the figures in brackets refer to the volume fraction of α at the appropriate temperature. At 850°C, the alloy contains only about 25% α and the flow curve is characteristic of a material undergoing restoration by dynamic recovery alone, i.e., the brass behaves almost as though it were wholly β. At this temperature, the different specimen types exhibit more or less the same flow behavior. The implication is that relatively small amounts (<30-40%) of the harder α-phase have little influence on the high-temperature σ(ε), with flow apparently concentrated mainly in the softer constituent. However, at lower temperatures (>50% α), and especially at 650°C, σ(ε) exhibits a clear maximum for both specimen types, which indicates that the α is deformed to at least the level of the critical strain for dynamic recrystallization. Furthermore, the two sorts of specimen are characterized by very different flow curves, the one where the α/β banding is initially parallel to the axis of compression lying highest and exhibiting a more pronounced stress drop. Metallographic examination reveals that the bands of the softer phase (β), in cases where these lie initially parallel to the

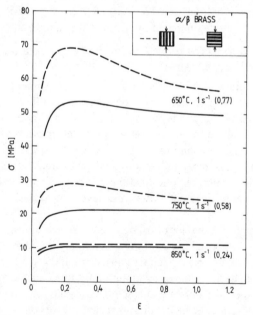

Fig. 23. Typical σ(ε) curves for α/β brass at various temperatures, illustrating the effects of having α/β bands oriented perpendicular and parallel to the axis of compression.

compression axis, are reoriented during deformation so as to be perpendicular to the latter, i.e., the softer constituent is redistributed in such a way as to facilitate the localization of deformation to it. The driving force for such a change is, of course, a reduction in flow stress (work of deformation). This reorientation of the microstructure is illustrated by optical micrographs in Fig. 24. Initially, both constituents in specimens with α/β bands parallel to the axis of compression are deformed. However, redistribution of the phases means that flow is more and more concentrated in the β as straining proceeds and there is a marked stress drop, much more pronounced than in the other type of specimen (α/β bands perpendicular to the axis of compression), where the softer phase is more favorably oriented from the beginning. At 650°C, the amount of α is so high that appreciable flow within this constituent is inevitable; dynamic recrystallization of the α can then take place, as evidenced by the peak in $\sigma(\varepsilon)$ even for specimens with the initial bands at right angles to the compression axis (Fig. 23), and by metallographic observation (Fig. 24). For testing at 750°C, however, the strain

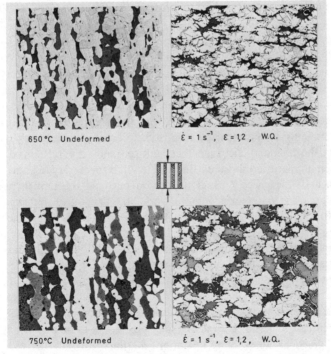

Fig. 24. Reorientation of α/β bands during hot compression of a dezincification-resistant duplex brass; the undeformed microstructure, formed on holding for 3 min at the appropriate temperature and then quenching, is shown on the left.

is taken up principally via deformation of the now larger volume fraction of β-phase, and σ(ε) for the last-mentioned specimen type exhibits no stress peak, i.e., the strain in α does not attain $ε_c$. For this temperature, the sharp stress drop in σ(ε), observed with specimens having α/β bands initially parallel to the axis of compression, derives purely from the reorientation of the β-phase.

It is not unfeasible that the redistribution of the constituents in α/β brass, discussed above, is assisted by the diffusional migration of α/β crystal boundaries, even during high-rate deformation. One piece of evidence is that the difference in flow behavior between the two types of specimen examined by Roberts and Otterberg[75] is strongly strain-rate dependent. Furthermore, it is well established that α⇄β can proceed very rapidly in brasses, because these two phases are in close compositional proximity and transformation thus requires relatively small amounts of diffusion. Clearly, the possibility of deformation-induced phase-boundary migration during deformation of coarse two-phase microstructures is an interesting one, and this subject merits further attention.

3.6. Summary of Current State of Knowledge Regarding Microstructural Development in Connection With Dynamic Recrystallization

The microstructural changes occurring as a result of dynamic recrystallization, and their dependence on the deformation parameters and initial grain size, are quite well understood, at least in a qualitative sense. However, additional experimental studies pertaining to the relationship between dynamically recrystallized grain size and stationary stress, i.e., Z, in materials of varying stacking-fault energy would be welcome. Furthermore, while the mechanism for the initial development of dynamic grains at pre-existing grain boundaries is fairly clear, the details pertaining to the repeated nucleation process, whereby new grains are formed at the boundary between recrystallized and unrecrystallized material, would seem to merit further study. Indeed, it will be necessary to know the strain dependence of the specific areal nucleation rate and the growth rate of the dynamic grains in order to develop a workable quantitative description of the recrystallization kinetics. Factors controlling the spread of $ε_c$ for grain boundaries of varying character must also be elucidated and considered in such a treatment.

The few available results indicate that the dynamic recrystallization kinetics characterizing cast material are much slower than for wrought specimens of equivalent grain size and tested under the same conditions. Because refinement of as-cast microstructures via recrystallization is a process of considerable significance for commercial hot working, further exploration of the recrystallization behavior in cast material, both static and dynamic, is to be commended.

In any practical hot working operation, the microstructure derived from dynamic recrystallization will, following the termination of deformation, be modified by metadynamic processes. The research effort which has hitherto been devoted to metadynamic recrystallization is very meager in relation to the importance of this phenomenon, and much more work is required. It is of particular significance to establish the nature of the metadynamic reaction in micro-structures previously subjected to varying amounts of dynamic recrystallization.

Several practical alloys are dual-phase materials under hot working conditions. The few studies which have been performed indicate that a redistribution of the constituent phases can occur during hot deformation in such a way as to facilitate the preferential localization of strain in the softer constituent. In such a case, the nature of the flow curve depends strongly on the initial distribution of phases in the specimen being tested and its relationship to the principal stress axes. Additional endeavor in this area would be welcome; of particular interest is whether or not the phase boundaries migrate in response to hot deformation, the driving force being the reduction in flow stress derived from a more favorable dispersion of the softer constituent.

4. Crack Development at Temperatures and Strain Rates Characteristic of Hot Working

4.1. Factors That Determine Hot Workability

Microcrack formation in association with hot deformation can be regarded as a dynamic structural change and will therefore be given some consideration in this review. The development of cracks during straining at hot working rates determines the *hot ductility* of a material, which in turn is an important factor dictating *hot workability,* i.e., the extent to which a given metal or alloy can be shaped without cracking in a specific hot working operation.

The hot ductility of a metal or alloy at a fixed temperature and strain rate can be defined in several ways; one is the reduction of area in a tensile test, ψ, which is related to the strain to fracture as $\varepsilon_f = \ln[100/(100 - \psi)]$. There exist a number of indications that ψ is a useful starting quantity for the evaluation of hot workability, i.e. the cracking propensity in a hot working operation.[76] Either ψ or some parameter like it can be regarded as a measure of *intrinsic hot ductility* and is, as such, influenced by microstructure and chemical composition as well as by temperature and strain rate. However, the hot workability of a material *in a given process* depends on a combination of factors, of which intrinsic hot ductility is only one. In addition, cognizance must be given to:

- The detailed geometry of the shape changes engendered by the working operation; of particular interest are those parts of the stock which are subjected to significant tensile strains.
- The surface condition of the stock, especially in regard to the presence of cracks or other defects which might propagate during working.
- The possible effects of the often appreciable frictional forces between stock and tools.

These additional *process-oriented* factors, which in practice will always influence hot workability to a greater or lesser extent, complicate the practical application of intrinsic (laboratory) hot ductility data. It is beyond the scope of the present report to go into the detailed evaluation of hot workability, and the remaining discussion in this chapter is therefore confined to a consideration of the factors that determine intrinsic hot ductility. However, the importance of the process-oriented contributions to the over-all hot workability cannot be ignored whenever one is interested in applying intrinsic data in order to predict the cracking propensity in any working operation.

4.2. Factors That Determine Intrinsic Hot Ductility

The discussion in this section and those following will be confined, in the main, to a consideration of the hot ductility of high-alloy steels. This reflects in part the direction of the author's interests, but also the fact that cracking during hot working is one of the most serious problems plaguing the commercial production of such qualities. In general, inferior ductility, within the temperature range for hot working, stems from one or more of the following phenomena:

- Development of intercrystalline cracks analogous to the wedge-type cracks formed in creep; this is the principal mechanism dictating poor hot ductility in highly alloyed austenitic stainless steels, e.g., 20Cr-25Ni-4Mo-1.5Cu or 17Cr-17Ni-4Mo(+N), and also certain copper alloys (Cu-10Ni, Cu-25Ni, Cu-20Zn-18Ni).

- Crack formation at the interface between δ-ferrite and austenite arising from the tendency for strain concentration in the softer ferrite constituent; this phenomenon is encountered in duplex stainless steels, of course, but it occurs even in austenitic grades in the cast condition, where δ-ferrite is often present as a result of segregation.

- Nucleation and coalescence of voids around hard particles in a matrix of high flow stress; examples of steels exhibiting limited hot ductility as a result of this mechanism are the free-machining stainless steels (because of sulfides) and cold work tool steels or high-speed steels (because of carbides).

- Formation of precipitates at grain boundaries; this is not really of importance at hot working strain rates but is known to be the cause of cracking problems during cooling of large ingots (because of the deformation effected by residual stresses) or during vertical continuous casting (when the strand is subjected to slow bending within the temperature range that permits formation of grain-boundary particles). This phenomenon has been studied in particular detail for microalloyed steels,[77-79] in which precipitation of AlN and Nb(C,N) give trouble; however, since this cracking mechanism lies somewhat outside the scope of the present report, it will not be considered further here.

In what follows, the principal characteristics of the first three cracking mechanisms will be described. The tensile reduction of area is used almost exclusively as a measure of hot ductility; as stated earlier, this quantity is thought to be of particular relevance to evaluation of hot workability in real working processes.[76]

4.2.1. Intercrystalline Wedge Cracking. Single-phase austenitic stainless steels exhibit a ductility trough at 950 to 1000°C, which is due to the development of intercrystalline wedge cracks. The minimum in ψ is especially prevalent in high-alloy steels with more than 3% Mo which have solidified via the formation of primary austenite. Several ψ-T diagrams characteristic of these steels are presented in Fig. 25.[80] Attention is drawn to the following common features:

• As the testing temperature is increased above 1050°C, the ductility improves and reaches a maximum at 1250 to 1300°C before incipient melting in grain-boundary regions causes a drastic impairment of ψ above the burning temperature. The improvement of ductility between 1050 and 1200°C derives from the increased facility of dynamic recrystallization as the deformation temperature is raised (i.e., as ε_c decreases); development of dynamically recrystallized grains at pre-existing grain boundaries (see Sec. 3) effectively prevents growth of wedge cracks. The region in stress-temperature space over which the occurrence of dynamic recrystallization ensures a high level

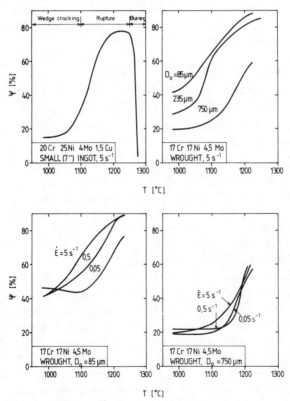

Fig. 25. Effects of temperature, grain size and strain rate on the hot ductility of highly alloyed austenitic stainless steels.

of ductility is usually denoted as "rupture" in fracture-mechanism maps.[81] (N.B. Dynamic recrystallization does not usually occur in commercial hot working of stainless steels; however, its occurrence in laboratory tensile tests is a measure of the likelihood of interpass static recrystallization taking place during, say, rolling, and producing the improvement in ductility that is described.)

- The ductility level in the trough decreases with increasing grain size.[80,82] For cast material, therefore, the principal factor dictating the poor hot ductility is large grain size; as long as no δ-ferrite is present, the normal segregation pattern in an ingot does not seem to exert much negative influence on ψ. The starting grain size also affects the temperature dependence of ψ because of its influence on the kinetics of dynamic recrystallization (Sec. 3.2).

- Varying the strain rate has little effect on the ductility level in the region of intercrystalline cracking; however, the behavior at higher temperatures is affected somewhat, with the improvement in ψ that corresponds to a given temperature increase being more pronounced as $\dot{\varepsilon}$ increases. The influence of strain rate is not well understood, although the behavior in the intercrystalline regime conforms to the Monkman-Grant[83] formalism often quoted in the context of creep fracture, i.e.:

$$\dot{\varepsilon} t_f = C_{MG} \tag{45}$$

where C_{MG} is a constant which can vary between 0.1 and 1 depending on D_0; obviously, for constant $\dot{\varepsilon}$, $C_{MG} = \varepsilon_f$.

Apart from the features illustrated in Fig. 25, the following additional factors are known to affect $\psi(T)$:

- The general level of the ψ-T curve is lowered due to the presence of alloying elements which increase the flow stress of the material. Examples of such additions to austenitic stainless steels are molybdenum (>3%) and nitrogen.[84] These elements are known to affect the rates of both dynamic and static recrystallization,[84] which means that the enhancement of ψ derived from these processes is delayed to higher temperatures in austenitic steels containing molybdenum and/or nitrogen.

- The minimum level of ψ in the ductility trough is very sensitive to the presence of impurities and trace elements which segregate to grain boundaries.[85,86] Many segregants — bismuth, lead, sulfur, antimony, tin and arsenic, in decreasing order of potency — exert deleterious influences on ductility, whereas others, such as boron, raise the ductility level within the regime of intercrystalline cracking. Typical effects for a 16Cr-9Ni-2Mo-base material ($\cong$AISI 316) are illustrated in Fig. 26.[86] Modern converter practice for stainless steel manufacture ensures that the levels of impurities which are harmful to hot workability can be maintained at low values. However, even then, boron is often added (up to ~40 ppm) to

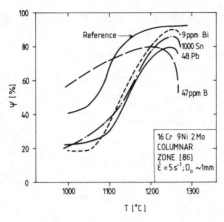

Fig. 26. Influence of impurities (Bi, Pb, Sn) and trace additions (B) on the ductility-temperature relationship for 16Cr-9Ni-2Mo stainless steel.[86]

provide a valuable ductility increase at lower temperatures; on the other hand, this addition carries the disadvantage of a reduced burning temperature. The operative mechanisms that dictate the effects of harmful impurities on hot ductility are discussed further in Sec. 4.3. The influence of boron is not really understood; this element also seems to be effective in improving hot ductility in cases where this is determined by microcracks formed in association with δ-ferrite, i.e., in as-cast lower-alloy austenitic steels like 18Cr-8Ni.

In many ways, it is surprising that intercrystalline wedge cracking can be the operative mechanism determining hot ductility under hot working conditions. After all, the development of wedge cracks requires the occurrence of grain-boundary sliding, which might not be expected to proceed to any great extent for times to fracture of ~0.1 s or less. However, effects of grain-boundary sliding can indeed be observed at hot working strain rates (see Fig. 29, in Sec. 4.3, below). As the test temperature is lowered, the ductility of highly alloyed austenitic steel starts to increase once again (<900°C) and the failure mode becomes ductile and transcrystalline (cavitation around inclusions). A fracture-mechanism map for 17Cr-17Ni-4.5Mo constructed on the basis of the observations of Carlsson et al.[80] is presented in Fig. 27. Some additional boundaries reported for AISI 316[81,87] are included in this figure for comparison. Note the wedge-cracking region, which extends all the way up to the dynamic fracture line for temperatures between 900 and 1050°C. The detailed mechanisms governing the development of intercrystalline wedge cracks during high-rate hot deformation are considered in Sec. 4.3.

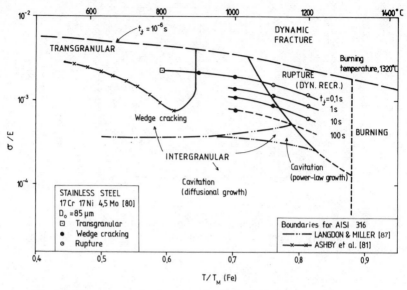

Fig. 27. Fracture-mechanism map for 17Cr-17Ni-4.5Mo stainless steel constructed on the basis of hot tensile tests,[80] and literature data for AISI 316.[81,87]

As regards the improvement of ductility engendered as a result of dynamic recrystallization, no quantitative data for the relationship between ψ and fraction recrystallized have been reported. However, Sandberg and Sandström[88] have measured ψ as a function of X during static recrystallization (two-step tensile testing) of austenitic stainless steels; in fact, this process is much more important for the gradual improvement of hot workability during, say, rolling of an ingot, where the pass strains are never sufficiently high to effect dynamic recrystallization. Their observations for 18Cr-15Ni-4.5Mo (+N) conform to the relationship

$$\psi = \psi_0 + (\psi_{max} - \psi_0) [1 - \exp(-\delta X)] \qquad (46)$$

where ψ_0 is the initial ductility level and ψ_{max} is the value corresponding to complete recrystallization, i.e., the peak of the ψ-T curve. The value of the constant δ is about 10, which means that the ductility improvement attains 90% of its maximum value before X reaches a value of 0.25. This reflects the fact that recrystallization is initiated at pre-existing grain boundaries and thus that the fraction recrystallized required to effectively impair the propagation of inter-crystalline cracks is comparatively low.

4.2.2. Cracking at Ferrite/Austenite Interfaces. Several authors (e.g., Ref 89) have demonstrated that ψ attains a minimum value at a critical amount of δ-ferrite (15 to 30%) during hot tensile testing of stainless steels with various nickel and chromium contents. For austenitic stainless steels, where ferrite is present in segregated ingots, the ductility is therefore improved through prolonged high-temperature homogenization, which reduces the volume fraction of the δ-constituent. The geometry of the ferrite is also important, and coarse, elongated stringers perpendicular to a principal tensile stress would seem to be especially detrimental.[89] On the other hand, austenite-ferrite mixtures in Fe-Cr-Ni alloys can, if sufficiently finely divided, exhibit superplastic characteristics.

The generally accepted explanation for the formation of primeval cracks at δ/γ interfaces under hot working conditions is the difference in softening behavior, and therefore in flow-stress level, between ferrite and austenite. Hence, strain is localized to the former, and the plastic inhomogeneity will lead to crack formation if the interface cohesion is not particularly high. Impurities which reduce interface cohesion will therefore be detrimental as far as hot workability is concerned, e.g., Gittins et al.[90] report that addition of 50 ppm Pb to a duplex 21Cr-4Ni stainless steel renders the material practically unworkable. As regards the temperature dependence of ψ, it will be clear that dynamic or static recrystallization of austenite will not engender any drastic improvement in ductility of ferrite-austenite mixtures, as it does when the matrix is completely austenitic. Nevertheless, ψ can be expected to increase with increasing T, partly because the flow stress is lower, which reduces the tractions at δ/γ interfaces, and partly as a consequence of microstructural changes with increasing temperature, e.g., the amount of δ in cast austenitic steels decreases (homogenization), or ferrite increases at the expense of austenite in duplex steels. The reader requiring a deeper insight into the hot ductility of stainless steels containing both ferrite and austenite is referred to the reviews by Ahlblom and Sandström[12] and by Järvinen.[91]

4.2.3. Void formation around hard particles, as stated earlier, is the cracking mechanism that is responsible for the poor hot ductility exhibited by certain tool steels and high-speed steels. The hot rolling yield from conventional production of such grades is often as low as 50%. These alloys have the common features that they undergo a eutectic solidification and that their ingot structures are characterized by networks of primary carbides. The hot ductility in the cast condition is then determined by the size of the eutectic cells (secondary-dendrite-arm spacing) and by the nature of the eutectic itself. The value of ψ increases with decreasing λ_2 (Fig. 28), because the microcracks created via the coalescence of voids within the eutectic colonies become shorter and require a greater strain before they themselves can coalesce and engender fracture.[91] Furthermore, for high-speed steels, the ductility associated with a plate eutectic is considerably inferior to that for a finely divided lamellar one; this is a consequence of the unfavorable shape of the particles, from the point of view of void formation, in the plate eutectic.[92]

During working, and especially in connection with hot rolling, the cells of primary carbides are stretched out into bands parallel to the rolling direction. The

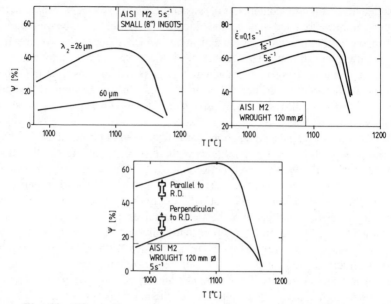

Fig. 28. Effects of temperature, strain rate, secondary-dendrite-arm spacing (cast material) and orientation of carbide bands relative to the tensile axis (wrought material) on hot ductility of M2 high-speed steel.

particles in such bands are so closely spaced that, once cavitation has occurred, under the action of a tensile stress, coalescence ensues over a very small interval of subsequent strain, which results in formation of long microcracks. Fracture then proceeds very rapidly if a tensile stress is applied at right angles to the carbide bands, i.e., transverse with respect to the rolling direction.[93] In consequence, wrought high-speed steel exhibits a characteristic anisotropy of hot ductility (Fig. 28), with ψ_T being considerably less than ψ_L. It can be noted that the transverse ductility might well *deteriorate*, as a result of working, to a level below that for cast material; this is quite different from the situation with austenitic steels, for which hot working of ingots always improves ductility. These effects can give rise to severe problems during hot rolling of high-speed steels, especially in situations where the stock is subjected to even quite moderate strains in the transverse direction.[76,93] After a large degree of hot working, e.g., rolling to rod, the bands of carbides are broken up to a sufficient extent that the transverse hot ductility is restored once again to about the same level as that in the longitudinal direction. It is interesting to note that high-speed steels manufactured via powder metallurgy, instead of by conventional melting and casting, are characterized by an excellent level of hot ductility with little or no anisotropy; this is a direct result of the dispersion of small, randomly distributed carbides which is attainable via powder processing.[93]

The characteristic temperature and strain-rate dependence of ψ for high-speed steels and tool steels is illustrated in Fig. 28. The ductility increases systematically with a decrease in the flow stress (increased temperature, reduced strain rate); hence, a better hot workability is expected for operations involving relatively slow deformation, e.g., forging. At temperatures exceeding ~1130°C, incipient melting within the eutectic regions or the bands of carbides causes the ductility to drop drastically over a very small temperature interval (burning).

The micromechanisms governing high-temperature fracture in high-speed steels and tool steels are discussed in greater detail in Sec. 4.4.

4.3. Micromechanisms of Intercrystalline Crack Development During Hot Working of Austenitic Stainless Steels

A detailed microstructural study of the evolution of intercrystalline cracks, during hot tensile testing of 17Cr-17Ni-4.5Mo in the temperature range 950 to 1050°C, has been undertaken by Carlsson, Lehtinen and Roberts.[80] The principal observations reported by them are summarized below:

- There is evidence for the occurrence of grain-boundary sliding during hot tensile testing (Fig. 29a), even at $\dot{\epsilon} = 10\ \text{s}^{-1}$.

- Grain-boundary wedge cracks are initiated at triple points after fairly low strains (<0.05). The development of these microcracks proceeds via the nucleation, growth and coalescence of cavities in the crack-tip region (Fig. 29b); this growth mechanism is further supported by the dimpled appearance of the macroscopically intergranular fracture.

- Cracks are usually found on grain boundaries which make an angle of less than 45° with the tensile axis, but are by no means exclusively confined to those boundaries lying exactly normal to the applied tensile stress.

- The number of cracks per grain (or the area fraction of cracks in a section) is not greatly affected by the hydrostatic stress component, e.g., by necking in a tensile test.

- Voids develop ahead of a microcrack as a result of grain-boundary sliding. The dearth of grain-boundary particles in the material, however, precludes their being the principal void-nucleation sites. During deformation, a local migration of grain boundaries occurs, giving rise to both distinct steps and variations of curvature; the steps invariably appear at intersections between subgrain and grain boundaries. Cavities formed ahead of a crack are often associated with bulges or steps in the boundary; a good example of a void, which appears to have been nucleated at a grain-boundary step as a result of sliding between adjacent grains, is shown in Fig. 29(c). Voids will not be stable unless the pressure due to their curvature is balanced by the applied stress, i.e., unless $r_v > 2S/\sigma$. With $\sigma \simeq 100$ MPa and $S = 0.8$ J·m^{-2}, which are typical for austenitic stainless steels at 1000°C, then we find that stable voids can form at steps with heights $\gtrsim 0.05$ μm, which seems quite tenable.

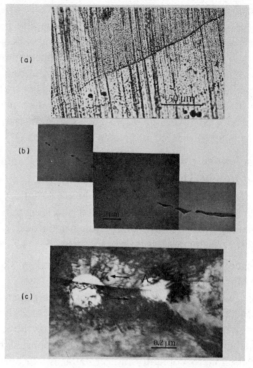

Fig. 29. Growth mechanism for intercrystalline wedge cracks in 17Cr-17Ni-4.5Mo austenitic stainless steel. (a) Evidence for grain-boundary sliding ($\varepsilon = 0.1$ at 1000 °C, 5 s^{-1}). (b) Crack growth via coalescence of cavities ($\varepsilon = 0.3$ at 1000°C, 5 s^{-1}; SEM). (c) Cavity nucleation at a grain-boundary step due to grain-boundary sliding as indicated by the arrows ($\varepsilon = 0.25$ at 1000°C, 5 s^{-1}; HVEM).

In attempting to develop a quantitative model for intercrystalline crack evolution during hot working, it is logical to apply the extensive body of theoretical work which has been directed toward the description of intergranular failure under creep conditions. However, models of the type proposed by Nix et al.[94] and by Vitek and Wilkinson,[95] which treat void coalescence with a wedge crack under the assumption that the elastic stress concentration at the tip of the crack is not relaxed via creep of the surrounding grains, are not likely to be applicable to hot tensile testing, where the intragranular strain will be appreciable (sufficient to effect dynamic recrystallization under appropriate conditions). Moreover, it is a reasonable assumption that at high strain rates, the contribution of the diffusional flow of matter to crack-tip cavity growth is negligible. For this reason, many models are not applicable to the description of hot ductility in high-rate tensile testing.

Edward and Ashby[96] have considered the situation in which grain-boundary cavities grow via power-law creep alone; their approach has been extended to include the effects of multiaxial stress states by Cocks and Ashby.[97] Carlsson, Lehtinen and Roberts[80] have proposed a modified form of these treatments, which provides a description of the growth of intercrystalline wedge cracks during high-rate tensile testing at constant $\dot{\varepsilon}$. The principal assumptions invoked are listed below:

• Wedge cracks are initiated at triple points due to grain-boundary sliding. These cracks grow via the incorporation of cavities generated locally in the crack-tip region.

• The presence of grain-boundary microcracks imparts a deformational inhomogeneity, with the local stress, and thus the local strain rate, being greater in the boundary regions than in the grain interiors, i.e., $\dot{\varepsilon}_g > \dot{\varepsilon}_i$, where both these quantities are equal to the externally applied strain rate in the absence of cracks. $\dot{\varepsilon}_i$ is defined by Eq 2, which is rewritten, in order to conform with the original presentation,[80] as

$$\dot{\varepsilon}_i = \dot{\varepsilon}_0 \exp (\sigma/\sigma_0). \qquad (47)$$

• Grain boundaries slide during deformation so that the increase in volume of a slab of material on either side of the boundary is taken up by the relative rigid-body displacement of adjoining grains. The rapidly deforming slab is constrained by the interiors of the surrounding grains to contract laterally at $\dot{\varepsilon}_i/2$ (see Fig. 30).

• The voids associated with wedge cracks are spherical at all stages of their development. This assumption is certainly not justified for simple tension,[97] but constitutes an acceptable approximation.

• Development of intercrystalline cracks is presumed to proceed at a constant stress level; this is probably not too bad an assumption, since the work-hardening rate has usually fallen off considerably by the time wedge cracks have nucleated.

If wedge cracks of initial length L_I are formed at time t_I after the commencement of straining, then the initial area fraction of cracks is

$$F_I = k_1 L_I D_0 N_A \qquad (48)$$

where N_A is the areal crack frequency and k_1 is a geometrical constant ($1/2\sqrt{2}$ for tetrakaidecahedral grains). Crack growth proceeds, as depicted in Fig. 30, via the incorporation of cavities of initial size r_0 at a distance ℓ ahead of the crack tip. The cavity grows rapidly because of the high local strain rate in the grain-boundary region, $\dot{\varepsilon}_g$. When the cavity radius is $\ell/2$, localized necking between the void and the microcrack can no longer be constrained by the surrounding material, and coalescence ensues over a strain interval which is almost

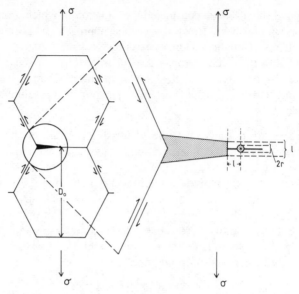

Fig. 30. Model for the extension of wedge cracks; for discussion, see text.

vanishingly small. At this point, a new cavity is envisaged to have nucleated ahead of the crack, and the unit process is repeated.

Following Edward and Ashby,[96] an upper bound for the strain rate in the grain-boundary regions is

$$\dot{\varepsilon}_g = \dot{\varepsilon}_0 \exp \frac{\sigma}{(\sigma_0(1 - F))} \tag{49}$$

In other words, the strain rate within the grain-boundary slab is faster because the stress acting locally on this region is greater by a factor of $1/(1 - F)$. If the thickness of the rapidly deforming (at $\dot{\varepsilon}_g$) region is ℓ, which is the diameter of each cavity when it coalesces with the main crack, then $\dot{\varepsilon}$, $\dot{\varepsilon}_g$ and $\dot{\varepsilon}_i$ are related as

$$\dot{\varepsilon}_i = \dot{\varepsilon} - \frac{\ell}{D_0}(\dot{\varepsilon}_g - \dot{\varepsilon}_i) \tag{50}$$

By invoking the third assumption in the list above and comparing expressions for the rate of change of cavity volume, it is easy to develop a relation for the time required for the unit process of the coalescence of a single void with a developing crack, i.e.:

$$dt = \frac{\ell}{\dot{\varepsilon}_i} \ln \left(\frac{H + 0.25\dot{\varepsilon}_i}{H + f_i\dot{\varepsilon}_i} \right) \tag{51}$$

where $H = (\dot{\varepsilon} - \dot{\varepsilon}_i) \cdot D_0/\ell$ and $f_i = (r_0/\ell)^2$. The corresponding increase in F is $1.5k_1\ell D_0 N_A$, and so the following expression for the rate of growth of wedge cracks is obtained:

$$\frac{dF}{dt} = \frac{3k_1\ell D_0 N_A}{2} \cdot \dot{\varepsilon}_i \left[\ln\left(1 + \frac{\dot{\varepsilon}_i}{\dot{\varepsilon} - \dot{\varepsilon}_i} \cdot \frac{\ell}{4D_0}\right) \right]^{-1} \qquad (52)$$

where $\dot{\varepsilon}_i$ is a function of F. The functional dependence is derived from Eq 47, 49 and 50 as

$$\dot{\varepsilon}_i = \dot{\varepsilon} - \dot{\varepsilon}_i \cdot \frac{\ell}{D_0} \left[\left(\frac{\dot{\varepsilon}_i}{\dot{\varepsilon}_0}\right)^{F/1-F} - 1 \right] \qquad (53)$$

In order to obtain an implicit dependence, $\dot{\varepsilon}_i$ in the brackets is replaced by $\dot{\varepsilon}$, giving

$$\dot{\varepsilon}_i = \dot{\varepsilon} \left\{ 1 + \frac{\ell}{D_0} \left[\left(\frac{\dot{\varepsilon}}{\dot{\varepsilon}_0}\right)^{F/1-F} - 1 \right] \right\}^{-1} \qquad (54)$$

which is a very good approximation for $\dot{\varepsilon}_i$ provided that $\dot{\varepsilon}_0 \ll \dot{\varepsilon}$ and $\dot{\varepsilon}_i$, as is nearly always the case. The numerical evaluation of t_f is considerably simpler if Eq 54 is used instead of Eq 53. Equation 52 can be integrated between the limits:

$$F = F_I \qquad at \qquad t = t_I$$

and

$$F = F_f \qquad at \qquad t = t_f$$

provided that some functional dependence of N_A on D_0 is known. The strain to fracture is evaluated as $\dot{\varepsilon}(t_I + t_f)$, which is reasonable as long as the ductility is relatively low (limited necking).

Experimental measurements on 17Cr-17Ni-4.5Mo stainless steel, via the metallographic examination of deformed tensile specimens, indicate that $N_A = k_2/D_0$, where for tetrakaidecahedral grains $k_2 = 8$ mm^{-1}.[80] Hence, from Eq 48, $F_I = 2\sqrt{2} L_I$, where L_I is intuitively expected to increase with increasing grain size; for example, if L_I is proportional to the wedge displacement, then $L_I = k_3 \varepsilon_{I,gbs} D_0$, where $\varepsilon_{I,gbs}$ is that portion of ε_I that is derived from grain-boundary sliding. F_f is troublesome to specify; some suggestions are given by Carlsson et al.[80] However, the computed ε_f is, in fact, rather insensitive to the exact value chosen for F_f between 0.2 and 1.

The predicted dependence of ε_f on initial grain size obtained via numerical integration of Eq 52 and 54 is shown in Fig. 31 (ε_I is approximated to zero). The values for the limits F_I and F_f at different D_0's are indicated in the diagram; these concur well with the lengths of individual wedge cracks found at small strains. The experimentally observed dependence of ε_f on D_0 is quite well reproduced by the theory, although the uncertainty regarding the exact values of the integration

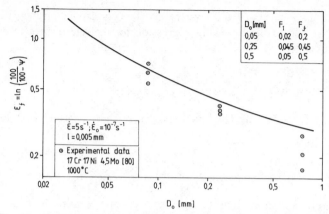

Fig. 31. Variation of fracture strain with initial grain size predicted by the model for intercrystalline crack extension; experimental data for 17Cr-17Ni-4.5Mo stainless steel[80] are included for comparison.

limits, particularly F_I, precludes a concerted attempt to replicate the observed behavior. Values of ε_f calculated using the more accurate expression, Eq 53, differ from those evaluated on the basis of Eq 54 by less than 1%.

The computed effects of varying $\dot{\varepsilon}$ and $\dot{\varepsilon}_0$ on the strain to fracture are demonstrated in Table 4. The following points can be noted:

- If all other factors are equal, ε_f is determined uniquely by $\dot{\varepsilon}/\dot{\varepsilon}_0$.
- A definite dependence of ε_f on strain rate is predicted; this is not in agreement with the behavior found experimentally, which conforms to the Monkman-Grant formalism (Eq 45; see Sec. 4.2.1). The anomaly is resolved if the amount of grain-boundary sliding decreases with increasing strain rate, i.e., L_I, and therefore F_I, are reduced as $\dot{\varepsilon}$ is raised in such a way that the dependence of ε_f on $\dot{\varepsilon}$ indicated in Table 4 is counteracted.

Table 4. Effects of Strain Rate and $\dot{\varepsilon}_0$ on Strain to Fracture Evaluated From Eq 52 and 54

D_0, μm	$\dot{\varepsilon}$, s^{-1}	$\dot{\varepsilon}_0$, s^{-1}	F_I	F_f	$\varepsilon_f - \varepsilon_I$
0.5	5	10^{-5}	0.05	0.5	0.564
0.5	5	10^{-7}	0.05	0.5	0.337
0.5	5	10^{-9}	0.05	0.5	0.223
0.5	0.05	10^{-7}	0.05	0.5	0.564
0.05	5	10^{-5}	0.02	0.2	1.245
0.05	5	10^{-7}	0.02	0.2	0.885
0.05	5	10^{-9}	0.02	0.2	0.685
0.05	0.05	10^{-7}	0.02	0.2	1.245

- A reduction in $\dot{\varepsilon}_0$, which corresponds to an increased flow-stress level, effects a corresponding decrease in ε_f, in qualitative accord with experimental experience (see Sec. 4.2.1).

In summary, it would appear that the treatment of wedge-crack growth presented in this section can account quite well for the dependence of tensile hot ductility on grain size and flow-stress level in austenitic stainless steels tested within the temperature regime of intercrystalline failure. However, more accurate information concerning the details of wedge-crack nucleation is required if further refinement of the theory — e.g., in order to rationalize the insensitivity of ductility to strain rate — is to be meaningful.

The approach presented above is capable of providing some indication as to the differences in shape of $\psi(T)$ curves associated with an alteration in $\dot{\varepsilon}$ (Fig. 25). For a given D_0, the temperature at which the ductility starts to improve, as a result of the increasing ease of dynamic recrystallization, is progressively shifted upward as the strain rate is reduced from 5 to 0.05 s^{-1}. This is also reflected in the areal density of wedge cracks, which increases as $\dot{\varepsilon}$ is diminished.[80] It would therefore appear that the occurrence of dynamic recrystallization is actually promoted by higher strain rates, which is the reverse of experimental experience from compression testing where ε_c is found to decrease with $\dot{\varepsilon}$;[49] one is forced to conclude that the inhomogeneity of deformation, when wedge cracks develop, drastically modifies the normal dependence of dynamic recrystallization kinetics on strain rate.

Combination of Eq 50 and 52 leads to the following expression for the local strain in the grain-boundary slab:

$$\varepsilon_g \doteq \frac{\dot{\varepsilon}D_0}{4\ell^2}\int_{F_I}^{F_f}\frac{1}{\dot{\varepsilon}_i}\left[1 - \left(1 - \frac{\ell}{D_0}\right)\frac{\dot{\varepsilon}_i}{\dot{\varepsilon}}\right]\cdot\ln\left(1 + \frac{\dot{\varepsilon}_i}{\dot{\varepsilon} - \dot{\varepsilon}_i}\frac{\ell}{4D_0}\right)\cdot dF \quad (55)$$

where $\dot{\varepsilon}_i$ is given by Eq 53 or 54. Because $F(\varepsilon)$ can be evaluated from Eq 52 and 54, it is possible to define ε_g as a function of the macroscopic tensile strain, ε. Such compilations (see Fig. 32) indicate that, at a given level of macroscopic strain, the degree of strain localization in the grain-boundary slab is considerably greater at higher $\dot{\varepsilon}$. Hence, for testing at a specified temperature, ε_c will be exceeded locally (in the vicinity of wedge cracks) at progressively lower values of ε as $\dot{\varepsilon}$ is raised; conversely, the ductility improvement derived from dynamic recrystallization starts to be apparent at progressively lower temperatures with increasing $\dot{\varepsilon}$. Both effects are found experimentally (Fig. 25). This calculation should be regarded as very preliminary, especially if it is necessary to postulate a strain-rate dependence of ε_I and/or L_I in order to obtain agreement with the Monkman-Grant equation (see above); in such a case, the separation between the curves for different strain rates in Fig. 32 will be diminished.

As a final point in this section, it is of interest to try to rationalize the influences of impurities and trace elements on the hot ductility of stainless steels

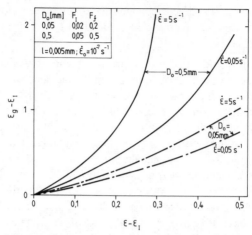

Fig. 32. Calculated local strain in the vicinity of grain boundaries during growth of intercrystalline cracks, as a function of macroscopic strain for two strain rates and two initial grain sizes.

(Sec. 4.2.1) in terms of the postulated model for wedge-crack growth. In this context, an important contribution has been made by McLean,[98] who considers the influence of impurity-atom segregation to grain boundaries on creep cavitation. He proposes that segregants can assist the nucleation of cavities either by reducing grain-boundary cohesion (i.e., nucleation occurs at lower stresses) or via a reduction of the energy of the internal surfaces of voids, which decreases the critical size, $r^* = 2S/\sigma$. In terms of the model sketched in Fig. 30, we can thus expect that r_0 is lower for a segregated boundary, but even ℓ will be reduced because of the increased cavity-nucleation rate. The net effect of the reduced ℓ, associated with a segregated boundary, is in fact to lower dF/dt (Eq 52) and thereby give rise to an increased fracture strain, which is obviously erroneous. This discrepancy might be rationalized in terms of the number of voids developing ahead of a wedge crack at any instant in time. The situation envisaged in the treatment outlined in this section and depicted in Fig. 30, i.e., the growth and coalescence of voids singly, is almost certainly oversimplified. In reality, a number of cavities at different stages of growth will exist ahead of the crack (e.g., see Fig. 29b) and, in a more realistic treatment, Eq 52, describing the rate of growth of wedge cracks, will have to be modified. Obviously, dF/dt will increase as the number of simultaneously developing cavities in front of the crack increases. On this basis, the reduced ε_f in austenitic steels containing certain grain-boundary segregants is understandable.

Another feasible line of reasoning, which can account for the effects of impurities such as bismuth, lead, arsenic and sulfur in reducing ε_f, is that these promote grain-boundary sliding and therefore increase L_I (F_I). The positive effect of boron on the hot ductility of austenitic stainless steels, within the range of intercrystalline wedge cracking, might then be explained via the converse argument, i.e., that this addition limits grain-boundary sliding.

4.4. Cracking Micromechanisms During Hot Deformation of High-Speed Steels and Tool Steels

The principal factors that determine hot ductility for these grades have been outlined in Sec. 4.2.3. Apart from temperature and strain rate, ψ depends strongly on the size of the eutectic cells formed during solidification. Furthermore, the hot ductility of partly worked material exhibits pronounced anisotropy, and ψ is much lower for specimens oriented with the tensile axis perpendicular to the rolling direction (i.e., perpendicular to the bands of primary carbides) than when the tensile axis lies parallel to the rolling direction.

For temperatures below the range for burning ($\lesssim 1100°C$ for high-speed steel), the level of ductility is raised by increasing the temperature or lowering the strain rate, i.e., via a reduced flow stress (Fig. 28). This behavior is summarized in Fig. 33, in which ε_f is given as a function of nominal fracture stress (maximum load during tensile testing $\div$ original area); the data refer to partially worked (120-mm $\square$) AISI M2 high-speed steel (0.85C-4Cr-5Mo-6W-2V). It is seen that measurements from different temperatures and strain rates fall on a unique line; furthermore, the slopes of the lines for ψ_L and ψ_T are the same. A similar behavior

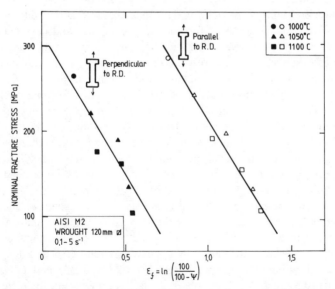

Fig. 33. Strain to fracture from tensile testing of wrought M2 high-speed steel plotted as a function of flow-stress level for two orientations of the tensile axis with respect to the rolling direction.

is also found for a 12%–Cr cold work tool steel (AISI D6);[93] however, in this instance, the banding of carbides in partly processed stock is not as pronounced as with high-speed steel, and the lines for ψ_L and ψ_T are not as widely separated.

The total fracture strain (ε_f) characterizing the high-rate hot tensile testing of high-alloy steels containing fairly coarse, hard particles is composed of two distinct contributions — the nucleation strain for voids (ε_{nuc}) and the strain required for the voids to grow, coalesce and cause fracture (ε_{gr}). Carlsson and Roberts[93] argue that ε_{nuc} is, under hot working conditions, determined by the level of *applied* stress. This is quite at variance with the situation prevailing during cold deformation, where void formation is controlled by internal stresses in the immediate vicinity of hard inclusions; such stresses are, however, likely to be very small at high temperatures due to very rapid recovery within the arrays of the geometrically necessary dislocations which are created at the particles. On this basis, one can expect ε_{nuc} to increase with increasing T or decreasing $\dot{\varepsilon}$, the exact relationship depending on the cohesive strength of the carbide/matrix interface and its possible variation with temperature. Conversely, ε_{gr} is independent of T and $\dot{\varepsilon}$ but is strongly influenced by the particle dispersion; hence, it can be anticipated that this quantity is different for specimens whose tensile axes are oriented at right angles to and parallel to bands of carbides. This interpretation casts some light on the behavior shown in Fig. 33; the slopes of the lines for ψ_L and ψ_T derive from the stress dependence of ε_{nuc} whereas their separation is due to the differences in ε_{gr} for the two types of specimens.

The close proximity of particles in the bands implies that the additional strain required, over and above ε_{nuc}, for coalescence of the voids associated with adjacent carbides is very small.[93] If the strings are oriented at right angles to the tensile stress, then the long rows of voids develop rather quickly into microcracks as deformation proceeds, and fracture ensues within a small strain interval following cavity nucleation, i.e., ε_{gr} is very small and, accordingly, ε_f is only slightly larger than ε_{nuc}. For the other type of specimen, the adjacent voids, which are formed around particles within bands lying parallel to the tensile axis, coalesce fairly rapidly, but no long cracks are generated, i.e., void growth in the direction perpendicular to the tensile stress is very restricted. Instead, fracture of such a specimen is controlled by the *very short* portions of the original eutectic network lying perpendicular, or nearly so, to the tensile axis. The primeval microcracks so formed are short and relatively widely dispersed, and considerable additional strain is required for these to grow, coalesce and give rise to final fracture; in other words, ε_{gr} is large and $\varepsilon_f \gg \varepsilon_{nuc}$. The picture outlined above rationalizes the difference in ε_f between transverse and longitudinal specimens of wrought high-speed steel (Fig. 28 and 33) and is supported unequivocally by metallographic observation.[93] Furthermore, the difference in ε_{gr} between the two types of specimens would be expected to depend on the details of the carbide dispersion and not on T and $\dot{\varepsilon}$ (i.e., flow stress), which is exactly what is found experimentally (Fig. 33).[93]

4.5. Crack Development During Hot Working: Summary of Current State of Knowledge

Crack formation during hot working requires tensile stresses. Unfortunately, tensile stresses can never be completely eliminated via process-technical precautions such as edge rolling, box-pass rolling and so on. One is therefore forced to try to improve the intrinsic hot ductility of the stock, and in some cases this has been done. For example, Swedish austenitic stainless steels produced by modern converter processing are characterized by very low levels of those residuals which are deleterious to ductility; the introduction of this practice has resulted in a considerable improvement in yield from ingot rolling of such grades.

At the present time, our understanding of the factors governing intrinsic hot ductility is incomplete, even at a qualitative level. The over-all cracking mechanisms are relatively well established, but many microscale details require further elucidation. When failure derives from wedge cracking, for example, the growth of cracks proceeds via the incorporation of grain-boundary cavities, but the exact mechanism whereby these are nucleated is still uncertain, although a suggestion has been given in Sec. 4.3. Furthermore, the role played by impurities and trace additions (like B) in the development of intercrystalline cracks warrants additional study, as does the improvement in ductility derived from static and dynamic recrystallization.

For duplex alloys comprising one constituent with a relatively low flow stress and one with a high flow stress, the treatments that consider crack development in terms of the differences in strength and rate of softening between the two phases are not very sophisticated (see Järvinen[91]) and should be improved. Again, the mechanisms pertaining to the influences of residuals and other trace additions on ductility remain unexplored.

The micromechanisms of fracture in high-speed steels and tool steels under hot working conditions, where voids are formed in association with hard carbides, have been studied only superficially. Important questions which remain to be elucidated are the relative significance of the applied stress and internal stresses developed around particles in the initiation of cavities, and the precise influence of the size distribution and dispersion of carbides on the strain to fracture and on the anisotropy of this quantity in partially wrought material.

Acknowledgments

A good deal of the information given in this overview was taken from work carried out at the Swedish Institute for Metals Research during the period from 1975 to 1982 — work which has not previously been presented in the open literature. This endeavor has been financed jointly by the Scandinavian steel- and copper-producing industry and the Swedish Board for Technical Development. Thanks are due to Mrs. Hilkka Andersson for painstakingly typing the manuscript.

References

1. R. C. Gifkins, "The Influence of Thallium on the Creep of Lead," *J. Inst. Metals,* Vol 81, 1952-53, p 417.
2. R. C. Gifkins, "The Effect of Prior Strain on the Creep of High-Purity Lead," *J. Austr. Inst. Metals,* Vol 2, 1957, p 66.
3. R. C. Gifkins, "The Effect of Stress on the Recrystallization of Lead During Creep," *J. Inst. Metals,* Vol 86, 1957-58, p 15.
4. R. C. Gifkins, "Recrystallization of Lead During Creep," *J. Inst. Metals,* Vol 87, 1958-59, p 255.
5. R. C. Gifkins, "Oxygen as an Impurity in High-Purity Lead," *Acta Met.,* Vol 6, 1958, p 132.
6. D. Hardwick, C. M. Sellars and W. J. McG. Tegart, "The Occurrence of Recrystallization During High-Temperature Creep," *J. Inst. Metals,* Vol 90, 1961-62, p 21.
7. J. J. Jonas, C. M. Sellars and W. J. McG. Tegart, "Strength and Structure Under Hot-Working Conditions," *Met. Rev.,* Vol 14, 1969, p 1.
8. H. J. McQueen and J. J. Jonas, "Recovery and Recrystallization During High-Temperature Deformation," in *Plastic Deformation of Metals,* edited by R. J. Arsenault, Academic Press, New York, 1975, p 393-493.
9. C. M. Sellars, "Recrystallization of Metals During Hot Deformation," *Philos, Trans. Roy. Soc.,* Vol 288, 1978, p 147.
10. H. Mecking and G. Gottstein, "Recovery and Recrystallization During Deformation," in *Recrystallization of Metallic Materials,* edited by F. Haessner, Riederer Verlag, Stuttgart, 1979.
11. C. M. Sellars, "Dynamic recrystallization," *Metals Forum,* Vol 4, 1981, p 75.
12. B. Ahlblom and R. Sandström, "Hot Workability of Stainless Steels: Influence of Deformation Parameters, Microstructural Components and Restoration Processes," *Int. Met. Rev.,* Vol 27, 1982, p 1.
13. K. -E. Hensger, "Dynamische Realstrukturprozesse bei Warmumformung," *Freiberger Forschungshefte,* No. B232, 1982.
14. F. Garofalo, *Fundamentals of Creep and Creep Rupture in Metals,* Macmillan, New York, 1965.
15. C. Zener and J. H. Hollomon, "Effect of Strain Rate Upon the Plastic Flow of Steel," *J. Appl. Phys.,* Vol 15, 1944, p 22.
16. O. D. Sherby and C. R. Burke, "Mechanical Behaviour of Crystalline Solids at Elevated Temperature," *Prog. Mater. Sci.,* Vol 13, 1967, p 325.
17. B. Ahlblom, "A Metallographic Study of Dynamic Recrystallization in an 18/9 Austenitic Stainless Steel," Swedish Institute for Metals Research Report No. 1208, 1977.
18. F. Garofalo, O. Richmond, W. F. Domis and F. von Gemmingen, "Strain-Time, Rate-Stress and Rate-Temperature Relations During Large Deformations in Creep," in *Proc. Joint Int. Conf. on Creep,* Institution of Mechanical Engineers, London, 1963, p 1-31.
19. S. K. Samanta, "Resistance to Dynamic Compression of Steel at Elevated Temperatures and High Strain Rates," Licentiat thesis, Royal Institute of Technology, Stockholm, 1966.
20. E. Hakulinen and W. Roberts, "Classification and Utilization of Flow-Stress Data for Steels Under Hot-Working Conditions," to appear in the report series of the Swedish Institute for Metals Research, 1982.

21. H. Mecking and K. Lücke, "Quantitative Analyse der Bereich III-Verfestigung von Silber-Einkristallen," *Acta Met.*, Vol 17, 1969, p 279.
22. K. Lücke and H. Mecking, "Dynamic Recovery," in *Inhomogeneity of Plastic Deformation*, edited by R. E. Reed-Hill, ASM, Metals Park, OH, 1973, p 223-250.
23. R. W. Bailey, "Note on the Softening of Strain-Hardened Metals and Its Relation to Creep," *J. Inst. Metals*, Vol 35, 1926, p 27.
24. E. Orowan, "The Creep of Metals," *J. West Scotland Iron Steel Inst.*, Vol 54, 1946-47, p 45.
25. U. F. Kocks, "Laws for Work Hardening and Low-Temperature Creep," *J. Eng. Mater. Tech.* (ASME H), Vol 98, 1976, p 76.
26. H. Mecking, U. F. Kocks and H. Fischer, "Hardening, Recovery and Creep in f.c.c. Mono-and Polycrystals," in *Proc. 4th Int. Conf. on Strength of Metals and Alloys*, Nancy, 1976, Vol 1, p 334.
27. H. Mecking, "Description of Hardening Curves of f.c.c. Single and Polycrystals," in *Work Hardening in Tension and Fatigue*, edited by A. W. Thompson, AIME, New York, 1977, p 67-88.
28. U. F. Kocks and H. Mecking, Discussion of Ref 30, *J. Eng. Mater. Tech.* (ASME H), Vol 98, 1976, p 121.
29. U. F. Kocks and H. Mecking, "A Mechanism for Static and Dynamic Recovery," in *Proc. 5th Int. Conf. on Strength of Metals and Alloys*, Aachen, 1979, Vol 1, p 345.
30. P. Öström and R. Lageneborg, "A Recovery–Athermal Glide Creep Model," *J. Eng. Mater. Tech.* (ASME H), Vol 98, 1976, p 114.
31. P. Öström and R. Lagneborg, "A Dislocation Link-Length Model for Creep," *Res Mechanica*, Vol 1, 1980, p 159.
32. H. Friedel, *Dislocations*, Pergamon Press, Oxford, 1964, p 239.
33. E. Voce, "The Relationship Between Stress and Strain for Homogeneous Deformation," *J. Inst. Metals*, Vol 74, 1948, p 537.
34. C. N. Ahlguist and W. D. Nix, "The Measurement of Internal Stresses During Creep of Al and Al–Mg Alloys," *Acta Met.*, Vol 19, 1971, p 373.
35. J. P. Poirier, "Is Power-Law Creep Diffusion Controlled?", *Acta Met.*, Vol 26, 1978, p 629.
36. H. J. McQueen, W. A. Wong and J. J. Jonas, "The Deformation of Aluminium at High Temperatures and Strain Rates," *Canad. J. Phys.*, Vol 45, 1967, p 1225.
37. C. M. Sellars and W. J. McG. Tegart, "La Relation Entre la Résistance et la Structure Dans le Déformation à Chaud," *Mém. Sci. Rev. Met.*, Vol 63, 1966, p 731.
38. J. Cotner and W. J. McG. Tegart, "High-Temperature Deformation of Aluminium-Magnesium Alloys at High Strain Rates," *J. Inst. Metals*, Vol 97, 1969, p 73.
39. D. J. Abson and J. J. Jonas, "Substructure Strengthening in Zirconium-Tin Alloys," *J. Nucl. Mater.*, Vol 42, 1972, p 73.
40. J. H. Gittus, "Theoretical Equation For Steady-State Dislocation Creep: Effects of Jog Drag and Cell Formation," *Phil. Mag.*, Vol 34, 1976, p 401.
41. D. Kuhlmann-Wilsdorf and J. H. Van der Merwe, "Theory of Dislocation Cell Sizes in Deformed Metals," *Mater. Sci. Eng.*, Vol 55, 1982, p 79.
42. D. Kuhlmann-Wilsdorf, "Unified Theory of Stages II and III of Work Hardening of Pure f.c.c. Metal Crystals," in *Work Hardening*, edited by J. P. Hirth and J. Weertman, Gordon and Breach, New York, 1968, p 97-139.
43. G. Glover and C. M. Sellars, "Recovery and Recrystallization During High-Temperature Deformation of α-Iron," *Met. Trans.*, Vol 4, 1973, p 765.
44. M. J. Luton and C. M. Sellars, "Dynamic Recrystallization in Nickel and Nickel-Iron Alloys During High-Temperature Deformation," *Acta Met.*, Vol 17, 1969, p 1033.

45. H. J. McQueen and S. Bergerson, "Dynamic Recrystallization of Copper During Hot Torsion," *Met. Sci. J.*, Vol 6, 1972, p 25.

46. R. Bromley and C. M. Sellars, "High-Temperature Deformation of Copper and Copper-Aluminium Alloys," in *Proc. 3rd Int. Conf. on Strength of Metals and Alloys*, Cambridge, Vol 1, 1973, p 300.

47. W. Roberts, "Studies Pertaining to Austenite Recrystallization in Association With Controlled Rolling of Nb-Microalloyed HSLA Steels," Swedish Institute for Metals Research Report No. 1211, 1977.

48. A. Le Bon, J. Rofes-Vernis and C. Rossard, "Recristallisation et Précipitation Provoquées par la Déformation à Chaud: Cas D'un Acier de Construction Soudable au Niobium," *Mém. Sci. Rev. Met.*, Vol 70, 1973, p 577.

49. B. Ahlblom, Tekn. Dr. thesis, Royal Institute of Technology, Stockholm, 1977.

50. C. Rossard, "Contribution à L'etude de la Déformation Plastique à Chaud des Aciers," *Metaux-Corrosion-Ind.*, Vol 35, 1960, p 102.

51. J. -E. Carlsson and W. Roberts, unpublished research, Swedish Institute for Metals Research.

52. J. P. Sah, G. J. Richardson and C. M. Sellars, "Grain-size Effects During Dynamic Recrystallization of Nickel," *Met. Sci.*, Vol 8, 1974, p 325.

53. R. W. Pethen and R. W. K. Honeycombe, Reprints of *Int. Conf. on Recrystallization in the Control of Microstructure*, Iron and Steel Institute/Institute of Metals, London, 1973.

54. W. Roberts, H. Bodén and B. Ahlblom, "Dynamic Recrystallization Kinetics," *Met. Sci.*, Vol 13, 1979, p 195.

55. G. Gottstein, D. Zabardjadi and H. Mecking, "Investigations of Dynamic Recrystallization During Tensile Deformation of Copper Single Crystals," in *Proc. 4th Int. Conf. on Strength of Metals and Alloys*, Nancy, 1976, Vol 3, p 1126.

56. R. Sandström and R. Lagneborg, "A Controlling Factor for Dynamic Recrystallization," *Scripta Met.*, Vol 9, 1975, p 59.

57. W. Roberts and B. Ahlblom, "A Nucleation Criterion for Dynamic Recrystallization During Hot Working," *Acta Met.*, Vol 26, 1978, p 801.

58. J. E. Bailey and P. B. Hirsch, "The Recrystallization Process in Some Polycrystalline Metals," *Proc. Roy. Soc.*, Vol A267, 1962, p 11.

59. J. P. Sah, G. J. Richardson and C. M. Sellars, "Quantitative Correlation Between High-Temperature Strength and the Kinetics of Dynamic Recrystallization," *Indian J. Tech.*, Vol 11, 1973, p 445.

60. H. P. Stüwe and B. Ortner, "Recrystallization in Hot Working and Creep," *Met. Sci.*, Vol 8, 1974, p 161.

61. R. Sandström and R. Lagneborg, "A Model for Hot Working Occurring by Recrystallization," *Acta. Met.*, Vol 23, 1975, p 387.

62. S. Sakui, T. Sakai and K. Takeishi, "Effects of Strain, Strain Rate and Temperature on the Hot-Worked Structure of a 0.16%-Carbon Steel," *Tetsu-to-Hagane*, Vol 62, 1976, p 856.

63. S. Sakui and T. Sakai, "Deformation Behaviours of a 0.16%-Carbon Steel in the Austenite Range," *Tetsu-to-Hagane*, Vol 63, 1977, p 285.

64. S. Sakui, T. Sakai and K. Takeishi, "Hot Deformation of Austenite in a Plain Carbon Steel," *Trans. Iron Steel Inst. Japan*, Vol 17, 1977, p 718.

65. J. J. Jonas and T. Sakai, "The Transition From Multiple to Single-Peak Recrystallization During High-Temperature Deformation," in Proc. 24th Colloquium on Metallurgy, *Les Traitements Thermomecaniques*, Saclay, 1981, Inst. Nat. Sci. et Tech. Nucleaires, p 35-56.

66. T. Sakai, M. G. Akben and J. J. Jonas, "The Role of Dynamic Recrystallization in Producing Grain Refinement and Grain Coarsening in Microalloyed Steels," in *Proc.*

Int. Conf. on Thermomechanical Processing of Microalloyed Austenite, AIME, Warrendale, PA, 1982, p 237-252.

67. J. W. Cahn, "The Kinetics of Grain-Boundary-Nucleated Reactions," *Acta Met.*, Vol 4, 1956, p 449.

68. A. T. Cole and G. J. Richardson, "The Effect of Deformation Conditions on the Hot Ductility and Density of Austenitic Stainless Steel," in *Hot Working and Forming Processes*, edited by G. J. Davies and C. M. Sellars, Metals Society, London, 1980, p 128-132.

69. R. Otterberg, H. Modin and W. Roberts, "Mikrostrukturens Utveckling Under Varmbearbetning av en α-Mässing och en α/β-Mässing," Swedish Institute for Metals Research Report No. 1360, 1979.

70. C. Ouchi and T. Okita, "Dynamic Recrystallization Behaviour of Austenite in Nb-Bearing HSLA Steels and Stainless Steel," *Trans. Iron Steel Inst. Japan*, Vol 22, 1982, p 545.

71. R. A. P. Djaic and J. J. Jonas, "Recrystallization of High-Carbon Steel Between Intervals of High-Temperature Deformation," *Met. Trans.*, Vol 4, 1973, p 621.

72. C. Rossard, "Mechanical and Structural Behaviour Under Hot Working Conditions," in *Proc. 3rd Int. Conf. on Strength of Metals and Alloys*, Cambridge, 1973, Vol 2, p 175.

73. B. Ahlblom and W. Roberts, "Possibilities for Microstructural Control During Hot Working of Austenitic Stainless Steels," in Proceedings from *Micon 78* (ASTM Special Publication 672), Philadelphia, 1979, p 296-305.

74. R. A. Petkovic, M. J. Luton and J. J. Jonas, "Recovery and Recrystallization of Polycrystalline Copper After Hot Working," *Acta Met.*, Vol 27, 1979, p 1633.

75. W. Roberts and R. Otterberg, "Microstructure Evolution in Association With Hot Working of a Dezincification-Resistant α/β-Brass," Swedish Institute for Metals Research Report No. 1362, 1979.

76. W. Roberts, "Practical Application of Laboratory Hot-Ductility Data for High-Alloy Steels," Swedish Institute for Metals Research Report No. 1691, 1982.

77. B. Mintz and J. M. Arrowsmith, "Influence of Microalloying Additions on the Hot Ductility of Steels," in *Hot Working and Forming Processes*, edited by G. J. Davies and C. M. Sellars, Metals Society, London, 1980, p 99-104.

78. G. D. Funnell, "Observations on Effect of AlN on Hot Ductility of Steel," ibid., p 104-108.

79. J. R. Wilcox and R. W. K. Honeycombe, "Influence of Prior Precipitation on Hot Ductility of C-Mn-Nb-Al Steels," ibid., p 108-112.

80. J. -E. Carlsson, B. Lehtinen and W. Roberts, "Brottmekanismer Vid varm-bearbetning av Austenitiska Rostfria Stål," Swedish Institute for Metals Research Report No. 1505, 1980.

81. R. J. Fields, T. Weerasooriya and M. F. Ashby, "Fracture-Mechanism Maps for Pure Iron, Two Austenitic Steels and One Ferritic Steel," *Met. Trans.*, Vol 11A, 1980, p 333.

82. K. A. Bywater and T. Gladman, "The Influence of Composition and Microstructure on Hot Workability of Austenitic Stainless Steels," *Met. Tech.*, Vol 3, 1976, p 358.

83. F. C. Monkman and N. J. Grant, "An Empirical Relationship for Rupture Life and Minimum Creep Rate in Creep-Rupture Tests," *Proc. ASTM.*, Vol 56, 1956, p 593.

84. L. -Å. Norström, "Inverkan av Mo och N på Varmduktiliteten hos Austenitiskt Rostfritt Stål," presented at Jernkontoret meeting on *Hot Ductility*, 1976 (available on request).

85. L. G. Ljungström, "The Influence of Trace Elements on the Hot Ductility of 17Cr-13Ni-Mo Steel," *Scand. J. Met.*, Vol 6, 1977, p 176.

86. B. Ahlblom, B. Lehtinen and R. Sandström, "Mekanismer för Spårämnens Inverkan

på Varmduktiliteten hos Austenitiska Rostfria Stål," Swedish Institute for Metals Research Report No. 1426, 1980.

87. D. A. Miller and T. G. Langdon, "Creep Fracture Maps for 316 Stainless Steel," *Met. Trans.,* Vol 10A, 1979, p 1635.

88. A. Sandberg and R. Sandström, "Static Recrystallization and Hot Ductility of Mo- and N-Alloyed Austenitic Stainless Steels in Association With Two- and Multi-step Deformations," Swedish Institute for Metals Research Report No. 1625, 1981.

89. J. H. Decroix, A. M. Neveu and R. J. Castro, "Some Observations on the Hot Workability of Stainless Steels in the As-Cast Condition," in *Deformation Under Hot-Working Conditions,* Iron & Steel Institute, London, 1968, p 135-144.

90. A. Gittins, L. G. Hinton and W. J. McG. Tegart, "Ductility of Steels in Hot Working," *Manufact. Eng. Trans.,* Vol 2, 1973, p 199.

91. A. Järvinen, "Plastisk Deformation av Material Med Flerfasstruktur," Jernkontoret Research Report No. D47, 1972.

92. W. Roberts and O. Sandberg, "Silicon and Nitrogen Alloying of High-Speed Steels and Its Influence on Microstructure and Properties of Grades AISI M2 and M7," Swedish Institute for Metals Research Report No. 1654, 1982.

93. J. -E. Carlsson and W. Roberts, "Klyvning vid Varmvalsning av Snabbstål," Swedish Institute for Metals Research Report No. 1583, 1981.

94. W. D. Nix, D. K. Matlock and R. J. Dimelfi, "A Model for Creep Fracture Based on the Plastic Growth of Cavities at the Tips of Grain-Boundary Wedge Cracks," *Acta Met.,* Vol 25, 1977, p 495.

95. V. Vitek and D. S. Wilkinson, "Intergranular Crack Growth at High Temperatures," in *Proc. 5th Int. Conf. on Strength of Metals and Alloys,* Aachen, 1979, Vol 1, p 327.

96. G. H. Edward and M. F. Ashby, "Intergranular Fracture During Power-Law Creep," *Acta Met.,* Vol 27, 1979, p 1505.

97. A. C. F. Cocks and M. F. Ashby, "Intergranular Fracture During Power-Law Creep Under Multiaxial Stresses," *Met. Sci.,* Vol 14, 1980, p 395.

98. D. McLean, "Influence of Segregation on Creep Cavitation," *Met. Forum,* Vol 4, 1981, p 44.

A New Approach to Dynamic Recrystallization

J. J. JONAS and T. SAKAI*

Abstract

The critical strain criterion $\varepsilon_p = \varepsilon_x$ for the transition from cyclic to single-peak recrystallization is shown to be invalid for the high-temperature deformation of fcc metals in tension and compression, and therefore probably also in rolling, forging and extrusion. The role of the strain and strain-rate gradients present in solid torsion bars in raising the apparent torsion peak strain ε_p above the ε_p values obtained from homogeneous tension or compression testing is clarified. A similar, and larger, effect is shown to cause discrepancies in the torsion values of the recrystallization strain ε_x. An alternative criterion for the transition is described, based on the grain-size measurements performed by Sakai and co-workers on a 0.16% C steel deformed in the γ range of temperatures. Their observations indicate that *cyclic* flow curves are associated with grain *coarsening* and that *single-peak* flow curves are associated with grain *refinement*. The critical condition is $D_0 = 2D_s$, where D_0 and D_s are the initial and stable grain sizes, respectively. These results indicate that single-peak behavior is caused by the "necklace" or "cascade" recrystallization of coarse-grain materials, which produces a large spread in the nucleation strain ε_c and, accordingly, a highly unsynchronized form of local recrystallization. The growth process (and consequently the grain size) in this case appears to be deformation-limited. By contrast, recrystallization is nearly completely synchronized in fine-grain materials, because the high density of grain nuclei leads to a small spread in the nucleation strain. During successive cycles of grain coarsening, the stress oscillations are gradually damped out as the final stable grain size is approached. Furthermore, the final grain size under these conditions is determined by impingement and is thus nucleation-controlled rather than growth-controlled.

The nature of the dynamic recrystallization flow curves associated with an increase or decrease in strain rate after deformation into the steady-state region is described. After an increase in strain rate, the flow curve displays a *single* peak, whereas, after a strain-rate decrease, *multiple* peaks are observed. The critical condition at which the shape of the stress-

*On leave from, respectively: Dept. of Metallurgical Engineering, McGill University, 3450 University Street, Montreal, Canada H3A 2A7; and Dept. of Mechanical Engineering, University of Electro-Communications, Chofu-shi, Tokyo 182, Japan.

strain curve changes from the multiple to the single-peak type is $D_{s1} = D_{s2}$, where D_{s1} and D_{s2} are the stable dynamically recrystallized grain sizes before and after the change in strain rate, respectively. This differs from the critical condition for the annealed structure, i.e., $D_0 = 2D_s$, referred to above. These results are interpreted in terms of a grain-size-based model for dynamic recrystallization.

The polycrystal observations are compared with those obtained on single crystals and on very coarse-grained polycrystals. It is shown that nucleation in single crystals (and in deformation bands in coarse polycrystals) requires much higher initiation stresses than does nucleation at grain boundaries of polycrystals, and is therefore likely to involve a different mechanism.

A New Approach to Dynamic Recrystallization

The occurrence of dynamic recrystallization during the creep of metals has been recognized for over 25 years.[1-7] It has also received close study for its role in the periodic acceleration of glaciers.[8,9] Whether or not it occurred during high-temperature deformation of industrial alloys under metal-processing conditions was subject, however, to dispute. The view held by some[10-13] was that softening *during* deformation was always by dynamic recovery and that, when recrystallized grains were observed in hot worked materials, they were always the result of static recrystallization during the interval (however short) between working and quenching. This controversy has happily been resolved with a fairly general consensus regarding its unequivocal occurrence[14-23] in fcc materials of low to medium stacking-fault energy under suitable conditions of temperature and strain rate. Dynamic recrystallization has even been reported in single crystals of these materials,[24-27] although the details of this process differ somewhat, as will be seen below, from those of the polycrystalline case.

The classical approach to dynamic recrystallization was primarily concerned with accounting for the transition from cyclic to single-peak recrystallization.[17-23] As can be seen in Fig. 1(a) and 1(b),[28,29] the flow curves pass through this transition as the strain rate is increased or the temperature decreased. Less attention was paid to the characteristics of the microstructural processes taking place within the material. More recently, these processes have been studied in some detail — principally by Sakai and co-workers, initially in Japan,[30-33] and subsequently in Canada.[34-38] These investigations have shown that a single peak in flow stress signifies that grain *refinement* is occurring, whereas multiple peaks indicate that discrete cycles of grain *coarsening* are taking place. It is the purpose of this review to synthesize recent work and, in particular, to consider the conditions under which dynamic recrystallization can produce grain refinement during industrial metal processing.

Classical Approach to Dynamic Recrystallization

The classical approach to the transition from cyclic to single-peak flow is based on the difference between the stress dependence of two characteristic strains

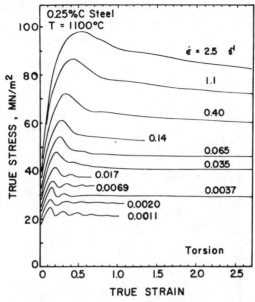

Fig. 1(a). Influence of strain rate on flow curves derived from hot torsion data at 1100°C for a 0.25% C steel. After Rossard and Blain.[28]

associated with the flow curve. These are illustrated and defined in Fig. 2. Although several slightly different versions of these quantities have been defined in the literature, we will confine our discussion to just two of them: ε_p (the strain to the peak stress or maximum in the flow curve) and ε_x. Under single-peak conditions, the latter can be defined as $\varepsilon_s - \varepsilon_p$, where ε_s is the minimum strain required to attain steady-state flow (see Fig. 2). Under cyclic conditions, ε_x is again equal to $\varepsilon_s - \varepsilon_p$; however, ε_s can be either the strain at which the mean value of the flow stress is attained during the second cycle of work hardening (Fig. 2) or the strain associated with the first minimum in the flow stress.[17] According to this model, recrystallization is periodic when $\varepsilon_x < \varepsilon_p$; that is, when the *growth* of new grains occurs more rapidly (over smaller strains) than their nucleation (see Fig. 3). Conversely, recrystallization is sustained or "continuous" when $\varepsilon_x > \varepsilon_p$; that is, when the strains associated with the *nucleation* of new grains are smaller than those associated with their growth.[17]

The Luton and Sellars analysis involved computer simulations which predicted a transition from cyclic to single-peak flow when either the strain rate was increased or the temperature was decreased. It also reproduced the principal features of the flow curves *obtained in torsion* reasonably well, although the simulated cyclic curves did not gradually die out as they did in the experiments. The approach of Luton and Sellars suffered from a number of further limitations,[18] which led Sah, Richardson and Sellars to publish more refined models

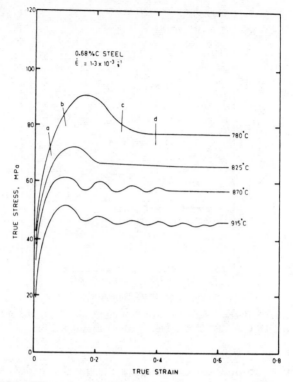

Fig. 1(b). Effect of temperature on flow curves determined in axisymmetric compression on a 0.68% C steel at a strain rate of 1.3×10^{-3} s^{-1} After Petkovic et al.[29]

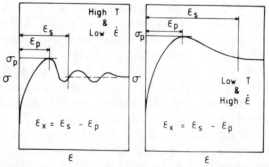

Fig. 2. Schematic definitions of the peak, recrystallization and steady-state strains, ε_p, ε_x and ε_s, respectively. (Left) Periodic recrystallization; $\varepsilon_p > \varepsilon_x$. (Right) Single-peak flow; $\varepsilon_p < \varepsilon_x$. After Luton and Sellars.[17]

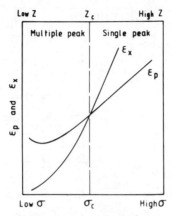

Fig. 3. Stress dependences of the peak and recrystallization strains ε_p and ε_x observed in torsion tests.[17] Single-peak flow occurs when $\sigma > \sigma_c$ (i.e., when the temperature-corrected strain rate Z > the critical value Z_c). Conversely, multiple-peak flow is observed when $\sigma < \sigma_c$ and $Z < Z_c$. The critical stresses and normalized strain rates σ_c and Z_c correspond to the experimental conditions under which $\varepsilon_p = \varepsilon_x$.

in 1973[19] and 1974.[20] The salient features of the Sah et al. treatment will not be reviewed here, because they have already been dealt with elsewhere.[34] Furthermore, because it was originally proposed to account for observations obtained in torsion testing, it is not really applicable in unmodified form, as will be seen below, to the results derived from other methods of testing.

Conflicts Observed Between Torsion Results and Those Obtained From Other Modes of Testing

The extent to which the classical explanation for the transition from cyclic to single-peak flow does not apply to other modes of straining was first pointed out by Sakai and co-workers in Japan.[30–32] They carried out extensive investigations on a 0.16% C steel in the temperature range 940 to 1040°C and the strain-rate range 10^{-5} to 20 s^{-1}. Their experiments were conducted in *tension* on steel strips which were deformed in a vacuum and then rapidly quenched (2000°C/s) by means of hydrogen jets. The martensitic structures formed in this way enabled the shapes and sizes of the austenite grains present during testing to be established. Their detailed metallographic observations also led to a clarification of the roles of grain refinement and grain coarsening in the dynamic-recrystallization process. The principal findings of these investigators will now be reviewed briefly.

The ε_p and ε_x measurements of Sakai et al. determined in *tension* on the 0.16% C steel are presented in Fig. 4(b). For comparison purposes, equivalent

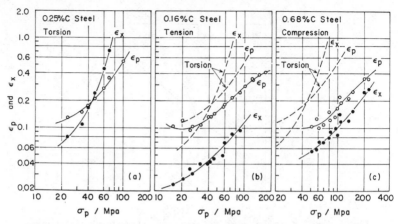

Fig. 4. Dependence of ε_p and ε_x on the peak stress σ_p. (a) Determined from the *torsion* flow curves of Fig. 1(a). Note intersection. (b) Determined from the *tension* data of Sakui et al.[30–32] Note that ε_p and ε_x do not intersect in this case. (c) Determined from the axisymmetric *compression* data of Petkovic et al.[29] Again, the ε_p and ε_x curves do not intersect.

quantities determined from the *torsion* experiments of Rossard and Blain[28] are illustrated in Fig. 4(a), and those from the *compression* data of Petkovic et al.[29] in Fig. 4(c). It is evident that the two characteristic strains pertaining to torsion intersect in the middle of the stress range (i.e., in mid Z range), as called for by Fig. 3. By contrast, neither the tension data (Fig. 4b) nor the compression data (Fig. 4c) *intersect at all* within the experimental range. Similar comparisons have since been carried out by Weiss et al.[37] between solid-bar torsion results on the one hand and data obtained in plane-strain compression as well as on *tubular* torsion specimens on the other. According to their analysis, the discrepancy shown in Fig. 4 applies to these other modes of testing as well, and is not limited to comparisons with the results of axisymmetric tensile or compressive straining.

When the results of the torsion tests are compared with those obtained from tension and compression, it is readily apparent that:

- The torsion ε_p is *somewhat* greater than the equivalent quantity measured in tension or compression (see Fig. 4).
- The torsion ε_x is *considerably* greater than the equivalent quantity measured in tension or compression.

If the torsion curves are taken as "correct," the lack of intersection in the tension and compression data arises because the "errors" in the uniaxial ε_x measurements are greater than those associated with the ε_p measurements. Alternatively, as will be demonstrated below, if the tension and compression data are taken as approximately correct, then the intersection that is generally present in torsion data can be regarded as unrepresentative of plastic deformation in general.

It is instead particular to torsion testing and can be attributed to some feature of this mode of testing that influences primarily the experimental value of ε_x and, to a lesser extent, ε_p. The particularity of the intersection to torsion also indicates that the transition from periodic to single-peak flow is related to some aspect of the recrystallization process other than the simple ratio of the "nucleation" strain to the "growth" strain.

The characteristics of torsion testing that are principally responsible for the apparent divergence in behavior were identified by Sakai and co-workers[30-32] as the radial strain and strain-rate gradient present in solid torsion bars. Other features of the various testing modes which also influence comparisons are their varying susceptibilities to strain localization during testing, as well as the slight divergences in the rate of work hardening and of texture development associated with them. The different factors will now be considered in turn.

Effects of Strain and Strain-Rate Gradients Present in Solid Torsion Bars

There are numerous ways of deriving stress/strain curves from the torque/twist data determined in the torsion testing of solid bars. In the method most frequently used,[39] the shear strain in the surface layer, γ_R, is given by:

$$\gamma_R = \theta R/L \tag{1}$$

where θ is the angle of twist, and R and L are the radius and gage length of the specimen, respectively. When the von Mises convention is applied, the effective strain $\bar{\varepsilon}$ in the surface layer is in turn expressed as:

$$\bar{\varepsilon} = \gamma_R/\sqrt{3} = \theta R/\sqrt{3}L \tag{2}$$

with similar expressions being employed for the shear strain rate at the surface, $\dot{\gamma}_R$, and the effective strain rate at the surface, $\dot{\bar{\varepsilon}}$.

The "surface" shear stress τ_R that is commonly plotted against γ_R is deduced from the measured torque or moment M through the relation:[39]

$$\tau_R = (M/2\pi R^3)(3 + m + n) \tag{3}$$

Here, $m = \partial \ln M/\partial \ln \dot{\theta}|_\theta$ is the (logarithmic) twist-rate sensitivity of the torque, and $n = \partial \ln M/\partial \ln \theta|_{\dot{\theta}}$ is the normalized torque-hardening rate. As before, the effective stress in the surface layer, $\bar{\sigma}$, can be obtained from τ_R with the aid of the von Mises assumption, leading to:

$$\bar{\sigma} = \sqrt{3}\tau_R \tag{4}$$

Although Eq 3 is exact for materials that obey simple work-hardening laws of the power type (i.e., for which m and n are constant), it is only an approximation when applied to materials being deformed at elevated temperatures, under which conditions the work-hardening rate is itself rate-dependent[40] (i.e., when the

coefficient n is no longer independent of strain rate and m is no longer independent of strain). Unfortunately, exact methods (e.g., that given in Ref 41) require
a more involved analysis as well as the use of several specimens of increasing
diameter for each condition to be investigated, and so are considerably less
convenient to use.

The approximations involved in the application of Eq 3 to high-temperature
flow lead to overestimation of ε_p, and to even greater overestimation of ε_x. These
errors arise because the material flow curve reaches a peak (n = 0) at a strain
which is rate-sensitive, followed by flow softening (n < 0) — characteristics
both of which are incompatible with the derivation of the equation. The manner
in which this arises can be seen by referring to Fig. 5(a), in which a cylindrical
torsion specimen is represented as consisting of concentric thin tubes for each of
which the strain rate $\dot{\gamma}$ is proportional to the radius r. Because the peak strain
increases less than linearly with strain rate, it is attained first in the outermost
shell, and then progressively later and later in each of the successively smaller
concentric shells. Now the moment developed in the torsion bar is, of course, the
sum of the moments developed in the individual shells; however, when some of
these are decreasing while others are increasing, the behavior of a particular shell
can be masked by the influence of the others. Thus, when the outer shell attains
the peak stress and begins to flow soften, the bar as a whole continues to show
hardening; as a result, the peak in the over-all couple is attained *later* than the
instant associated with τ_{max} in the outermost shell.

Effect on ε_p

The influence of this type of "averaging error" is depicted in Fig. 5(b). The
broken lines represent the behavior of two tensile specimens being deformed at
the strain rates corresponding to the outermost shells of two different torsion bars.

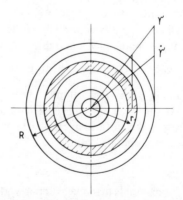

Fig. 5(a). Cross section of a solid torsion bar represented as a series of
concentric thin tubes or shells. The linear dependence of strain γ and
strain rate $\dot{\gamma}$ in each shell on the tube radius r is also shown.

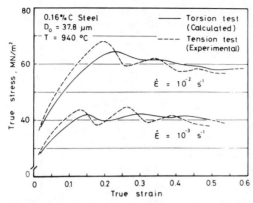

Fig. 5(b). Comparison of *actual* true stress–true strain curves determined in tension with *calculated* torsion curves. The latter were derived from tensile data determined at the strain rate appropriate to each cylindrical "shell" in the simulated torsion bar.[31,32] The torque contributions of the individual shells were then summed and the effective stress–effective strain curves were deduced by the method of Fields and Backofen.[39]

The full lines represent the *estimated* torsion behavior as built up from a series of imaginary concentric shells, each of which follows the behavior expected for such a shell from a tension test of appropriate strain rate. The equivalent strain $\bar{\varepsilon}$ in the simulated torsion test of Fig. 5(b) is indeed that associated with the outer radius R of the specimen: however, the effective stress $\bar{\sigma}$ is *not* that of the outermost shell, but represents instead a kind of "average stress" which reflects only the over-all torque exerted by the individual layers, of which some are undergoing hardening and others are experiencing softening. The point of interest is that the peak strain ε_p for the composite torsion specimen is larger than ε_p for the tensile specimen, *even though both flow curves are based on the same set of tensile data.* This construction accounts for a good part of the discrepancy between the ε_p's for torsion and tension illustrated in Fig. 4 and 5(c).[30–32] An extension of this argument also applies to the torsion/compression discrepancy, as well as to the differences between torsion and plane-strain compression.[37]

The averaging error described above can be eliminated either by using tubular torsion specimens or by employing the differential testing method of Canova et al.,[41] which enables the behavior of the outermost layer to be determined with accuracy. Even under these conditions, however, there remain differences between the flow curves established by means of torsion testing and those determined by other testing modes. The divergences that remain arise from two separate classes of physical phenomena.[42] The first effect is that associated with the differences in the textures produced at large strains by deformation along different strain paths. The regular testing methods produce four distinct textures at large deformations; these can be classified as follows:

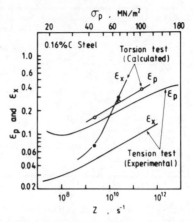

Fig. 5(c). Comparison of the ε_p-vs-σ_p and ε_x-vs-σ_p relations determined in the tension experiments of Ref 31 and 32 and from the simulated torsion curves for the composite bar of Fig. 5(a) as calculated from the same tension data.

1. Uniaxial tension (similar* to that of wire drawing, swaging, etc.).
2. Uniaxial compression (similar* to that of balanced biaxial tension).
3. Plane-strain tension and compression.
4. Torsion.

In the first three of these testing methods, the material directions associated with the axes of symmetry of the specimen remain fixed. In torsion, by contrast, the material direction at the surface that is originally longitudinal is gradually rotated toward coincidence with the transverse direction. Because of this difference, the first three textures have more planes of symmetry than the torsion texture, which has none. Thus, although fixed-end torsion is a plane-strain mode of deformation (the radial direction is equivalent to the width direction in rolling), it does not lead to the same textures as do conventional plane-strain tension and compression.

The importance of the texture differences can be assessed from the following relation:[42,43]

$$\sigma_Y(1) = \overline{M}(1) \cdot \overline{\tau}_C \qquad (5)$$

Here $\sigma_Y(1)$ is the uniaxial yield strength along the (1) or longitudinal axis, $\overline{M}(1)$ is the average Taylor factor associated with unconstrained loading along this direction at a given strain,[43] and $\overline{\tau}_C$ is the current mean value of the CRSS; it is

---

*In materials which have an initial preferred orientation which is not axisymmetric, the application of a uniaxial tensile stress does not lead to axisymmetric deformation, so that the strain paths associated with uniaxial loading and with wire drawing are not quite identical. Similar remarks apply to axisymmetric compression and balanced biaxial straining.

closely linked with the current dislocation configuration. In general, the four different textures produced by the four different testing methods described above will lead to four different values of $\overline{M}(l)$ at the same value of the accumulated glide or suitably defined equivalent strain.[42] Thus, even in the simple case where the current value of $\overline{\tau}_C$ is the same for all four test methods, the associated macroscopic "flow stress" is different (albeit only slightly) for the specimens deformed along the four distinct strain paths.

The situation is further complicated by the observation that the fundamental rate of work hardening (in terms of $\overline{\tau}_C$ versus $\overline{\gamma}_C$, the accumulated slip) is different, once again only slightly,[41–45] for deformation along different strain paths. Although more detailed data are required to confirm this trend, it appears that the rate of work hardening of the CRSS is lowest in torsion (simple shear), intermediate in conventional plane strain (i.e., pure shear), higher in axisymmetric compression, and highest of all in axisymmetric tension.[42] The effect of these two factors (i.e., work-hardening and texture-development differences) is that, even when the results of *tubular* torsion are compared with tension or compression, ε_p is larger for the former than for the latter two testing modes.

Effect on ε_x

The factors leading to a difference between the values of ε_p determined in torsion and in other testing modes (see Fig. 4) have now been considered in some detail. These factors continue to play a role in causing the ε_x values determined in solid-bar torsion to be larger than those established by the other modes of testing. In this case, however, there is an additional factor which enters into the "averaging" problem, and which causes the ε_x discrepancy to be considerably larger than the one associated with ε_p. The peak strain, it will be recalled, is reached first in the outermost shell of a solid-bar torsion specimen (see Fig. 5a), and then successively in the various internal layers in turn. The same progression applies to the recrystallization strain ε_x for the same physical reasons (i.e., that the dependence of ε_x on strain rate, and therefore on radius r, is considerably less than linear, whereas the strain rate in the various layers is in fact linear with r). There is, however, one essential difference. Whereas the time to reach ε_p in each layer is measured from t = 0, when $\overline{\varepsilon}$ = 0 in all the layers, the time to reach ε_x in each layer is measured from the respective time to reach ε_p, *which is different in each layer*. The increase in the spread of the times associated with ε_x in the various layers over that associated with ε_p has an effect on $\varepsilon_s = \varepsilon_p + \varepsilon_x$ which is similar but greater than the one described above for ε_p. It has the practical consequence that ε_s is reached *far later* in the innermost layers (and therefore in the solid bar as a whole) than in the outermost shell.* This appears to be the principal reason why the differences between the ε_x values determined

---

*This is also why stress oscillations in torsion tend to die out more rapidly than in compression, for example (see Fig. 1 and Fig. 5b).

in torsion and in tension and compression are so much greater (see Fig. 4 and 5c) than those that apply to ε_p. There are, in addition, effects attributable to the differences in: (a) texture development, (b) rate of work hardening and (c) susceptibility to strain localization. Finally, it should be added that the above factors affecting the results of torsion and axisymmetric testing also affect the results obtained in plane-strain tension and compression tests.[37]

Microstructural Changes Associated With Single- and Multiple-Peak Flow Curves

It was shown above that the $\varepsilon_p \gtrsim \varepsilon_x$ criterion for the transition from multiple- to single-peak flow is not applicable to axisymmetric tension and compression, and it also appears to be invalid for plane-strain tension and compression[37] and, by extension, for rolling as well as extrusion. The investigations of Sakai and co-workers[30–32] were therefore aimed at finding a substitute for the chararacteristic strain model, and one which ideally would be microstructurally based. They arrived at such a model by observing that single-peak flow is associated with grain *refinement,* and that multiple-peak flow, conversely, is generally associated with grain *coarsening.* They verified the dependence of the stable dynamic grain size D_s on the temperature-corrected strain rate Z,[14] and introduced the concept of the critical value of Z, Z_c, which distinguishes between periodic and single-peak flow (see Fig. 3). By determining the dependence of Z_c on the initial grain size D_0 and comparing it with the D_s–Z relationship, they established that the critical condition associated with the transition is $D_0 = 2D_s$, with the relations $D_0 > 2D_s$ and $D_0 < 2D_s$ leading to single- and multiple-peak flow, respectively. Because of their technological importance, the observations of Sakai et al.[30–32] relating to grain refinement and grain coarsening will now be described in more detail.

In their experiments, a 0.16% C steel was tested in tension at 890 and 940°C at a series of strain rates. The initial austenite grain size was varied over the range 25 to 70 μm. Specimens were rapidly quenched using the hydrogen jets at increasing strains at each testing strain rate. Because the tensile mode of testing prevented the attainment of truly large strains, the grain size at the end of each experiment was taken to be approximately equal to the "equilibrium" grain size D_s. Examples of the grain structures produced by this technique when the initial grain size was 32 μm are given in Fig. 6(a). It is evident that, at the highest strain rate (18 s^{-1}), grain *refinement* has occurred, although the process apparently did not go to completion as a result of the limited strain attainable (0.6) prior to quenching. At the other extreme in strain rate (2.6 × 10^{-4} s^{-1}), grain *coarsening* has taken place, the structure shown being the result (see Fig. 6b) of three successive cycles of work hardening followed by flow softening or recrystallization.

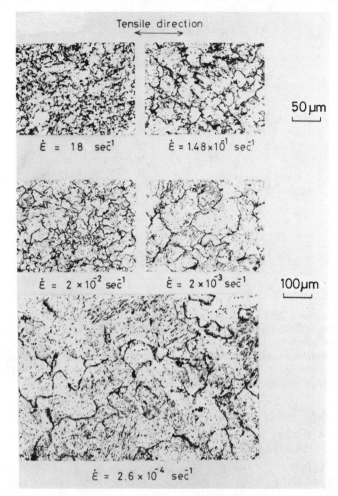

Fig. 6(a). Dependence of the steady-state austenite microstructure on testing strain rate in a 0.16% C steel deformed in tension at 940°C and H_2 quenched. The original mean grain size was about 32 μm; grain *refinement* took place at the three highest, and grain *coarsening* at the two lowest, strain rates.[30,32]

The detailed dependence of the mean grain size on strain at these two strain rates is illustrated in Fig. 6(b), in which the results obtained at three intermediate strain rates are also presented. Several features of this diagram bear comment.

- In the grain-coarsening region of behavior (lowest two strain rates), several stress peaks are observed (P_1, P_2, P_3) before steady-state flow is established. The recrystallization that follows the first peak does not succeed in produc-

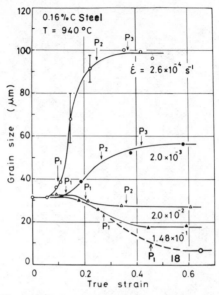

Fig. 6(b). Effect of strain and strain rate on the mean grain size of a 0.16% C steel deformed in tension at 940°C. P_i identifies the strain at the ith flow-stress peak. Note that *multiple peaks* are observed during grain *coarsening* (and when grain refinement produces less than a 2:1 reduction in grain size). Conversely, a *single peak* is observed when the grain *refinement* ratio is above 2:1.[30,32]

ing the final, stable grain size, which appears instead to be attained in stages. Conversely, once the final grain size has been reached, no further stress peaks are detected.

- At the highest two strain rates (18 and 1.48×10^{-1} s^{-1}), single-peak flow is observed (no P_2 or P_3) in association with grain-refinement ratios of 4.5 and 2.0.
- At the intermediate strain rate of 2.0×10^{-2} s^{-1}, the microstructure is refined *slightly* during deformation, the refinement ratio being less than 2:1. Despite the occurrence of grain refinement, a second flow-stress peak is detected, indicating that the flow curve remains periodic unless the refinement ratio exceeds 2:1. (A mechanistic explanation for the association of this limited amount of grain *refinement* with *periodic* recrystallization will be presented below.)

Dependence of Transition Condition on Initial Grain Size

The stable or equilibrium grain size D_s produced by dynamic recrystallization has long been recognized to correlate well with the steady-state stress σ_s.[17–23,46–48]

In a given alloy, the latter depends in turn only on the temperature and strain rate, i.e., on the independent-state variables, or, more concisely, on the temperature-corrected strain rate Z.[14,17–23] Thus it is possible to establish the dependence of D_s on Z as well as on σ_s. The D_s data given in Fig. 6 are replotted as a function of Z in Fig. 7(a), where the dependence on the peak stress σ_p is also shown. (The steady-state stress was not employed in this work, because steady-state flow could not be attained reliably in some of the tensile experiments. Nevertheless, as the ratio σ_p/σ_s is sensibly constant, or nearly so, in the usual grain-size range of most materials, it is considered that there is no fundamental difference between the use of D_s-σ_p and D_s-σ_s plots.) Also included in Fig. 7(a) is the datum point from an experiment carried out at 890°C. The relation illustrated in Fig. 7(a) is typical for materials undergoing dynamic recrystallization, and is similar to that reported for several kinds of steel.[46–48]

The investigations of Sakai and co-workers[30–32] revealed that, in addition to the previously observed correlation between D_s and Z (Fig. 7a), there is also a correlation between the critical value of the temperature-corrected strain rate Z_c associated with the transition from periodic to single-peak flow (Fig. 3) and the initial grain size D_0. Their results are reproduced here in Fig. 7(b), from which it can be seen that a finer initial grain size leads to a higher transition strain rate (or a lower transition temperature), whereas a coarser initial grain size leads in a similar manner to a lower transition strain rate (or a higher transition temperature). Of particular interest was their observation that the *form* of the dependence resembled that of the dependence of D_s on Z (Fig. 7a), although in the latter case Z is the independent variable whereas in Fig. 7(b) D_0 is the independent variable. The similarity in nature of the two relationships led to the possibility of describing

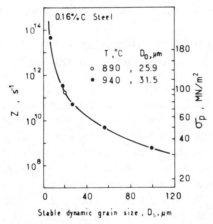

Fig. 7(a). The relation between the temperature-corrected strain rate Z (or peak stress σ_p) and the stable dynamic grain size D_s for a 0.16% C steel.[30,32]

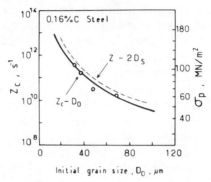

Fig. 7(b). Dependence of the critical Z_c on initial austenite grain size D_0 in a 0.16% C steel.[31,32] The broken line is the Z-2D$_s$ relation determined in Fig. 6(b) and 7(a). Note that the Z-2D$_s$ and Z_c-D$_0$ relations are nearly coincident.

them by a single curve, given suitable algebraic representations. The manner in which this was done is indicated in Fig. 7(b) by the broken line, which illustrates the dependence of *twice* the equilibrium grain size (i.e., 2D$_s$) on Z. The close correspondence between the 2D$_s$–Z and D$_0$–Z$_c$ relations can be readily seen. (In the original publications,[30–32] the correspondence was considered to occur at 1.8D$_s$, but here, for simplicity, the value 2D$_s$ is employed.)

The use of 2D$_s$ in Fig. 7(b) corresponds to the observation made above that the transition from oscillatory to single-peak behavior occurs when the grain-size-reduction ratio (i.e., original grain size divided by stable grain size) is 2:1. It also leads to the critical condition $D_0 = 2D_s$ for the transition, a relation that has a physical significance that will be considered more closely later.

Deformation Mechanism Map for Grain Refinement and Grain Coarsening

The significance of the similarity between the 2D$_s$-Z and D$_0$-Z$_c$ relations can be seen to better effect in Fig. 8, where the crosshatching distinguishes the grain-refinement from the grain-coarsening region. The solid line represents the experimentally determined locus $D_0 = 2D_s$. From this diagram, the outcomes of two different types of tests can be predicted: the so-called "vertical" and "horizontal" experiments. A third type of test, which combines horizontal and vertical components and involves strain-rate changes during straining, will be introduced later. The vertical experiments begin at vertically separated points on Fig. 8 and are carried out with a *fixed* initial grain size D$_0$ but at a series of increasing, or decreasing, values of Z. These correspond to the classical experiments of Fig. 1, in which either the strain rate is increased or the temperature is decreased. The transition from multiple- to single-peak flow occurs at the value

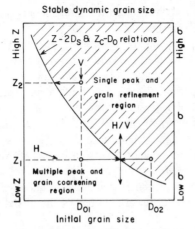

Fig. 8. A microstructural mechanism map for distinguishing between the occurrence of two types of dynamic recrystallization. The curve describing the Z-$2D_s$ and Z_c-D_0 relations separates the single-peak (grain-refinement) region from the multiple-peak (grain-coarsening) region. Three distinct types of experiments are represented: "vertical" (V) tests carried out over a range of strain rates and temperatures on material with a *fixed* initial grain size D_{01}; "horizontal" (H) tests carried out at a fixed temperature-compensated strain rate Z_1 with a *series* of initial grain sizes in the range D_{01} to D_{02}; combined "horizontal/vertical" (H/V) tests involving changes in strain rate or temperature after a period of steady-state deformation.[34-36]

of Z_c appropriate to D_0 indicated by the solid line, which functions in this case as a D_0-Z_c relation. As the critical parameter Z_c is also associated with the particular values of D_0 and D_s which satisfy the relation $D_0 = 2D_s$, the grain-refinement ratio all along the transition is $2:1$. At Z values *greater than* Z_c (i.e., under single stress-maximum conditions), the grain-refinement ratio is greater than $2:1$.

According to this model, once the peak has been attained, the flow stress keeps dropping while a progressively larger volume fraction of the material undergoes dynamic recrystallization. Concurrently, through the process of grain refinement, the unstable D_0 grain structure is gradually replaced by the stable D_s one (Fig. 6b). The current location of the experiment on the Z–D map moves horizontally *leftward* as a result of the refinement until the $D_0 = 2D_s$ locus is attained. Under Z conditions *well below* Z_c, cyclic grain coarsening is observed, accompanied by multiple stress peaks. In this case the grain coarsening corresponds to the horizontal movement of the experimental coordinate in the *rightward* direction until, once again, the $D_0 = 2D_s$ locus is reached. At Z values just below Z_c, a small amount of refinement takes place, even though the curves are still of the oscillatory type. A rationalization for this intermediate behavior will be introduced below,* together with a microstructural model for the existence of the critical refinement ratio $D_0/D_s = 2$.

The "horizontal" experiments, in contrast to the vertical ones described above, begin at horizontally distributed points on Fig. 8 and are carried out at a *fixed* temperature-corrected strain rate Z. The initial grain size, instead of Z, is varied in these tests, and is permitted to approach the same fixed value of D_s. When $D_0 < 2D_s$, each cycle of recrystallization produces grain coarsening, and the experimental point moves rightward until the $D_0 = 2D_s$ locus is reached, i.e., until the stable grain size appropriate to, say, Z_l (Fig. 8) is attained. Conversely, when $D_0 > 2D_s$, grain refinement takes place, and the horizontal coordinate moves leftward, coming to rest at the same abscissa as before. Such "horizontal" experiments, which have not been carried out as frequently in the past as have the "vertical" ones, provide a useful critical test of the model of Fig. 8.

Critical Tests of the Grain Refinement/ Grain Coarsening Model

One of the critical tests carried out to evaluate the grain-size-based transition criterion was performed on a microalloyed steel containing 0.05% C, 1.2% Mn, 0.115% V and 0.006% N.[35] In order to produce a range of initial grain sizes, individual specimens were austenitized at temperatures of 975, 1000, 1030, 1100 and 1260°C. This led to grain sizes of 65 to 300 μm, as indicated in Fig. 9. The specimens were hot compressed in a modified Instron testing machine at a constant true strain rate of $1.4 \times 10^{-3}s^{-1}$ and a temperature of 975°C. For this Z condition, D_s was about 50 μm, so that $2D_s = 100$ μm. It is apparent from Fig. 9 that when D_0 was less than $2D_s$, i.e., for the 65- and 80-μm material, multiple stress peaks were induced, which were not damped out in the 65-μm material by the time the experimental strain limit of 1.0 was reached. It should be noted, however, that the amplitude of the oscillations decreased with strain. In terms of Fig. 6, this means that the amplitude decreases as each successive wave of grain coarsening brings the mean grain size closer to D_s. For the coarse-grain materials, i.e., when D_0 was greater than $2D_s$, a single, marked flow-stress peak was observed, followed by a small, but distinct, second peak. This was accompanied by the occurrence of grain refinement, which came to an end when the stable mean grain size D_s was attained.

A similar series of tests leading to results that corroborate those of Fig. 9 was carried out on a 0.06C-1.43Mn plain carbon steel.[34] In this case, the experimental temperature was 900°C, and the strain rate was $1.4 \times 10^{-3}s^{-1}$, as for the microalloyed steel. The material was austenitized at a series of temperatures ranging from 900 to 1260°C and selected to produce initial grain sizes of 60 to 375 μm. As in the experiments of Fig. 9, multiple stress peaks were associated

---

*See the section dealing with Eq 7, below.

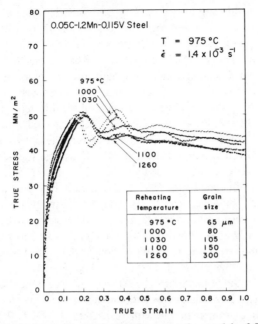

Fig. 9. Critical experiment designed to test the model of Fig. 8. A 0.05C-1.2Mn-0.115V steel was austenitized at a series of temperatures ranging from 975 to 1260°C and selected to produce initial austenite grain sizes of 65 to 300 μm, respectively (see inset). Testing was carried out in axisymmetric compression at 975°C and 1.4×10^{-3} s^{-1}, for which $2D_s \simeq 100$ μm. Cyclic σ/ε behavior was observed when $D_0 < 2D_s$; by contrast, single-peak behavior was obtained when $D_0 > 2D_s$.[35]

with the condition $D_0 < 2D_s$, with a total of four stress maxima being observed at the strain limit of 1.0. The amplitude of the oscillations decreased with strain, as each successive wave of coarsening brought the mean grain size closer to the final value D_s. For the condition $D_0 > 2D_s$, a single, marked flow-stress peak was displayed, after which grain refinement occurred. Although the shapes of the single-peak curves differed distinctly from those of the multiple-peak ones, a slight second peak was nevertheless detected, as in the flow curves of Fig. 9.

Literature Support for the Critical-Grain-Size Model

It is of interest that a transition of the type described above can be seen in the work of Sah et al.[20] published in 1974. In torsion tests on nickel carried out at 1000°C and 0.05 rpm, multiple peaks were observed in the flow curves pertaining to material with an initial grain size of 140 μm. By contrast, when specimens with grain sizes of 350 and 400 μm were deformed, only single peaks were detected. These results are fully consistent with their D_s value of 120 μm, which leads, according to the present criterion, to a critical D_0 of 240 μm.

Additional support for the critical-grain-size criterion can be gained from an investigation carried out by Ouchi and Okita.[47] They studied the recrystallization behavior of six basic steels, which they deformed in compression at 1000 and 1100°C. Specimens with five different initial grain sizes were prepared, ranging from 29 to 350 μm, and these were strained at three different mean strain rates. Sixteen of the 18 cases they looked at in detail agreed with the $D_0 = 2D_s$ condition for the transition in flow behavior. Nevertheless, two of the 18 did not: one of these involved a grain-refinement ratio of just less than 2:1, even though multiple peaks were *not* observed; the other concerned a multiple-peak test in which the grain-refinement ratio was apparently slightly greater than 2:1. The latter "exception," however, can be explained in terms of the recent results of Blaz et al.[38] concerning the dynamic recrystallization behavior of coarse-grain OFHC copper. Blaz et al. found that when the grain size was greater than about 100 μm, numerous dynamic-recrystallization nuclei were formed in the deformation bands within the grains, in addition to those associated in the normal way with the original grain boundaries. The effect of this phenomenon was to increase the nucleus density above that expected, and thus reduce the *effective* initial grain size of the coarse-grain material with respect to the $D_0 = 2D_s$ criterion. An interpretation of this type, based on the nucleating role of deformation bands in coarse-grain specimens, is consistent with the data of Ouchi and Okita, whose "exceptional" material had an initial grain size of 350 μm.

Finally, it should be added that support for the present model can be gained from the results of another recent investigation. Maki et al.[48,49] studied the recrystallization behavior of six laboratory steels and also found that the transition from oscillating to single-peak flow was associated with the initial grain size of the material. It appears from their data[49] that a critical condition $D_0 = A \cdot D_s$ fits their results, where the proportionality constant A is between one and two. Thus, although the grain-coarsening/grain-refinement model seems to be well supported by literature data, for wider validity the critical condition may be better described by the relation $D_s < D_0 < 2D_s$.

Effect of Alloying Additions on Grain Refinement and Grain Coarsening

The usual effect of the presence of alloying elements on the shape of the flow curve is depicted in Fig. 10.[35] Here it can be seen that, as the over-all concentration of alloying elements is increased, the oscillations evident in the case of the plain carbon material gradually disappear. The tests in Fig. 10(a), which were conducted at 1075°C, were all completed in about 20 s; thus no precipitation of nitrides or carbonitrides occurred during the experiments. The differences be-

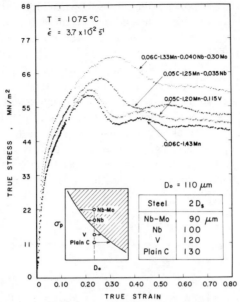

Fig. 10(a). Effect of microalloy addition on the shape of the flow curve. The plain carbon and 0.115% V steels undergo cyclic recrystallization because $D_0/2D_s < 1$. The Nb and Nb-Mo grades display single-peak behavior because $D_0/2D_s > 1$. The transition in the character of the curves can be ascribed to the effect of the solute elements on the stable grain size produced at a given temperature and strain rate.[35]

tween the curves are therefore attributable entirely to the influence of the alloying elements *in solution*.

In the case of Fig. 10(a), however, the rationalization based on the present model differs somewhat from the one proposed above for Fig. 9. In that experiment, although the *final* grain sizes were the same for all five tests, the *initial* ones were not, leading to relative grain sizes, $D_0/2D_s$, both larger and smaller than one. By contrast, in the present instance, it was the *initial* grain size which was held constant. The change in the $D_0/2D_s$ criterion was produced instead by varying the *stable* grain size D_s (at each fixed temperature and strain rate) through the influence of the microalloying elements in solution on the recrystallization process. For example, D_s values for the carbon and vanadium steels are about 65 and 60 μm, respectively, so that their relative grain sizes, $D_0/2D_s$, are 0.85 and 0.91, which are consistent with the occurrence of grain coarsening and stress oscillations. For the niobium and Nb-Mo steels, on the other hand, the D_s values are about 50 and 45 μm, respectively, so that the corresponding values of $D_0/2D_s$ are 1.10 and 1.22. Because these values are greater than one, grain refinement and single-peak behavior are indicated.

A similar explanation applies to the curves shown in Fig. 10(b). Here the initial grain size is again constant and the addition of alloying elements changes the character of the flow curves by reducing D_s. However, because these experiments were carried out at 900°C, and because each test took about 2000 s to complete, the precipitation of NbCN and of VN took place in the 0.035% Nb and 0.115% V steels, respectively.[35] Thus, the retardation of recrystallization by the solute effect was supplemented by a component of precipitate retardation. It is in fact the occurrence of carbonitride precipitation that is responsible for the considerable increase in the peak stress and strain in the HSLA steels over those associated with the plain carbon grade. This is why there is a larger difference in Fig. 10(b), between the behaviors of the carbon steel on the one hand and of the vanadium and niobium steels on the other than there is in Fig. 10(a), where no precipitation is taking place.

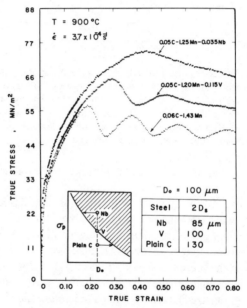

Fig. 10(b). Effect of dynamic precipitation on the shape of the flow curve. The plain carbon steel exhibits cyclic recrystallization because $D_0/2D_s = 0.77$. Conversely, the niobium steel, with $D_0/2D_s = 1.18$, displays single-peak behavior. The vanadium steel, with $D_0/2D_s \simeq 1$, is at the transition between the two types of flow. The greater differences among the three steels compared with the differences shown in Fig. 10(a) are attributable to the occurrence of dynamic precipitation during compression of the two HSLA materials.[35]

Dynamic Recrystallization Under Transient Conditions of Deformation

When the microstructural mechanism map of Fig. 8 was introduced, reference was made to *three* types of experiments:

1. The "vertical" tests, which correspond to a vertically distributed set of initial conditions (i.e., of temperature-corrected strain rates Z), and which are carried out on specimens of a fixed initial grain size D_0.

2. The "horizontal" tests, which correspond to a horizontally distributed set of initial grain sizes, and which are carried out at a fixed value of Z.

3. The combined "horizontal/vertical" tests, which include changes in both grain size and strain rate (or temperature), and which will now be described. This method of investigation is based on first establishing a stable dynamic recrystallization microstructure (horizontal movement on Fig. 8) and then carrying out strain-rate changes of the traditional type (vertical movement on Fig. 8). The nature of the transient flow curve that follows the change in Z is then studied, as are the microstructural adjustments that are necessary for adaptation to the new testing conditions (further horizontal movement on Fig. 8).

The effect of a strain-rate *increase* is depicted in Fig. 11 (left). Here it can be seen that multiple-peak flow at $\dot{\varepsilon}_1$ leading to D_{s1} is followed by single-peak flow at $\dot{\varepsilon}_2 = 26.4\dot{\varepsilon}_1$; after a new stable grain structure D_{s2} has been more or less established at $\dot{\varepsilon}_2$, multiple-peak flow is again displayed when the strain rate is reduced to its previous value, $\dot{\varepsilon}_1$.[36] These experiments were carried out in compression on a modified Instron testing machine equipped with appropriate computer interfacing. Prior to testing, each specimen was austenitized at 1000°C for 30 min directly in the test chamber and under an argon atmosphere. This heat treatment was applied to ensure the complete dissolution of the carbonitrides ($T_{sol} = 980$°C) and to stabilize the initial austenite grain size at about 80 μm. Because the tests were performed at the austenitization temperature, i.e., above the solution temperature of the carbonitrides, neither static nor dynamic precipitation took place during the experiments.

By performing the strain-rate changes at strains of 0.45 and 0.85 and then continuing testing to deformations of 1.4, a complete cycle on the map of Fig. 8 (see also inset in Fig. 11, left) was traced out, and the dynamic conditions prevailing after the completion of the prestraining interval were reattained. The results of some related experiments[36] are illustrated in Fig. 11 (right). In this case, the initial strain rate was the same, as was the interruption strain, but the second strain rate $\dot{\varepsilon}_2$ was varied over the range $1.36\dot{\varepsilon}_1 \leq \dot{\varepsilon}_2 \leq 38.95\dot{\varepsilon}_1$.

The occurrence of the single peak in Fig. 11 (left and right) was not entirely expected, because the "starting material" in these experiments was not an an-

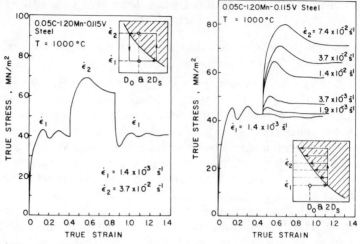

Fig. 11. Stress-strain curves for a 0.05C-1.20Mn-0.115V steel deformed at 1000°C. (Left) The strain rate was cycled from 1.4×10^{-3} to 3.7×10^{-2} and back to 1.4×10^{-3} s^{-1} at strain intervals of about 0.45. (Right) After a strain interval of 0.45 at an initial strain rate of 1.4×10^{-3} s^{-1}, the strain rate was increased to a new rate in the range 1.9×10^{-3} to 7.4×10^{-2} s^{-1}.[36]

nealed or dislocation-free specimen, as in Fig. 1, Fig. 5(b), and Fig. 9 and 10, but instead was a specimen in which dynamic recrystallization was already well under way and in which a dynamically stable grain (and dislocation) structure, referred to here as D_{s1}, had already been established. Because the new stable grain structure D_{s2} was finer than D_{s1}, it is evident that, even under strain-rate-change conditions, single-peak flow remains associated with grain refinement, as it is in annealed materials. Although temperature changes were not attempted in this particular investigation, the results of Fig. 11 suggest that a single flow-stress peak, accompanied by grain refinement, would probably be observed after a *decrease* in temperature following a period of steady-state flow at a higher temperature.

Effect of Decreases in Strain Rate

The effect of a strain-rate change which is the *reverse* of that effected in Fig. 11 (left) is presented in Fig. 12 (left). The broken lines in this diagram are for purposes of comparison and indicate the shapes of the flow curves when *annealed* materials are tested at $\dot{\varepsilon}_1$ and $\dot{\varepsilon}_2$. In this case, following a decrease in strain rate by a factor of 26.4, multiple peaks are observed. Because $D_{s2} < D_{s1}$, it is apparent that cycles of grain coarsening accompany oscillatory flow, as in the deformation of dislocation-free specimens. It should be noted that, after the strain-rate decrease, the transient flow curve shows a rapid drop, with a stress

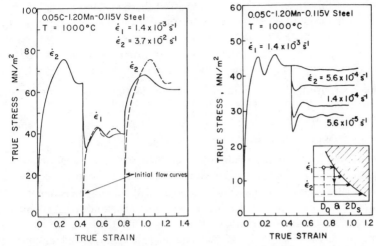

Fig. 12. Stress-strain curves for the vanadium steel of Fig. 11 deformed at 1000°C. (Left) The strain rate was cycled from 3.7×10^{-2} to 1.4×10^{-3} s^{-1} and then back again. The broken flow curves are for the annealed material. (Right) After a strain interval of 0.45 at an initial strain rate of 1.4×10^{-3} s^{-1}, the strain rate was decreased to a new rate in the range 5.6×10^{-5} to 5.6×10^{-4} s^{-1}.[36]

minimum after an additional strain of 0.026. Beyond the minimum, the curve is similar to that for the annealed structure, but the oscillations tend to die out more quickly, and the peak strain is slightly larger than in the virgin material. The flow curve accompanying the strain-rate increase, on the other hand, differs considerably from that for the annealed condition, in that both the peak stress and the peak strain are smaller than those associated with the annealed structure. These differences are related to the presence of a highly inhomogeneous dislocation substructure in the "starting material," and therefore to the much larger distribution of nucleation strain in the strain-rate-change specimens than in the annealed material. The strain-rate increase and decrease transients are in sharp contrast to those observed when the deformation is controlled by dynamic recovery,[36] a difference which can have an effect on the loads required to perform subsequent forming operations.

Somewhat similar remarks apply to the curves presented in Fig. 12 (right), in which the strain rate was decreased by ratios of 2.5 to 25. It should be noted, however, that when the ratio was decreased below 10, the dip in stress became less sharp, and the second and later stress peaks following the minimum disappeared. For consistency, although other interpretations are possible, the view is taken here that the first peak after the minimum is actually the *second* flow-stress peak after the change, so that, in terms of the classification system adopted here, the flow is of the oscillating type after *all* strain-rate decreases, however small.

According to this view, only the second half of the first "stress peak" is seen; it consists of the flow stress at the moment of the strain-rate decrease (the "peak" stress) together with the flow softening that follows until the minimum is reached.

Thus, it appears that under dynamic conditions (in opposition to the annealed behavior) only *two* (and not three) types of transient flow are observed.

1. After a strain-rate *increase*, there is single-peak flow accompanied by grain refinement (i.e., $D_{s1} > D_{s2}$).
2. After a strain-rate *decrease*, there is multiple-peak flow, accompanied by grain coarsening (i.e., $D_{s1} < D_{s2}$).

Cases 1 and 2 above can readily be identified with the categories of grain refinement ($D_0 \gg D_s$) and grain coarsening ($D_0 < D_s$), respectively, in annealed materials. However, the critical condition associated with the transition in microstructural and mechanical behavior under dynamic conditions is given by $D_{s1} = D_{s2}$;[36] this differs somewhat from the one that applies to previously undeformed materials ($D_0 = 2D_s$). As will be seen below, single- and multiple-peak curves are associated with two contrasting recrystallization mechanisms, the former with deformation-limited growth ("growth" control) and the latter with grain impingement ("nucleation" control).

Effects of Dynamic Recovery and Dynamic Recrystallization Under Conditions of Transient Deformation

When the strain rate is decreased abruptly during deformation under dynamic-recovery conditions, the relatively stable dislocation substructure formed at the higher strain rate, say $\dot{\varepsilon}_1$, is too dense for the new strain rate $\dot{\varepsilon}_2$ and gradually changes to a coarser one, along with a gradual decrease in flow stress.[50,51] Thus, as indicated in Fig. 13(a), the transient flow stress is always higher than that pertaining to the annealed material, or even to steady-state flow at $\dot{\varepsilon}_2$. Due to the substructural stability, this transient persists for a strain interval of 0.2 or more.[50] By contrast, the structure of the dynamically recrystallized grains at $\dot{\varepsilon}_1$ is highly unstable when the strain rate is decreased to $\dot{\varepsilon}_2$. This is because the mobility of the high-angle boundaries is much higher than that of the collections of dislocations responsible for the flow softening in the lower part of Fig. 13(a). The high boundary mobility, in turn, ensures that the transient in the flow curve persists for a much shorter interval, e.g., 0.02 to 0.04 strain (see upper part of Fig. 13a), after which the flow curve follows almost the same path as that for the annealed material.

The net effect of the factors described above is that, when the strain rate is decreasing (see Fig. 13b) and dynamic *recovery* is rate-controlling, the flow stress is *higher* than expected from the steady-state values. Conversely, under conditions of dynamic *recrystallization*, the flow stress is *below* the steady-state value, and therefore considerably lower than that associated with dynamic recov-

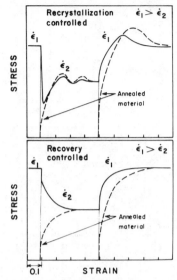

Fig. 13(a). Comparison of the transient flow curves observed after cyclic changes in strain rate under conditions of dynamic recrystallization (upper curves) and dynamic recovery (lower curves). The broken lines indicate the behavior of the annealed material.

ery. The picture changes somewhat when the strain rate is increasing. In this case, as long as dynamic recovery is rate-controlling, the transient flow stress is below the steady-state level (see Fig. 13). However, this is true also under dynamic-recrystallization conditions, so that a parallel exists between the effects of the two mechanisms when the strain rate is increasing (or the temperature decreasing). This parallel is absent, and their effects are opposite, when the strain rate is decreasing (or the temperature is increasing). In summary then, because of the relatively rapid motion of the grain boundaries responsible for dynamic recrystallization, this mechanism acts much like static recrystallization in keeping both the dislocation density and the flow stress below the values expected from steady-state considerations. By contrast, the relative stability of the dynamic-recovery substructures makes possible the inheritance of hard structures which can raise the flow stress above expected levels.[36]

Physical Basis for the Critical-Grain-Size Criterion

The physical basis for the microstructural mechanism map of Fig. 8 pertaining to the annealed structure will now be considered. First to be taken up is the question of how it is possible for the *nature* of the flow curve to change as sharply as it does when the temperature and strain rate are both held constant, and only

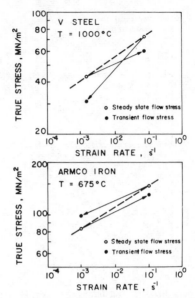

Fig. 13(b). Comparison of flow stresses developed during gradual strain-rate increases and decreases with steady-state behavior (broken lines). The dynamic-recrystallization behavior was derived from tests on the vanadium steel of Fig. 11 and 12 carried out at 1000°C (upper curves).[36] The dynamic-recovery behavior is based on tests on Armco iron carried out at 675°C (lower curves).[50]

the initial grain size is changed (horizontal series of tests in Fig. 8). Under these conditions, the flow stresses and therefore the stored energies promoting recrystallization change only slightly with initial grain size because of the constancy of Z. At this point, for simplicity, it is assumed that the nucleation of new grains occurs only at or near existing grain boundaries, so that the potential nucleus density is determined principally by the specific grain-boundary area.[17–20,30–36,48,52] According to this view, the possibility of nucleating new grains at deformation bands, twins or inclusions is assumed to be considerably smaller than at the grain boundaries.*

When nucleation occurs mainly at the grain boundaries, the relation between the nucleus density N_0 and the initial grain size D_0 is given by:[36]

---

*For completeness, however, it should be added that dynamic nucleation has been observed to occur within deformation bands at grain sizes greater than 100 μm in copper,[38] although in this case the twin boundaries did not play a significant role in the germination process. The modifications to the present model required to include nucleation in deformation bands, and, by extension, at inclusions and twin boundaries, are discussed elsewhere,[38] and will not be considered in detail in this chapter.

$$N_0 = 2\frac{P_0}{D_0} \tag{6}$$

where P_0 is the probability per unit surface area of an initial grain-boundary site being activated. When N_0, the nucleus density in the annealed material, is lower than N_s, the equilibrium density for steady-state flow (coarse-grain materials), grain refinement occurs until the appropriate density N_s is attained. Because nucleation occurs at the *existing* boundaries, the "necklace" mechanism generally operates (see Fig. 14a), the limit for which is a necklace containing a single strand.[34-36] Thus, the critical-grain-size condition

$$D_0 = 2D_s \tag{7}$$

is associated with the transition nucleus-density condition (see Fig. 14b):[36,53]

$$N_0 = N_s \tag{8}$$

Conversely, when $N_0 > N_s$ (fine-grain materials), grain coarsening occurs until the appropriate density N_s is again attained. The grain size during "necklace" recrystallization can be considered to be growth-controlled, ceasing when the

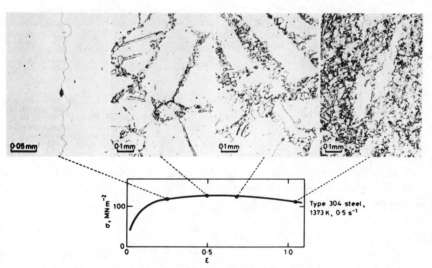

Fig. 14(a). Microstructures illustrating the progress of dynamic recrystallization in type 304 austenitic stainless steel with a coarse initial grain size. The specimens were deformed in compression at 1373 K and 0.5 s^{-1} to increasing strains and rapidly quenched (after *Roberts et al.*[52]). The occurrence of necklace recrystallization, leading to grain refinement, is clearly visible. In this case, $D_0 \gg 2D_s$, and nucleation at the original grain boundaries is followed by deformation-limited growth.

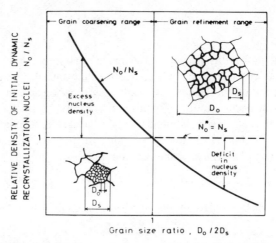

Fig. 14(b). Relation between the normalized initial nucleus density N_0/N_s and the normalized grain size $D_0/2D_s$ (see Eq 6 and 9). When $D_0/2D_s < 1$, $N_0/N_s > 1$, and grain *coarsening* occurs simultaneously with nucleus loss; in this way, N_0 approaches the stable value N_s from above. Conversely, when $D_0/2D_s > 1$, $N_0/N_s < 1$, and grain *refinement* takes place concurrently with the production of additional nuclei; under these conditions, N_0 approaches N_s from below. The *local* nucleus density N^*_0 in coarse-grain materials (broken line; $D_0/2D_s > 1$) is the *stable* nucleus density N_s because of the deformation-limited growth taking place during necklace recrystallization.[34,53]

driving force is destroyed by the concurrent deformation. The grain size during coarsening, on the other hand, can be considered to be impingement, and therefore nucleation-controlled (see Fig. 14b).[36,53]

Extension to Dynamic Conditions

If the same equations are now applied to the dynamically recrystallized structure, the relation between the stable nucleus density N_s and the equilibrium dynamic grain size D_s is given by:[36]

$$N_s = 2\frac{P_s}{D_s} \qquad (9)$$

where P_s is again the probability of activation of a grain-boundary site per unit area. Combining Eq 6 to 9, it is evident that

$$P_0 = 2P_s \qquad (10)$$

This signifies that the probability of nucleation in the dynamic grain structure is *less* than in the annealed one. The decreased probability is directly linked to the more homogeneous substructure distribution in the annealed material while it is undergoing deformation. By contrast, the presence of recently recrystallized regions during dynamic recrystallization, which contain no dislocations or sub-grains, and therefore in which no nucleation is possible, reduces the mean value of P_s to about half the value of P_0 in the virgin material.* It follows from the above that, when a change is made from one set of dynamic conditions to another, the critical condition at which no transient is observed will be:

$$N_{s1} = N_{s2} \tag{11}$$

which corresponds in turn to

$$D_{s1} = D_{s2} \tag{12}$$

as already noted above. In this approach, of course,

$$P_{s1} = P_{s2} = P_0/2 \tag{13}$$

Relation Between Nucleus Density and the Spread in Nucleation Strain

As shown earlier,[34-36] the character of the flow curves depends sensitively on the relation between the nucleation strain, ε_c, and the spread in the nucleation strain, $\Delta\varepsilon_c$. This is because, when $\Delta\varepsilon_c$, is small, nucleation occurs in all regions at about the time that the critical strain ε_c is attained in the most favorably oriented region. This signifies that the entirety of the material is essentially at the same stage of the recrystallization process at a given moment. Under these conditions, the progress of recrystallization is highly synchronized and the macroscopic flow curve readily reflects the local cycles of work hardening and recrystallization.

The ratio of ε_c to $\Delta\varepsilon_c$ depends in turn on the nucleus-density ratio N_0/N_s (see Fig. 14b). In fine-grain materials, $D_0 < 2D_s$, so that $N_0/N_s \gg 1$ and there are many potential nucleation sites within every future stable grain. As a result, $\Delta\varepsilon_c/\varepsilon_c$ is small, and nucleation occurs substantially simultaneously in the different parts of the material. After the first cycle of recrystallization is complete, work hardening begins again, leading to the second peak of the oscillating curve. However, the second cycle cannot be observed unless the condition suggested above is satisfied as written in the more precise form:[34,53]

$$\varepsilon_{c2} > \Delta\varepsilon_{c1} \tag{14}$$

Here ε_{c1} is the first nucleation strain measured from the beginning of straining; $\Delta\varepsilon_{c1}$ is the observed spread in this strain, i.e., the difference between the nucle-

---

*The ratio P_0/P_s, which is equal to the value of the constant A in $D_0 = A \cdot D_s$, is equal to 2 in a simple necklace model. However, the magnitude of A can be affected, not only by nucleation at deformation bands, inclusions and twin boundaries, but also by the detailed grain-size distribution and texture. Thus, in cases where P_0/P_s is slightly more or slightly less than 2 (see Ref 31, 32, 38, 47 and 49, for example), A in $D_0 = A \cdot D_s$ will also be slightly more or less than 2, respectively.

ation strains associated with the earliest and latest regions to begin recrystallizing; and ε_{c2} is the second nucleation strain measured from the moment that the first grains associated with the first cycle begin to grow (see Fig. 15). According to this model, several recrystallization cycles are needed to attain D_s because, at fine D_0, the nucleus density is initially too high, so that too fine an intermediate grain size, D'_s, is produced. The number of cycles thus depends on the disparity between D_0 and $2D_s$. With each cycle of recrystallization, D'_s and $\Delta\varepsilon_c$ increase, until finally the cycles disappear when $\Delta\varepsilon_c \approx \varepsilon_c$ for the next cycle and $D'_s = D_s$.

The conditions in the coarse-grain material stand in sharp contrast to those indicated above, particularly with regard to the nucleus-density ratio N_0/N_s,

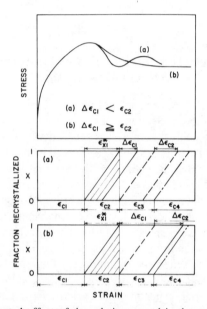

Fig. 15. Expected effect of the relative spread in the nucleation strain $\Delta\varepsilon_c/\varepsilon_c$ on the shape of the flow curve. (a) When $\Delta\varepsilon_c/\varepsilon_c$ is small (i.e., when recrystallization is nucleated approximately concurrently throughout the material, as in fine-grain specimens), distinct cycles of softening and hardening can be seen. These are gradually damped out because $\Delta\varepsilon_c$ increases as grain coarsening takes place. (b) When $\Delta\varepsilon_c/\varepsilon_c$ is large (i.e., when there is a considerable spread in the nucleation strain in different volume fractions of the material, as in coarse-grain specimens), only a single cycle of flow softening occurs, which persists until necklace recrystallization is complete.[34,53] (Because the various characteristic strains, ε_c, ε^*_x, $\Delta\varepsilon_c$, etc., all depend on grain size, as does ε_p, the schematic "strain" employed as the abscissa must, strictly speaking, be considered the *normalized* strain $\varepsilon/\varepsilon_p$.)

which in this case is *less than one*. Here only *one part* of a single D_0 boundary is associated with a particular future grain, a condition which leads to a real scarcity of potential nucleation sites. (The grain *interiors* take no part at all in the nucleation of the first strands of new grains.) Because of the necklace mechanism, the nucleation strain ε_{c1} pertaining to each successive strand of the necklace can be expected to cover a large range, from the smallest value associated with the first strand to the greatest value associated with the last strand. Thus, $\Delta\varepsilon_{c1}$ is large and generally exceeds ε_{c2}, the second nucleation strain, as shown in Fig. 15. In this case, $\Delta\varepsilon_{c1} > \varepsilon_{c2}$, and the different volume fractions of the specimen are at different stages of the recrystallization process, i.e., recrystallization is very heterogeneous and asynchronized at any particular moment. According to this view, flow softening occurs until the least favorably oriented grain interiors have participated in the recrystallization process. The model also indicates that when the grain-size distribution is bimodal or particularly broad, and when $D_0 \approx 2D_s$, a combination of "impingement" recrystallization (in the fine-grain-size fraction) and "deformation-limited growth" recrystallization (in the coarse-grain-size fraction) can occur. Because the impingement or nucleation-limited type of recrystallization is relatively well synchronized in the material, the volume fraction in which it takes place can be responsible for the small second peak frequently observed under refinement conditions when $D_0 \approx 2D_s$.

It should be added that, in the construction of Fig. 15, ε^*_{x1} was set, for simplicity, equal to ε_{c2}:

$$\varepsilon^*_{x1} = \varepsilon_{c2} \tag{15}$$

Furthermore, the macroscopic recrystallization strain ε_{x1} can be defined as:[34]

$$\varepsilon_{x1} = \varepsilon^*_{x1} + \Delta\varepsilon_{c1} \tag{16}$$

Thus, the present *microstructural* critical condition:[34]

$$\Delta\varepsilon_{c1} = \varepsilon_{c2} \tag{17}$$

can be expressed, by the addition of Eq 15, as:

$$\varepsilon^*_{x1} + \Delta\varepsilon_{c1} = 2\varepsilon_{c2} \tag{18}$$

or, by employing Eq 16, as:

$$\varepsilon_{x1} = 2\varepsilon_{c2} \tag{19}$$

Thus, it is evident that the present relative-grain-size-based critical condition, Eq 17, is broadly similar to the one derived by earlier workers (Eq 19) on the basis of computer simulations, and *in the absence of grain-size considerations*.[19,20,54]

Extension of the Relative Grain Size Model to Dynamic Conditions

When the strain rate is increased or decreased after a period of steady-state flow, the transient in the flow curve is associated with a gradual change in the dynamic grain structure from a mean size D_{s1} at Z_1 to a new equilibrium size D_{s2} at Z_2. This phenomenon can also be analyzed in terms of the two strain factors, ε_c and $\Delta\varepsilon_c$. As in the case of the annealed material, the ratio of these two strains determines the nature of the flow curve, as well as the type of recrystallization that takes place,[36] as discussed above. When the strain rate is decreased from $\dot{\varepsilon}_1$ to $\dot{\varepsilon}_2$, the spread in the nucleation strain, $\Delta\varepsilon_c$, decreases because the current grain size is now too fine and the potential nucleus density, N_{s1}, is too high. However, $\Delta\varepsilon_c$ gradually increases with strain as the current grain size gradually approaches its equilibrium value D_{s2}. By contrast, ε_c is much less sensitive to current grain size, and depends largely on the instantaneous value of the temperature-compensated strain rate Z. When the strain rate is decreased, it decreases suddenly, and vice versa. The availability of the abnormally high density of active nuclei after a strain-rate *decrease* produces the sharp drop in flow stress, leading to a minimum, which is a particular feature of the transient flow curves. In the interval between the rapid stress decrease and the subsequent attainment of steady-state flow, $\Delta\varepsilon_c < \varepsilon_c$, signifying that the recrystallization process is fairly well synchronized throughout the material, leading to multiple-peak flow and to grain coarsening.*

When the strain rate is *increased,* on the other hand, the ratio $\Delta\varepsilon_c/\varepsilon_c$ increases to values above 1, signifying that the recrystallization process is essentially desynchronized and that it becomes spatially inhomogeneous. The desynchronization of the recrystallization process can be attributed in turn to the relative scarcity of nuclei in the inappropriately coarse-grain material. Thus, single-peak flow and grain refinement are observed although, in contrast to the behavior of annealed materials, there is no necklace recrystallization under these conditions. Instead, there is a local and inhomogeneous type of "necklacing" associated with nucleation on the current grain boundaries and with the gradual propagation of recrystallization into the regions that do not contain grain boundaries at the moment of the change.

Dynamic Recrystallization in Single Crystals

Numerous investigations have been completed concerning the phenomenon of dynamic recrystallization in single crystals. Several fcc metals have been studied, including silver, gold, copper, nickel, and alloys of these metals.[24–27,55,56] Some

---

*The dependences of ε_c and $\Delta\varepsilon_c$ on strain rate and relative grain size are discussed in more detail in Ref 36 (see Fig. 9).

features of this work are of interest with respect to the present review, although they generally have not been considered in evaluation of the results of polycrystal experiments because neither the significance of the single-crystal data nor their relation to the behavior of industrial materials has been clearly evident. Nevertheless, certain aspects of the single-crystal observations can be put into focus and discussed in the framework of the present grain-coarsening/grain-refinement model, and such a synthesis will now be attempted.

There are three basic differences between the single-crystal and polycrystalline observations which will be highlighted here. The first is that the critical condition for nucleation in single crystals is generally taken to be the attainment of a certain minimum (resolved shear) *stress,* whereas in polycrystals it is the attainment of a critical peak *strain,* which displays a more systematic behavior than the peak *stress.* The former depends on the temperature and strain rate, and varies slightly with the initial orientation; it is taken to be a measure of the critical dislocation substructure required to initiate recrystallization.[24–26] The peak strain ε_p studied under polycrystalline conditions also varies with temperature and strain rate, as well as with other factors such as solute concentration, precipitate volume fraction and mean size, etc. Of particular interest here is the dependence of ε_p on grain size,[20,30–35,38] which indicates that ε_p increases with initial grain size D_0.

The second basic difference concerns the *values* of the peak stresses observed in single-crystal and polycrystalline materials. As can be seen in Fig. 16, the initiation stress in a typical single crystal oriented for multiple slip can be two or more times as high (depending on orientation[27]) as that in a polycrystal of the same material and purity, tested at the same temperature and strain rate. The third

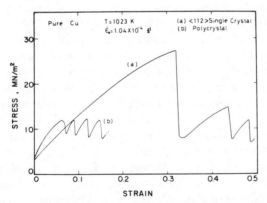

Fig. 16. Flow curves for copper deformed in tension at 1023 K and 10^{-4} s^{-1}: (a) $\langle 112 \rangle$ single crystal; (b) polycrystalline specimen (Kikuchi[27]). The single-crystal curve is plotted in terms of the average *normal* rather than resolved *shear* stresses and strains (Schmid factor m = 0.408).

difference is the obvious one that, whereas nucleation in polycrystals is largely restricted to the original grain boundaries, nucleation in single crystals occurs in the grain interiors.

These three factors are intimately interconnected, because they all concern the relevant nucleation mechanism and whether the final grain size is "nucleation" or "growth" controlled. With regard to the higher nucleation stresses in single crystals, it can be seen in Fig. 16 that, *once a single cycle of recrystallization has taken place,* the initiation stress for further cycles drops to the polycrystal level. Thus it appears that, once polycrystals have been introduced into a single crystal, further nucleation occurs at the boundaries of the polycrystalline region because it can take place at a much lower critical stress. The higher single-crystal critical stress thus corresponds to a kind of "homogeneous" nucleation, whereas the lower polycrystal critical stress concerns a type of "heterogeneous" nucleation that occurs at the previously existing grain boundaries.

Such an interpretation is consistent with the observations described briefly above regarding dynamic recrystallization in very coarse-grain copper poly-crystals.[38] In these experiments, which represent a kind of bridge between single-crystal and polycrystalline deformation, the peak stress was indeed ob-served to *increase with grain size* when the grain size exceeded about 100 μm. (It was approximately independent of grain size in the fine-grain specimens, in which nucleation was largely restricted to the prior grain boundaries.) In this work, extrapolation of the nucleation stress to "infinite" grain size, i.e., to $1/D_0 = 0$, led to estimates of the single-crystal nucleation stress which are in the lower ranges of such observations. In the coarse-grain material, a larger and larger proportion of the nuclei were formed at *deformation bands* within the grains rather than at the grain boundaries as the initial grain size approached the maximum investigated (815 μm). Thus, the extrapolated $1/D_0 = 0$ critical stress can be considered as the stress required to nucleate new grains *within deformation bands* in a suitably constrained single crystal. If nucleation is associated with other structural features, which appears to be the case in crystals of ⟨111⟩ or ⟨112⟩ orientations, for example,[27] then the critical initiation stress is even higher.

A further related question is whether the single-crystal behavior is basically multiple or single peak in the present context (despite the appearance of Fig. 16), and whether grain coarsening or grain refinement is taking place. In this sense, the data seem somewhat contradictory, because apparent multiple-peak flow appears to be combined with evident grain refinement, an association which is inconsistent with the present analysis. Nevertheless, the observations can be rationalized when the highly inhomogeneous deformation taking place under single-crystal testing conditions is taken into account. A detailed study of the data suggests clearly that experiments conducted using a suitable servocontrolled testing machine, *which maintains the strain rate at a constant value at a particu-lar location within the specimen,*[57] would produce single-peak flow curves of the expected shape, in keeping with the observed occurrence of grain refinement. This interpretation is indeed supported by the single-peak behavior displayed by copper single crystals when deformed in compression,[58] a deformation mode which leads to more homogeneous flow. If this interpretation is correct, then the recrystallization process in single crystals is *growth* controlled (i.e., deformation limited) and not *nucleation* controlled, a conclusion which is in opposition to the one reached in other recent investigations.[24-26]

Potentially Fruitful Areas for Further Investigation

It is evident from the single-crystal investigations described above that the role of nucleation in grain interiors (as opposed to that at grain boundaries) needs more study. Matters requiring clarification are the relative importance of nucleation at twins, deformation bands, kink bands, inclusions, etc. (i.e., in regions of high lattice curvature). Of particular interest in this regard would be experiments on *constrained* (or partly constrained) single crystals, which would be more representative of the deformation of polycrystals. Also of relevance is an assessment of the comparative importance in polycrystals of nucleation in the grain interiors, at sites such as those just mentioned, as opposed to the more common nucleation at or near grain boundaries.

A relatively unexplored area that should lead to fruitful results is the determination of the textures produced by dynamic recrystallization. One reason for the formation of preferred orientations is that the first grains to undergo dynamic recrystallization in a necklacing sequence (i.e., those in which nucleation occurs most easily), for example, lose their places in the distribution of orientations, so that the contributions of the remaining grains to the texture are reinforced. The dynamically formed textures which are characterized will have to be carefully distinguished from the related textures produced by: (*a*) deformation (dynamic recovery); (*b*) classical static recrystallization; and (*c*) postdynamic or metadynamic recrystallization. Such distinctions will require the use of very rapid quenching techniques and a reasonable knowledge of the kinetics of the various softening mechanisms. It is evident that if the textures produced by these four common and important mechanisms can be identified, then the conditions of industrial processing can be suitably altered or controlled to produce the textures that lead to the best combinations of material properties.

Although the $D_0 = 2D_s$ criterion for the transition from multiple- to single-peak flow seems well established, its validity should be checked for a wider series of alloys. Such investigations can also help to clarify the role of the grain-impingement mechanism during coarsening and that of deformation-limited growth during grain refinement. In a similar way, although the $D_{s1} = D_{s2}$ critical condition for transient deformation appears to be of general applicability, further work is required to determine the effects of such factors as the breadth of the grain-size distribution as well as the detailed evolution of the local critical strains and their standard deviations. Some of this research could involve the computer simulation of flow curves, with suitable physical expressions for nucleation and growth that take into account whether impingement or deformation-limited growth is the controlling process.

Much attention has been paid in this chapter to the characteristics of dynamic recrystallization. However, when this mechanism operates *during* deformation, it is evident that the material will be subject to metadynamic recrystallization as soon as deformation *ceases*. Nevertheless, the relation between dynamic and metadynamic recrystallization has received only limited attention to date. Research is therefore needed to determine, for example, the characteristics of metadynamic recrystallization following the grain-impingement/grain-coarsening type of dynamic mechanism on the one hand and following the necklace-formation/grain-refinement type on the other.

Practical Implications of Dynamic Recrystallization

The first extensive investigation of dynamic recrystallization under hot working conditions concerned plain carbon steels.[28] Nevertheless, this softening mechanism is unlikely to play a major role in the hot rolling of austenite, because static recrystallization generally intervenes after one or more rolling passes and removes the deformation substructure before the high strains needed to initiate dynamic recrystallization can be accumulated. As a result, dynamic recrystallization plays its most important role in the processes in which large strains are imparted in a single operation without unloading. The two most important cases of this type are planetary hot rolling and extrusion. Dynamic recrystallization may also be initiated in intermittent processes such as hot rolling and forging, as long as the unloading intervals are very short and the recrystallization kinetics are slow enough so that static recrystallization does not have the opportunity to intervene.[59] Examples of such processes are modern automatic bar mills and continuous forging lines.

As indicated above, under conventional rolling conditions, dynamic recrystallization is less likely to occur. Its appearance is nevertheless favored by relatively high temperatures, low strain rates, large pass reductions, and low concentrations of alloying elements.[59] One final application should be mentioned concerning materials which resist static recrystallization but which must nevertheless undergo microstructural refinement: e.g., certain stainless steels, superalloys and semiconductor materials. In such cases, dynamic recrystallization can be induced by deforming to sufficiently large strains and the original microstructure can be replaced in this way. In such cases, the principles outlined above can indicate the processing paths most likely to prove successful.

Conclusions

1. The critical-strain model for explaining the transition from periodic to single-peak flow in high-temperature austenite does not apply to tension or compression, or to other relatively homogeneous modes of deformation such as rolling and upset or continuous forging. This is because of the strain and strain-rate gradients that are present during torsion testing of solid bars — the type of testing that provided the data on which the model was based. As a result of the gradient, the peak strain ε_p is attained progressively later as the radius of an elemental shell in the solid bar is decreased. In this way, the peak torque (and therefore the peak stress determined experimentally) is attained *later* (and at a larger effective strain) than in the outermost layer of the bar, or than in a tensile test performed at the effective strain rate associated with the outermost layer. This effect is compounded for the recrystallization strain, ε_x, so that the differences between the values of ε_x determined in torsion and in compression are even greater than the differences between the torsion and compression values of ε_p. The net effect is that ε_p-ε_x curves do not in general intersect in uniaxial or plane-strain tension or compression testing. [In some cases of compression, an intersection *is* observed, but at stresses considerably higher than that corresponding to the transition from grain-coarsening (periodic) to grain-refining (single peak) flow.]

2. Periodic flow curves are associated with grain coarsening; the stress oscillations cease when the successive waves of coarsening finally produce the equilibrium grain size. By contrast, single-peak flow curves are observed when the grain-refinement ratio is at least 2:1. The criterion that distinguishes between the two types of behavior is $D_0 = 2D_s$, where D_0 and D_s are the initial and equilibrium grain sizes, respectively. Under industrial conditions of metal processing, because of the coarse grain size produced on reheating and the lower subsequent processing temperatures, single-peak behavior is generally expected, accompanied by grain refinement. The flow softening that follows the single-peak stress maximum comes to an end when refinement has affected the entire volume of the material, leading to the stable grain size D_s.

3. The dependence of twice the stable grain size ($2D_s$) on the temperature-corrected strain rate Z parallels that of the transition parameter Z_c on the initial grain size D_0. This correspondence is responsible for the transition criterion $D_0 = 2D_s$, according to which single-peak behavior is observed when $D_0 > 2D_s$ and stress oscillations are observed when $D_0 < 2D_s$. The critical-grain-size criterion has been verified in several materials, including niobium and vanadium microalloyed steels, which were tested under both solute and precipitate retardation conditions. It is evident from these results that "necklace" or "cascade" recrystallization (in which grain growth is halted by concurrent deformation) is

associated with single-peak behavior. Conversely, when grain growth is terminated by boundary impingement instead, as under grain-coarsening conditions, the flow curve is cyclic.

4. An analysis of these two types of behavior indicates that when necklace nucleation is taking place in coarse-grain materials there is a relative scarcity of nucleation sites. This leads in turn to a large spread in the nucleation strain and therefore to a form of dynamic recrystallization which is highly unsynchronized. In fine-grain materials, on the other hand, there is a high density of nuclei, as a result of which the spread in nucleation strain is small and all regions of the material are at approximately the same stage of the recrystallization process.

5. In very coarse-grain materials (grain size greater than 100 μm), nucleation can take place in deformation bands within the grains, in addition to the normal sites at the original grain boundaries. Because of the availability of such sites, the nucleus density is higher than that associated with the current grain size. The occurrence of nucleation at deformation bands thus reduces the *effective* initial grain size with respect to the critical transition condition for dynamic recrystallization. Similar considerations probably apply to the effects of twin boundaries and inclusions when these features contribute to the nucleation process.

6. The peak stresses associated with dynamic recrystallization in single crystals are about twice those observed in polycrystals; thus the nucleation processes involved in these two phenomena are quite distinct. In single crystals, the critical stress is a measure of the substructural density (and of the differences in substructural density) required to nucleate new grains "homogeneously," i.e., in the absence of existing grain boundaries; this is higher than the critical stress in polycrystals, where "heterogeneous" nucleation is possible. The recrystallization behavior of very coarse-grain polycrystals is intermediate between those of polycrystals and single crystals. Because of the low density of "heterogeneous" grain-boundary sites and the availability of "homogeneous" sites in the deformation bands, the peak stress is higher than in fine-grain specimens, but lower than in single crystals, of the same material.

7. When the strain rate is increased after a period of steady-state deformation, the flow curve displays a *single* peak. Both the peak stress and peak strain determined under these conditions are smaller than those associated with the annealed structure. When the strain rate is decreased, on the other hand, the flow curve displays *multiple* peaks, and the values of peak strain are greater than for the annealed structure.

8. The strain-rate-change results indicate that the nature of the transient flow curve changes abruptly according to whether the strain rate is increased or decreased after a period of steady-state flow. The critical condition at which the curve changes from the multiple-peak (grain-coarsening) to the single-peak (grain-refinement) type is $D_{s1} = D_{s2}$, where D_{s1} and D_{s2} are the equilibrium

dynamic grain sizes before and after the strain-rate change, respectively. This differs from the critical condition for the annealed structure, i.e., $D_0 = 2D_s$, where D_0 and D_s are the initial and stable grain sizes, because the probability of nucleation in the annealed structure is about twice that in the dynamic grain structure.

9. The transients displayed after a strain-rate change (inverse type) differ clearly from those observed when dynamic recovery is rate-controlling (normal transient). The latter transients involve a relatively *stable* dislocation substructure which changes only gradually with strain. By contrast, the microstructure of dynamically recrystallizing grains is relatively *unstable,* apparently because of the much greater mobility of the high-angle as opposed to the low-angle boundaries. As a result, when dynamic recrystallization is initiated under industrial conditions of processing, the dislocation density at any subsequent stage of the process is likely to be lower than when only dynamic (and static) recovery is involved.

Acknowledgments

The authors are indebted to Drs. M. G. Akben, G. Gottstein, U. F. Kocks, H. Mecking, W. Roberts and C. M. Sellars for stimulating discussions, and to Dr. S. Kikuchi of Kyoto University for making available the results of his single-crystal work (Fig. 16) prior to publication. They also acknowledge with gratitude the financial support received from the following sources: the Natural Sciences and Engineering Research Council of Canada; the Department of Energy, Mines and Resources, Canada; the Quebec Ministry of Education (FCAC program) (JJJ); the Ministry of Education of Japan; and the Hot Deformation Committee of the Iron and Steel Institute of Japan (TS). The authors express their thanks to McGill University (JJJ) and to the University of Electro-Communications, Tokyo (TS), for granting periods of sabbatical leave, and to the Centre de Mise en Forme des Matériaux (Director: Dr. J. L. Chenot), Ecole Nationale Supérieure des Mines de Paris, for providing manuscript-preparation facilities.

References

1. J. N. Greenwood and H. K. Worner, "Types of Creep Curves Obtained With 829 Lead and Its Dilute Alloys," *J. Inst. Met.,* Vol 64, 1939, p 135-158.
2. H. Hirst, "Deformation of Single Crystals of Lead by Creep," Part I: "Recrystallization and Failure," *Proc. Austr. Inst. Min. Met.,* No. 118, 1940, p 101-109; Part III: "Effect of Various Factors on the Creep Rate," ibid., No. 120, 1940, p 777-806; "Deformation by Creep of a Specimen of Lead Consisting of Five Large Crystals," ibid., No. 121, 1941, p 29-47.
3. E. N. da C. Andrade, "Creep of Metals and Recrystallization," *Nature,* Vol 162, 1948, p 410; "The Flow of Metals," *J. Iron Steel Inst.,* Vol 171, 1952, p 217-228.

4. R. C. Gifkins, "The Influence of Thallium on the Creep of Lead," *J. Inst. Met.,* Vol 81, 1952–1953, p 417-425.
5. R. C. Gifkins, "The Effect of Stress on the Recrystallization of Lead During Creep," *J. Inst. Met.,* Vol 86, 1957-1958, p 15-16.
6. R. C. Gifkins, "Recrystallization of lead during creep," *J. Inst. Met.,* Vol 87, 1958–1959, p 255-261.
7. R. C. Gifkins, "Oxygen as an Impurity in High-Purity Lead," *Acta Met.,* Vol 6, 1958, p 132-133.
8. S. Steinemann, "Experimental Investigations of the Plasticity of Ice" ("Experimentelle Untersuchungen zur Plastizität von Eis"), *Beiträge Geol. Schweiz (Hydrologie),* Vol 10, 1958, p 1-72.
9. J. J. Jonas and F. Müller, "Deformation of Ice Under High Internal Shear Stresses," *Can. J. Earth Sci.,* Vol 6, 1969, p 963-968.
10. H. P. Stüwe, "Dynamische Erhollung bei der Warmverformung," *Acta Met.,* Vol 13, 1965, p 1337-1342.
11. H. P. Stüwe, "Do Metals Recrystallize During Hot Working?," *Deformation Under Hot Working Conditions,* edited by C. M. Sellars and W. J. McG. Tegart, Iron Steel Inst., London, 1968, p 1-6.
12. H. J. McQueen, "Hot Working and Recrystallization of Face Centered Cubic Metals," *Trans. Japan Inst. Met.,* Vol 9 (Suppl.), 1968, p 170-177.
13. H. J. McQueen, "Deformation Mechanisms in Hot Working," *J. Met.,* Vol 20, 1968, p 31-38.
14. H. J. McQueen and S. Bergerson, "Dynamic Recrystallization of Copper During Hot Torsion," *Met. Sci.,* Vol 6, 1972, p 25-29.
15. H. P. Stüwe and B. Ortner, "Recrystallization in Hot Working and Creep," *Met. Sci.,* Vol 8, 1974, p 161-167.
16. J. J. Jonas, C. M. Sellars and W. J. McG. Tegart, "Strength and Structure Under Hot-Working Conditions," *Met. Rev.,* Vol 14, 1969, p 1-24.
17. M. J. Luton and C. M. Sellars, "Dynamic Recrystallization in Nickel and Nickel-Iron Alloys During High Temperature Deformation," *Acta Met.,* Vol 17, 1969, p 1033-1043.
18. G. Glover and C. M. Sellars, "Recovery and Recrystallization During High Temperature Deformation of α-Iron," *Met. Trans.,* Vol 4, 1973, p 765-775.
19. J. P. Sah, G. J. Richardson and C. M. Sellars, "Quantitative Correlation Between High Temperature Strength and the Kinetics of Dynamic Recrystallization," *Indian J. Tech.,* Vol 11, 1973, p 445-452.
20. J. P. Sah, G. J. Richardson and C. M. Sellars, "Grain-Size Effects During Dynamic Recrystallization of Nickel," *Met. Sci.,* Vol 8, 1974, p 325-331.
21. H. J. McQueen and J. J. Jonas, "Recovery and Recrystallization During High Temperature Deformation," in *Treatises on Materials Science and Technology,* Vol 6, *Plastic Deformation of Materials,* edited by R. J. Arsenault, Academic Press, New York, 1975, p 393-493.
22. C. M. Sellars, "Recrystallization of Metals During Hot Deformation," *Phil. Trans. R. Soc. Lond. A.,* Vol 288, 1978, p 147-158.
23. C. M. Sellars, "Dynamic Recrystallization," *Metalurgia I. Oldewnictwo* (Metallurgy and Foundry, Poland), Vol 5, 1979, p 377-404.
24. H. Mecking and G. Gottstein, "Recovery and Recrystallization During Deformation," in *Recrystallization of Metallic Materials,* edited by F. Haessner, Dr. Riederer Verlag, Stuttgart, 1978, p 195-222.
25. G. Gottstein, D. Zabardjadi and H. Mecking, "Dynamic Recrystallization in Tension-Deformed Copper Single Crystals," *Met. Sci.,* Vol 13, 1979, p 223-227.

26. G. Gottstein and U. F. Kocks, "Dynamic Recrystallization and Dynamic Recovery in ⟨111⟩ Single Crystals of Nickel and Copper," *Acta Met.*, Vol 31, 1983, p 175-188.

27. S. Kikuchi, Kyoto University, private communication.

28. C. Rossard and P. Blain, "Evolution of Steel Microstructure During Plastic Deformation at High Temperature" ("Evolution de la Structure de L'acier Sous L'effet de la Déformation plastique à Chaud"), *Mém. Sci. Rev. Mét.*, Vol 56, 1959, p 285-299.

29. R. A. Petkovic, M. J. Luton and J. J. Jonas, "Recovery and Recrystallization of Carbon Steel Between Intervals of Hot Working," *Can. Met. Quart.*, Vol 14, 1975, p 137-145.

30. S. Sakui, T. Sakai and K. Takeishi, "Effects of Strain, Strain Rate, and Temperature on the Hot Worked Structure of a 0.16% Carbon Steel," *Tetsu-to-Hagane*, Vol 62, 1976, p 856-865.

31. S. Sakui and T. Sakai, "Deformation Behavior of a 0.16% Carbon Steel in the Austenite Range," *Tetsu-to-Hagane*, Vol 63, 1977, p 285-293.

32. S. Sakui, T. Sakai and K. Takeishi, "Hot Deformation of Austenite in a Plain Carbon Steel," *Trans. Iron Steel Inst. Japan*, Vol 17, 1977, p 718-725.

33. T. Sakai, "Some Problems Involving Dynamic Recrystallization," in *Hot Deformation and Fracture*, edited by I. Tamura, Joint Society of Iron Steel Basic Research, Tokyo, 1981, p 34-52.

34. J. J. Jonas and T. Sakai, "The Transition From Multiple to Single Peak Recrystallization During High Temperature Deformation," in *Les Traitements Thermomécaniques*, 24ème Colloque de Métallurgie, INSTN (Saclay), 1981, p 35-56.

35. T. Sakai, M. G. Akben and J. J. Jonas, "The Role of Dynamic Recrystallization in Producing Grain Refinement and Grain Coarsening in Microalloyed Steels," in *Thermomechanical Processing of Microalloyed Austenite*, edited by P. J. Wray and A. J. DeArdo, AIME, 1982, p 237-252.

36. T. Sakai, M. G. Akben and J. J. Jonas, "Dynamic Recrystallization During the Transient Deformation of a Vanadium Microalloyed Steel," *Acta Met.*, Vol 31, 1983, p 631-642.

37. I. Weiss, T. Sakai and J. J. Jonas, "Effect of Test Method on the Transition From Multiple to Single Peak Dynamic Recrystallization," *Met. Sci.*, Vol 18, 1984, in press.

38. L. Blaz, T. Sakai and J. J. Jonas, "Effect of Initial Grain Size on the Dynamic Recrystallization of Copper," *Met. Sci.*, Vol 17, 1983, in press.

39. D. S. Fields, Jr., and W. A. Backofen, "Determination of Strain Hardening Characteristics by Torsion Testing," *Proc. ASTM*, Vol 57, 1957, p 1259-1272.

40. U. F. Kocks, J. J. Jonas and H. Mecking, "Development of Strain Rate Gradients," *Acta Met.*, Vol 27, 1979, p 419-432.

41. G. R. Canova, S. Shrivastava, J. J. Jonas and C. G'sell, "The Use of Torsion Testing to Assess Material Formability," in *Formability of Metallic Materials–2000 A.D.*, ASTM STP 753, edited by J. R. Newby and B. A. Niemeier, ASTM, 1982, p 189-210.

42. J. J. Jonas, G. R. Canova, S. C. Shrivastava and N. Christodoulou, "Sources of the Discrepancy Between the Flow Curves Determined in Torsion and in Axisymmetric Tension and Compression Testing," in *Plasticity of Metals at Finite Strain: Theory, Experiment and Computation*, edited by E. H. Lee and R. L. Mallett, Stanford Univ., 1982, p 206-222.

43. J. Gil-Sevillano, P. Van Houtte and E. Aernoudt, "Large Strain Work Hardening and Textures," *Prog. Mater. Sci.*, Vol 25, 1981, p 69-412.

44. S. S. Hecker and M. G. Stout, "Strain Hardening of Heavily Cold Worked Metals," in *Deformation, Processing, and Structure*, 1982 Materials Science Seminar,

ASM, 1984.

45. S. C. Shrivastava, J. J. Jonas and G. R. Canova, "Equivalent Strain in Large Deformation Torsion Testing: Theoretical and Practical Considerations," *J. Mech. Phys. Solids*, Vol 30, 1982, p 75-90.

46. T. Nakamura and M. Ueki, "The High Temperature Torsional Deformation of a 0.06%C Mild Steel," *Trans. Iron Steel Inst. Japan*, Vol 15, 1975, p 185-193.

47. C. Ouchi and T. Okita, "Dynamic Recrystallization Behavior of Austenite in Nb-Bearing High Strength Low Alloy Steels and Stainless Steel, *Trans. Iron Steel Inst. Japan*, Vol 22, 1982, p 543-551.

48. T. Maki, K. Akasaka and I. Tamura, "Dynamic Recrystallization Behavior of Austenite in Several Steels," in *Thermomechanical Processing of Microalloyed Austenite*, edited by P. J. Wray and A. J. DeArdo, AIME, 1982, p 217-234.

49. T. Maki, K. Akasaka and I. Tamura, Discussion of Ref 35 in *Thermomechanical Processing of Microalloyed Austenite*, edited by P. J. Wray and A. J. DeArdo, AIME, 1982, p 251-252.

50. J. P. A. Immarigeon and J. J. Jonas, "Flow Stress and Substructural Change During the Transient Deformation of Armco Iron and Silicon Steel," *Acta Met.*, Vol 19, 1971, p 1053-1061.

51. T. Takeuchi, "Load-Elongation Curves and Creep Curves of Pure Single Crystals," *J. Phys. Soc. Japan*, Vol 32, 1972, p 677-693.

52. W. Roberts, H. Boden and B. Ahlblom, "Dynamic Recrystallization Kinetics," *Met. Sci.*, Vol 13, 1979, p 195-205.

53. T. Sakai and J. J. Jonas, "Dynamic Recrystallization: Mechanical and Microstructural Considerations, An Overview," *Acta Met.*, Vol 32, 1984, in press.

54. R. Sandstrom and R. Lagneborg, "A Model for Hot Working Occurring by Recrystallization," *Acta Met.*, Vol 23, 1975, p 387-398.

55. A. Wantzen, P. Karduck and G. Gottstein, "Deformation Substructure in Dynamically Recrystallized Copper Single Crystals," in *Strength of Metals and Alloys*, ICSMA5, edited by P. Haasen et al., Aachen, 1979, p 517-522.

56. P. J. T. Stuitje and G. Gottstein, "Dynamische Recrystallisation in Zugverformten Silbereinkristallen," *Z. Metallk.*, Vol 279, 1980, p 279-285.

57. N. Christodoulou, J. J. Jonas and G. R. Canova, "Determination of First and Second Order Work Hardening and Rate Sensitivity Coefficients for OFHC Copper and 99.99 Per Cent Aluminum," in *Mechanical Testing for Deformation Model Development*, ASTM STP 765, edited by R. W. Rohde and J. C. Swearengen, ASTM, 1982, p 51-66.

58. A. J. DeArdo, University of Pittsburgh, private communication.

59. A. H. Ucisik, I. Weiss, H. J. McQueen and J. J. Jonas, "Multistage Hot Deformation With Decreasing Temperature of Two Plain Carbon and Two HSLA Steels," *Can. Met. Quart.*, Vol 19, 1980, p 351-358.

Discussion of Paper by J. J. Jonas and T. Sakai

J. P. IMMARIGEON

Structures and Materials Laboratory
National Aeronautical Establishment
National Research Council of Canada

The model for dynamic recrystallization presented by Jonas and Sakai (J and S) is generally consistent with an earlier phenomenological concept developed by Immarigeon and Floyd (I and F) to model the forging behavior of powder-processed nickel-base superalloys (*Met. Trans.,* July 1981, Vol 12A, p 1177-1185). Both models are in the form of microstructure-mechanism maps which allow peak stresses and steady-state grain sizes and stresses to be predicted as functions of strain rate and temperature. Both models also indicate that the response of alloys which recrystallize during hot working is governed by the size of the grains initially present in the material relative to the size of the grains present under steady-state conditions. These models differ, however, in purpose and applicability. While the J and S model provides a new criterion for the transition from cyclic to continuous recrystallization, the I and F model was developed to provide a rationale for the occurrence of either flow softening or flow hardening in dual-phase alloys which recrystallize during hot working at temperatures at which they are in a two-phase metastable condition.

Inherent in the I and F model is the assumption that deformation at hot working temperatures takes place both along grain boundaries by sliding (g.b.s.) and within the grain interiors by the motion of grain dislocations. Under such conditions and when the contribution from g.b.s. becomes significant, flow strength becomes dependent on grain size, tending to decrease with a decrease in grain size. The degree of this dependence is also expected to increase with a decrease in grain size and/or a decrease in strain rate at a given temperature.

The I and F model has been tested and found to be consistent with the flow behavior of several dual-phase alloys which recrystallize to a very fine grain size during hot compression where g.b.s. effects become predominant. In contrast, dislocation-density-dependent intragranular flow may be expected to predominate in coarse-grain alloys, such as the steels considered by J and S. A limited contribution from g.b.s. is, however, possible, as noted by W. Robert (discussion, this symposium), and at the relatively low strain rates considered by the authors the contribution from g.b.s. may increase to a significant degree. This should perhaps be considered in interpreting the flow behavior of their steel.

An ideal phenomenology for hot working should consider the whole range of grain sizes and quantify the corresponding relative contributions of g.b.s. and intragranular flow. In fine-grain materials, where grain size is stabilized by the presence of precipitates and where recrystallization may be induced by straining, the role of g.b.s. must be considered, because its effects are not negligible.

Reply to J. P. Immarigeon by Jonas and Sakai

The results of Dr. Immarigeon and his co-workers regarding grain coarsening and grain refinement under superplastic (SP) forming conditions certainly have some common features with ours, despite the differences in the basic mechanisms. Of particular interest are their observations of grain *refinement* at their higher strain rates. Because their starting material was derived from powder, the refinement was probably unlike the break-up of the fibrous and elongated structures produced by rolling or extrusion, a process which is frequently reported in the superplasticity literature.[1] If both grain coarsening *and* true grain refinement can occur under appropriate SP conditions, then dynamic recrystallization may indeed play a role, however modified, in superplastic phenomena. As a result, more research is called for to determine whether refinement can be produced in other alloys under conditions similar to those of I and F. A testable hypothesis that might explain the relative absence of reports of true grain refinement (not break-up) in the current SP literature could be based on the solute concentrations in each of the two phases. Thus, when highly alloyed materials are being tested, such as the INCO 713LC of I and F (approximately 25% various solutes and 75% nickel), the stable grain size D_s at a given temperature and strain rate is very small. By contrast, in experimental alloys based on pure metals such as lead and tin, or aluminum and zinc, the much lower solute concentration in each phase can diminish the solute retardation of grain-boundary motion, producing much larger values of D_s. If this reasoning is correct, experiments at similar homologous temperatures could lead, apparently paradoxically, to refinement in some materials (the highly alloyed ones), and to coarsening in others (the relatively pure ones).

An important difference between our observations and those of I and F concerns the shapes of the transients that precede steady-state flow. When $D_0 > D_s$, we detect oscillations in flow stress associated with *discrete* cycles of recrystallization, whereas they observe gradual flow hardening associated with *continuous* grain growth. Finally, because of the relative lack of work-hardening transients under SP conditions, the grain size is close to being a state parameter, as indicated by I and F, leading to a much simpler description of the flow stress than is possible under hot working conditions, alas.

Reference

1. "Superplastic Forming of Structural Alloys," Proceedings of International Symposium held at San Diego, CA, June 1982, edited by N. Paton and H. Hamilton, American Institute of Metallurgical Engineers, Warrendale, PA, 1982.

The Role of Dynamically Recovered Substructure in Dynamic Recrystallization

(A discussion of the paper by J. J. Jonas and T. Sakai)

H. J. McQueen
Professor of Mechanical Engineering
Concordia University
Montreal, Canada

Introduction

The new phenomenological description and model of dynamic recrystallization presented by Jonas and Sakai[1] constitutes an important advance in the understanding of high-temperature dynamic restoration mechanisms. Although their presentation is thorough in respect to its principal theme, there are several additional insights which I think would provide a more complete understanding of the phenomenon. An important aspect, which was hardly mentioned although probably was in the minds of the authors, was consideration of the substructure which forms before and during dynamic recrystallization and plays a definitive role in determining grain size. A second comment examines the relationship of the new theory to that of Sellars and colleagues.

Influence of Substructure on Recrystallization

While the paper by Sakai and Jonas[1] examines the influence of grain size on nucleation, it is almost completely silent on the role of substructure. In fact, the substructure is important in establishing the relationship between Z and $2D_s$ in Fig. 8 (in Ref 1), as is explained below. Furthermore, for a constant initial grain size, it influences the density and type of nucleation as Z varies.

It has been shown that the steady-state flow stress σ_s depends on the subgrain size d_s during steady-state deformation in Al, αFe and Zr:[2-6]

$$\sigma_s = \sigma_y + B(d_s^{-p})$$

231

and that d_s depends on the temperature T, strain rate $\dot{\varepsilon}$ and activation energy ΔH:

$$d_s^{-q} = C + D \log Z$$

$$Z = \dot{\varepsilon} \exp (+ \Delta H/RT)$$

where p and q usually have values between 0.7 and 1.5. In addition, it has been shown that these equations also hold for substructures in stainless steel developed by dynamic recovery, as the sole mechanism at strains lower than ε_c and also in the dynamically recrystallized grains during the steady state.[7,8] The fact that this dependence holds both before and after the peak demonstrates that flow stress depends on substructure. It has also been shown that the dynamically re-crystallized grain sizes D_s in copper and nickel and their alloys are related to σ and Z by relationships analogous to the two above.[2,7,9-14] One can therefore conclude that the dynamically recrystallized grain size is dependent on the sub-structural dimensions.[7,8,15,16] This has been indicated by transmission electron microscopy where (*a*) the serrations in the grain boundaries have been shown to be of the same dimensions as the subgrains[2,12] and (*b*) subgrains of high misori-entation have been observed in copper and brass hot worked either to just before the peak or into the steady state.[13,14]

The dependence of $2D_s$ on Z in Fig. 8 (in Ref 1) is the result of the density of recrystallization nuclei being dependent on the density of subgrains that become bounded on part of their surface by a high-angle boundary. If one considers a constant value Z_1, then regardless of the initial grain size (>20 μm, i.e., not superplastic), the subgrain size, d_1, both prior to nucleation and in steady-state deformation, is constant. The number of nuclei that form in the necklace along the original grain boundaries is dependent on the density of subgrains along the grain boundaries. The substructure both in front of and forming behind the migrating boundary defines the extent of growth of a necklace normal to the grain boundary before another necklace begins to nucleate.[15,16] When the grain size is smaller than the characteristic steady-state grain size D_{s1}, there is a large grain-boundary area available for nucleation sites, but only a few become actuated dependent on the substructure. As long as the grain size is an order of magnitude larger than the subgrains, the role of the latter in the formation of nuclei is independent of grain size (although it is granted that the grain size affects the rate of recrystallization and hence ε_c). Furthermore, this substructure ensures that the steady-state grain size finally becomes D_{s1} regardless of the occurrence of several stages of recrystallization (multiple peaks). Finally, the limitation of growth by concurrent deformation is related to the rate at which the substructure forms inside a growing nucleus.[15,16] A significant difference between static and dy-namic recrystallization is that, whereas the strain defines the grain size in the former, it is Z which defines it in the latter (partly because of its influences on ε_c).

Sakai and Jonas report that, in single crystals and in very large grains (>100 μm), there is nucleation within the grains. The proposal that the grain size be decreased to an effective value fails to come to grips with the fact that this is a new mechanism dependent entirely on the substructure. They fail to consider the effects of high strain rates (10 to 100 s^{-1}), which lead to the formation of small cells some of which attain high misorientations because the critical strains are on the order of 2 to 4.[13,14] As in single crystals, nuclei have every likelihood of being created inside the grains under such circumstances. There is the additional factor that the small cells are less likely to induce suitable grain-boundary bulging, thus not proportionately increasing grain-boundary nucleation.

There is also insufficient discussion of nucleation and the growth of new grains during the continual dynamic recrystallization of steady-state deformation;[8,14] the relevant substructure and mechanistic model are illustrated in Fig. 1 and 2. There cannot be necklace formation as in Fig. 14(b) (in Ref 1) because each new grain is the size of the old. Under these conditions the occurrence of nuclei is primarily dependent on the substructure cell size and the related cell-wall density, although the final size is partially controlled by strain-limited growth.[15,16] Additional evidence supporting the nucleation from cells is the much lower density of twins in dynamically recrystallized grains than in normally recrystallized ones where the twins form in the growth stage.[14] One can also imagine a model of continual dynamic recrystallization which is quite different from the static kind where the grain boundaries migrate into material of roughly uniform dislocation density until they impinge on another recrystallized region. In the former case, different segments of the advancing boundary of a new grain are in contact with regions of quite different density dependent on the strain since those regions recrystallized. Segments adjacent to low-strain regions are held up first, whereas those against high-strain regions advance more rapidly, which results in their bulging out, creating new nuclei (Fig. 3).

Fig. 1. Electron micrograph of the dynamically recovered and recrystallized structure in copper deformed into the steady-state regime at 800°C and 11.1 s^{-1} (magnification, 8000×). The dynamically recrystallized grains as determined by diffraction are outlined.[14]

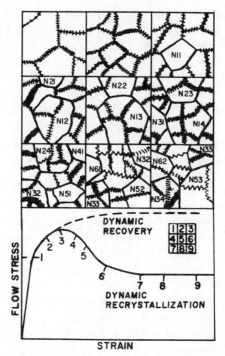

Fig. 2. The substructure formation,[1,2] the formation within a grain, and the growth and deformation of a dynamic nucleus (3-5) are shown. During steady-state deformation, nuclei are repeatedly forming from subgrains in different regions,[6-9] including those which have previously recrystallized.[8]

For the reasons just discussed, the stress-strain curves resulting from a change in strain rate during steady-state deformation[1] cannot be compared to the nucleation of the first dynamic grains in an almost static annealed starting material. The dynamic nuclei created just before the change continue to grow, and regions of varying strain density pre-exist to give rise to new nuclei. Increasing the strain rate raises the rate of dislocation accumulation and the pre-existing nuclei cannot provide enough softening. In addition, the creation of new nuclei even from predisposed regions is made increasingly more difficult by the greatly increased strain rate. On the other hand, a decrease in strain rate allows the existing nuclei to grow more easily (the lower the rate the more similar it is to metadynamic recrystallization[2]). The lower strain rate enhances nucleation in the predisposed regions with the result that there is an avalanche of recrystallization which gives rise to a grain structure similar to that in a statically annealed material with grains much larger than the dynamic grains at either the original or the new strain rate. Under such circumstances, the rules of nucleation described by Jonas and Sakai[1] apply.

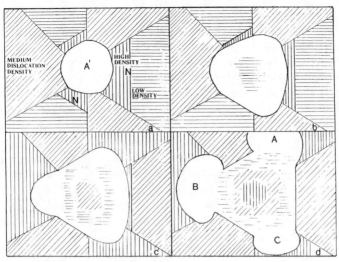

N-Boundries Halted by Strain Limited Growth
A',A,B,C - New Nuclei

Fig. 3. Nucleation during dynamic recrystallization as a result of a migrating boundary which, retarded by low strain in some regions, bulges into others. This is an alternative to the nucleation from subgrains illustrated in Fig. 2. N = boundaries halted by strain-limited growth; A', A, B, C = new nuclei.

The Relative Magnitudes of ε_p, ε_x, ε_c and ε_r

Jonas and Sakai overemphasize the differences between their theory and that of Sellars and colleagues.[10,11] Their criterion for a single peak ($\Delta\varepsilon_{c1} > \varepsilon_{c2}$), as expressed in Fig. 15 (in Ref 1), is substantially the same as Sellar's principle that for a single peak the strain for substantial recrystallization must be greater than the critical strain. This does not deny that Sakai has clearly shown that the crossover of the mechanical parameters ε_p and ε_x (Fig. 1, 2 and 3 in Ref 1) does not occur in tension and, in torsion, is a fortuitous result of the strain gradient. Sellars' metallographic work, which had determined that dynamic recrystallization occurs by successive necklaces around the edges of the initial grains, showed that $\varepsilon_c < \varepsilon_p$ and $\varepsilon_r \neq \varepsilon_x$.[10,11] The equivalences of terminology are that

$$\varepsilon_c = \varepsilon_{c1} \quad \text{and} \quad \varepsilon_r = (\Delta\varepsilon_{c1} + \varepsilon_{x1}^*)$$

This latter quantity is the normal definition of completion, i.e., when recrystallized grains have entirely replaced the initial structure. Thus the Sellars' criterion for single-peak flow curves in Sakai-Jonas terminology is

$$\Delta\varepsilon_{c1} + \varepsilon_{x1}^* > \varepsilon_{c1}$$

As compared with $\Delta\varepsilon_{c1} > \varepsilon_{c2}$, both terms in the Sellars' statement are larger than those in the more precise statement of Sakai and Jonas. In reality the quantities ε_{c2} and $\Delta\varepsilon_c$ are difficult to determine precisely whereas ε_{c1} and ε_r can be determined by normal techniques.

The demonstration of Sakai and Jonas[1] that $\varepsilon_x > \varepsilon_p$ is not a *criterion* for single-peak flow curves should not be taken to mean that such a condition may not exist in certain alloys even in tension or compression tests. Certainly for the carbon and HSLA steels reported[1] ε_x is much less than ε_p when single-peak curves first appear on increasing Z. However, for alloys in which boundary mobility is reduced by substitutional solute or particles, ε_x is likely to be increased relative to ε_p; an example of this appears in Fig. 14(a) (in Ref 1), which shows the very slow recrystallization of stainless steel.[16] For Waspaloy at 1038°C, where the γ' is in solution, ε_x, being $5 \times \varepsilon_p$ in torsion, is still likely to exceed ε_p in tension; in tests from 1100 to 980°C with $D_0 > 2D_s$, single peaks were observed[17] (see Fig. 4). Similarly for M2 tool steel, ε_x is much greater than ε_p throughout the entire hot working range from 1100 to 900°C; carbides, which

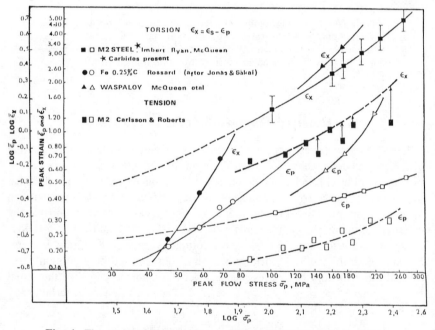

Fig. 4. The superiority of ε_x over ε_p exhibited by the 0.25% C steel in torsion is equivalent to an inferiority of ε_x relative to ε_p in tension.[1] The great superiority of ε_x over ε_p for M2 steel in torsion[18] is, however, related to a superiority of ε_x in tension as well.[19] Waspaloy, above the solvus, also exhibits retardation of recrystallization with an extended ε_x.[17]

amount to 17% at 1040°C, retard the growth of recrystallization grains, although it is possible that their coalescence accounts in part for the softening and the large ε_x,[18] Fig. 4. Confirmation that large $\varepsilon_x(\varepsilon_x \gg \varepsilon_p)$ in torsion is indicative of large $\varepsilon_x(\varepsilon_x > \varepsilon_p)$ in tension comes from the results on M2 of Carlsson and Roberts,[19] referred to by Roberts in these proceedings.

Conclusions

In conclusion, the demonstration by Jonas and Sakai[1] that the occurrence of grain refinement during dynamic recrystallization is the definitive condition for a single peak in the flow curve is a great advance in the understanding of that mechanism. However, consideration of the role of substructure in the dynamic recrystallization mechanism clarifies the description provided as follows:

1. The dependence of subgrain size on Z gives rise to the steady-state grain-size dependence on Z.
2. The substructure present at the critical strain determines the density of nuclei on the grain boundaries and also in succeeding stages of necklacing where that phenomenon occurs.
3. As D_0 becomes very large or as the strain rate becomes very high, the substructure gives rise to nucleation within the grains.
4. The substructure and the recently formed nuclei existing just before a change in strain rate determine the behavior following that change and provide a good framework for its explanation.

References

1. J. J. Jonas and T. Sakai, "A New Approach to Dynamic Recrystallization," in *Deformation, Processing, and Structure,* edited by G. Krauss, ASM, Metals Park, OH, 1984.
2. H. J. McQueen and J. J. Jonas, "Recovery and Recrystallization During High Temperature Deformation," in *Treatise on Materials Science and Technology,* Vol 6, *Plastic Deformation of Materials,* Academic Press, New York, 1975, p 393-493.
3. H. J. McQueen and J. E. Hockett, "Microstructure of Aluminum Compressed at Various Rates and Temperatures," *Met. Trans.,* Vol 1, 1970, p 2997-3004.
4. J. J. Jonas, D. R. Axelrad and J. L. Uvira, "On Substructure Strengthening and High Temperature Deformation of Cubic Metals," *Trans. Japan Inst. Metals,* Vol 9 (suppl.), 1968, p 257-267.
5. D. J. Abson and J. J. Jonas, Substructure Strengthening in Zr and Zr–Sn Alloys," *J. Nucl. Mater.,* Vol 42, 1972, p 73-85.
6. H. J. McQueen, "The Discovery of Dynamic Restoration Mechanisms in Hot Working," *Metals Forum* (Australia), Vol 4, 1981, p 81-91.
7. L. Fritzmeier, M. Luton and H. J. McQueen, "Dislocation Substructure in Dynamic Recovery and Recrystallization of Hot Worked Austenitic Stainless Steel," in *Strength of Metals and Alloys* (ICSMA 5 Aachen), Pergamon Press, Frankfurt, Vol 1, 1979, p 95-100.

8. H. J. McQueen, "Dynamic Recovery and Its Relation to Other Restoration Mechanisms," *Metallurgia i Odlewnictwo* (Metallurgy and Foundry, Poland), Vol 5, 1979, p 421-470.
9. R. Bromley and C. M. Sellars, "High Temperature Deformation of Cu and Cu-Al Alloys," in *Microstructure and Design of Alloys* (ICSMA 3, Cambridge), Inst. of Metals, London, Vol 1, 1974, p 380-385.
10. C. M. Sellars, "Recrystallization of Metals During Hot Deformation," *Phil. Trans. R. Soc. Lond.*, Vol A288, 1978, p 147-158.
11. C. M. Sellars, "Dynamic Recrystallization," *Metallurgia i Odlewnictwo*, Vol 5, 1979, p 377-404.
12. F. J. Humphreys, "Inhomogeneous Deformation of Al Alloys at Elevated Temperatures," in *Strength of Metals and Alloys*, Pergamon Press, Sydney, Vol 1, 1982, p 625-630.
13. H. J. McQueen, "Hot Working and Recrystallization of Face Centered Cubic Metals," *Trans. Japan Inst. Metals,* Vol 9 (suppl.), 1968, p 170-177.
14. H. J. McQueen and S. Bergerson, "Dynamic Recrystallization of Copper During Hot Torsion," *Metal Science*, Vol 6, 1972, p 25-29.
15. R. Sandstrom and R. Lagneborg, *Acta Met.*, Vol 23, 1975, p 387-398.
16. W. Roberts, H. Boden and B. Ahlblom, "Dynamic Recrystallization Kinetics," *Metal Science,* Vol 13, 1979, p 195-205.
17. H. J. McQueen, G. Gurewitz and S. Fulop, "Influence of Dynamic Restoration Mechanisms on the Hot Workability of Waspaloy," *High Temp. Tech.*, 1983 (in press).
18. C. Imbert, N. D. Ryan and H. J. McQueen, "Hot Workability Under Large Strain and Multistage Conditions of Tool Steels," *Met. Trans.*, 1983 (submitted).
19. J. -E. Carlsson and W. Roberts, "End Cracking in Hot Working of High-Speed Steel," Swedish Institute of Metals Research, Report No. 1583, 1982.

Reply to Discussion by H. J. McQueen

McQueen introduces a number of interesting issues relating to dynamic recrystallization which can be categorized under the headings that follow. We will comment briefly on these matters below.

1. The importance of the d_s versus Z relation, where d_s is the mean subgrain size; the relation of d_s to flow stress, nucleus density, etc.
2. The nucleation and growth of new dynamic grains during steady-state dynamic recrystallization on the one hand and after strain-rate changes on the other.
3. The particular dynamic recrystallization behavior of materials which are: (*a*) highly alloyed; (*b*) of very coarse grain size; or (*c*) tested at high strain rates.
4. The relation between our critical *mechanism-based* condition for the transition between multiple- and single-peak flow and the various criteria of Sellars and co-workers, which are either mechanically or microstructurally based.

Relation Between Our Critical Condition
($D_0/2D_s = N_s/N_0 = 1$) and Those of Sellars and Co-workers
($\varepsilon_p = \varepsilon_x$; $2\varepsilon_{c1} = \varepsilon_{x1}$)

We have discussed this matter in an earlier paper (Ref 34), and we will only summarize our opinions here. The original Sellars strain-based criterion ($\varepsilon_p = \varepsilon_x$) was a *mechanical* one, based purely on the shape of the flow curve. The point of the analysis described above is that the criterion applies *only to solid-bar torsion*, and not to *tubular* torsion or to plane-strain tension and compression, axisymmetric tension and compression, etc., because of the peculiarities of the usual methods employed to deduce the experimental flow curve (e.g., that of Ref 39) from the measured torque.* When test methods other than solid-bar torsion are used, the (peak or steady-state) stress associated with the transition from multiple- to single-peak flow is normally *well below* the stress at which $\varepsilon_p = \varepsilon_x$. The available high-temperature mechanical data obtained using the various test methods listed above are analyzed and compared in Ref 37, from which the large difference between the "transition stress" and the stress associated with the critical strain condition $\varepsilon_p = \varepsilon_x$ can be readily seen. Thus it is our view that the condition $\varepsilon_p = \varepsilon_x$ does not have any predictive ability (except for solid-bar torsion) with regard to the type of recrystallization mechanism operating (necklace formation versus impingement) or the shape of the flow curve (single versus multiple peak, respectively).

The original $\varepsilon_p = \varepsilon_x$ criterion was clearly a *mechanical* one, and thus different in concept from the various *microstructural* criteria we have discussed above. Nevertheless, Sellars and co-workers also deduced a microstructural condition from the results of their computer simulations, which concerned the behavior of *tubular* specimens. From their calculations, they arrived at the criterion (expressed in our notation):

$$\varepsilon_{x1} = 2\varepsilon_{c1} \tag{a}$$

which is equivalent to:

$$\Delta\varepsilon_{c1} + \varepsilon_{x1}^* = 2\varepsilon_{c1} \tag{b}$$

and is similar to our Eq 19. It should be noted that the RHS of the above two relations is *twice* that of the relation given by McQueen (i.e., RHS = $2\varepsilon_{c1}$ as opposed to ε_{c1}).

The various critical strains discussed above relate to certain microstructural events, and are therefore difficult to measure, as McQueen points out. Moreover,

---

*Such problems can, however, be avoided by the use of the technique described in Ref 41, which involves testing a series of specimens of increasing diameter.

because they are known only after the fact, their main use is restricted to analysis and interpretation. They are not therefore really suitable for predictive purposes, a shortcoming shared with the $\varepsilon_p = \varepsilon_x$ criterion. In terms of distinguishing between the two types of flow curve and of recrystallization, our view is that the grain-size-based condition $D_0/2D_s = 1$ is best. The terms involved in this relation can be readily measured; furthermore, the "relative grain size" $D_0/2D_s$ on which our treatment is based can be given a clear physical significance because it is related to the relative nucleus density by the simple expression:

$$D_0/2D_s = N_s/N_0 \qquad\qquad (c)$$

The importance of this relation, and the manner in which it is linked to the sub-boundary density in the vicinity of the grain boundaries, is discussed in more detail below.

Importance of the d_s-Versus-Z Relation

We have not discussed the d_s-versus-Z relation for two reasons. The first is simply that almost all reviews and original research papers dealing with high-temperature deformation published since 1969 have devoted considerable space to this dependence, and, in our "New Approach to Dynamic Recrystallization," we wanted to present the alternative merits of the new Z_c-versus-D_0 relation which, in combination with the $2D_s$-versus-Z dependence, leads to the new $D_0 = 2D_s$ criterion. The second reason is that d_s essentially factors out of the important expressions governing the flow behavior during either flow softening or multiple-peak flow (i.e., until steady-state flow is attained, when indeed both D_s and d_s depend on Z). This can be seen from the following.

1. During stress oscillations, or during flow softening after a single peak, d_s remains *constant;* nevertheless, the flow stress *fluctuates.* Thus the flow stress cannot "depend on the substructure" as stated by McQueen. Instead, σ and d_s both "depend" on the independent variables $\dot{\varepsilon}$ and T (and the stress on the strain as well).

2. In coarse-grain ($D_0/2D_s \gg 1$) materials experiencing the first strand of necklace nucleation (see Fig. 14a), the subgrains in the grain *interiors* play *no role at all in the recrystallization process.* Here, for simplicity, we leave aside nucleation at twins, deformation bands, inclusions, etc. — topics discussed by us elsewhere.[38] As a result, the relative nucleus density $N_0/N_s = 2D_s/D_0$ is very low. Conversely, in fine-grain ($D_0/2D_s \ll 1$) materials being deformed at the same value of Z, and *therefore containing subgrains of the same mean size d_s,* the relative nucleus density, as given by $N_0/N_s = 2D_s/D_0$, is initially very high, until it is reduced during successive oscillations to the steady-state level. The overabundance of nuclei (with respect to the final, stable configuration) can be attributed to the overabundance of grain-boundary sites (with respect to the stable value

based on D_s). In the two contrasting microstructures, *the subgrain size d_s is the same,* yet the flow curves diverge sharply, and the detailed mechanisms of recrystallization are also different. Thus, the transition in flow behavior (the principal subject of our paper) is in no way affected by the value of d_s. Nor is, for example, the number of peaks observed under stress-oscillation conditions, which also depends on the value of $D_0/2D_s$. The probable reason for this independence can be seen from what follows.

3. The nucleus density N_0 in the vicinity of the *original* grain boundaries can be linked to the mean subgrain size d_s through the probability term P_0 (see Eq 6 to 13 in the paper) as follows:

$$P_0 = k_0/d_s^2 \qquad (d)$$

Here k_0 is a nondimensional constant. Similarly, the probability of nucleation in the vicinity of the *dynamic* grain boundaries is given by:

$$P_s = k_s/d_s^2 \qquad (e)$$

where k_s is another constant. Note that the probability P_0 observed experimentally is apparently twice as great as P_s, suggesting that $k_0/k_s \simeq 2$. This can be attributed to the relative *homogeneity* of the substructure in the starting material as opposed to the relative *inhomogeneity* of the substructure in the dynamically recrystallizing material. (In the latter, the recently recrystallized regions containing *no* subgrains cannot contribute to the nucleation process.) Thus, the nucleus density in the starting material relative to the stable density (relative nucleus density) is given by:

$$N_0/N_s = (2P_0/D_0)/(2P_s/D_s) \qquad (f)$$

$$= (P_0/P_s)(D_s/D_0) \qquad (g)$$

At this point, Eq d and e can be substituted, leading to:

$$N_0/N_s = (k_0/k_s)(D_s/D_0) \simeq 2D_s/D_0 \qquad (h)$$

Three features of Eq h are of direct interest here. The first is that, although the probabilities of nucleation P_0 and P_s both depend on d_s, the latter is *factored out* when the *relative* probability is the term of importance. This is essentially why d_s does not play a direct role in determining whether grain coarsening or grain refinement will take place. It is also why we cannot agree with McQueen's contention that "the role of the subgrains in the formation of nuclei is independent of grain size." The second is that the important parts of Eq g and h are the *grain* (not the subgrain) sizes precisely because the relative nucleus density depends largely on the *grain-boundary* as opposed to the sub-boundary densities. The type of recrystallization and shape of flow curve depend on the relative *grain* (and not subgrain) sizes, because dynamic recrystallization is initiated principally along the existing grain boundaries. The third feature is that, in terms of the above analysis, the critical condition $D_0 = 2D_s$ can also be expressed as $D_0 = (k_0/k_s)D_s$. For this reason, the experimental values of the grain-size ratio D_0/D_s, *which are not always exactly 2*, can be taken as indicative of the relative probabilities of nucleation in the starting (annealed) and dynamic microstructures, and therefore of the degree of inhomogeneity present in the dynamic recrystallization substructure under steady-state conditions.

Nucleation and Growth During Steady-State and Strain-Rate-Change Conditions

McQueen has pointed out, quite correctly, that the classical necklace mechanism (Fig. 14a) cannot operate as such either during steady-state flow or under strain-rate-increase conditions. This is because of the inhomogeneity of the dynamic substructure we have been discussing immediately above. The phenomenon can nevertheless be visualized in our terms by considering that both the "pearls" in the strands of the classical model and the large nucleus-free regions in the grain interiors are uniformly distributed under dynamic conditions, rather than segregated, as in the annealed starting material. "Necklacing" can still be thought of as occurring in the sense that large volume fractions of the material (uniformly distributed in this case) are currently unable to participate in the nucleation process, in this case because of the recent occurrence of dynamic recrystallization (and therefore an absence of dislocations) within them. Of greater importance is the question of whether the quantitative description we have presented (Eq 6 to 13) is valid under dynamic conditions. It is our view that the *relative nucleus density* is particularly useful for this purpose; by analogy with Eq h, above, this property can be expressed as follows:

$$N_{s1}/N_{s2} = D_{s2}/D_{s1} \cdot (d_{s2}/d_{s1})^2 \tag{i}$$

Thus, when the strain rate is changed after attaining a steady state of flow, the behavior that follows depends on whether there is a current nucleus *scarcity* (leading to grain refinement and single-peak flow, as in annealed materials) or a current nucleus *overabundance* (accompanied by grain coarsening and multiple-peak flow, again as in annealed starting materials). The concept of a link among (a) flow-curve shape; (b) grain coarsening or grain refinement; and (c) relative nucleus density (with respect to the current steady-state configuration) therefore seems to apply to dynamic as well as to annealed "starting" microstructures. In a similar manner, coarsening is likely to be associated with impingement-limited growth and refinement with deformation-limited growth.

Dynamic Recrystallization Under Unusual Conditions

It remains for us to consider the behavior of materials which are highly alloyed, of very coarse grain size, or tested at high strain rates. McQueen has reported that, under *torsion-testing* conditions, the recrystallization strain ε_x can exceed the peak strain ε_p by a large margin, an inequality which he feels is likely to be preserved even under tensile-testing conditions, if such large strains could be attained in this mode. However, as outlined in the first part of our reply, and as shown in more detail in Ref 37, the condition $\varepsilon_p = \varepsilon_x$ in tension testing (and in other testing modes except for solid-bar torsion), is generally satisfied at *much higher stresses* (and temperature-corrected strain rates) than that associated with the transition from oscillating to single-peak flow. If $\varepsilon_x > \varepsilon_p$, then grain refinement must of necessity be occurring under the conditions he described (as long as dynamic recrystallization continues to be nucleated at the grain boundaries). While these examples are interesting, we nevertheless feel that the continued use of the $\varepsilon_p = \varepsilon_x$ criterion can only obscure the metallurgical processes that are actually taking place. The real issue is whether the $\varepsilon_p = \varepsilon_x$ criterion pertaining to these materials when tested by a relatively homogeneous mode of straining correlates even approximately with the transition in recrystallization and flow-curve behavior. We think this is highly unlikely.

Of greater importance is whether a quantitative description remains possible under exceptional conditions that correlates with the grain coarsening/grain refinement transition on the one hand and with the change in flow-curve shape on the other. For this purpose, let us consider an idealized case quite unlike our present model in which *no nucleation takes place at the grain boundaries at all!*

Suppose, for example, that nucleation occurs only at (*a*) inclusions (say, in a highly alloyed material); (*b*) deformation bands (as in single crystals and coarse-grain materials[38]); or (*c*) twin boundaries (e.g., at very high strain rates). It is at least possible that the relative-nucleus-density approach described above continues to apply under these circumstances, although in a modified form. We could speculate, for example, that the conditions

$$N_0/N_s \lessgtr 1 \tag{j}$$

apply, respectively, to grain coarsening (multiple peaks) and grain refinement (a single peak), only in this case the initial and stable densities are expressed in terms of (*a*) inclusion, (*b*) deformation-band and (*c*) twin-boundary densities.* Because of the nature of the roles they play in initiating dynamic recrystallization, it is conceivable that the initial densities are almost invariably lower than the dynamic densities, so that only single-peak flow can be observed under these conditions. Although this line of reasoning is highly speculative, it relates to materials and conditions of industrial interest, and may therefore be worthy of further investigation.

---

*Or in terms of some appropriate segment of the relevant population distributions (e.g., inclusions within a specific size range, etc.).

Static Recrystallization and Precipitation During Hot Rolling of Microalloyed Steels

C. M. SELLARS

Department of Metallurgy
University of Sheffield, England

Effective controlled rolling of microalloyed steel plates to obtain fine ferrite grain structures requires refinement of the austenite grain size by recrystallization during roughing and the development of a deformed and unrecrystallized austenite structure during finish rolling.

The rate of recrystallization and the resulting grain size are sensitive both to rolling variables, particularly pass reduction and temperature, and to metallurgical variables. In microalloyed steels the metallurgical variables are grain size, microalloy content in solution, and the state of precipitation of carbonitrides. This paper reviews the interactions of rolling variables with precipitation and recrystallization of microalloyed austenites, determined from experimental single-deformation studies carried out mainly on niobium steels, and then considers the way in which these interactions determine the evolution of microstructure during multipass controlled rolling.

1. Introduction

High-strength low-alloy steels based on carbon-manganese steel compositions with microalloy additions (generally less than 0.1%) of niobium, vanadium, titanium or a combination of these elements now form an important family of steels designed to meet the needs of an increasingly wide spectrum of markets. These steels are produced in various product forms, including plate, sheet, bar and sections, which may be used in the as-rolled condition or after subsequent heat treatment. The development of these steels and their basic metallurgy have been the subjects of a great deal of research in the past two decades, and much of this work has been presented at a series of conferences.[1-9] The physical metallurgy of these steels has also been the subject of several reviews, the latest of which, by DeArdo, Gray and Meyer,[10] gives an excellent over-all perspective with particular emphasis on niobium steels.

This paper is concerned primarily with the plate steels used in the as-controlled-rolled condition and concentrates on the interactions between composition and processing variables which affect the development of microstructure in the austenite. The principles of controlled rolling to obtain fine ferrite or bainite structures after transformation are well established and may be summarized as follows:

1. Reheating to a suitably high temperature to take all or most of the microalloy additions into solution in the austenite.
2. Roughing at a relatively high temperature to obtain grain refinement by repeated static recrystallization between passes.
3. Holding at intermediate temperatures where partial recrystallization between passes would lead to mixed austenite grain structures if rolling were continued.
4. Finishing at low temperatures in the austenite where recrystallization is sufficiently retarded so that it does not occur at all between passes and so that highly dislocated, elongated ("pancaked") austenite grains are developed to provide a high density of nucleation sites for the ferrite transformation during cooling after rolling.

There are several modifications of this basic scheme, including rolling down into the two-phase temperature range or even into the ferrite range, to obtain additional benefits from dislocation strengthening of the ferrite and from texture effects.

The intention in this paper is to examine the physical metallurgy governing the kinetics of recrystallization and precipitation of microalloy carbonitrides and their interactions and then to apply this knowledge, which has been gained mainly from the results of single-deformation experiments, to the microstructural changes taking place during the various stages of the basic scheme of controlled rolling.

2. Solution of Microalloy Elements

All the microalloy elements form stable cubic carbides and nitrides, the solubilities of which in austenite have been extensively investigated. These are normally represented by the solubility product as a function of temperature:

$$\log(M)(X)^x = A - \frac{B}{T} \tag{1}$$

where M is the content (wt %) of Nb, V, Ti or Al; X is the content (wt %) of C or N; x may be 1 or less, depending on the concentration of vacancies in the compound crystal structure; A and B are constants; and T is absolute temperature. The relative stabilities of the various carbides and nitrides are shown graphically in Fig. 1,[11] from which it can be seen that complete solution can only be obtained for most compounds in steels of normal compositional levels at high temperatures in the austenite range. The exceptions are VC, which is less stable and is taken into solution at relatively low temperatures, and TiN, which is more stable and is virtually insoluble in the austenite except at very high temperatures.

The carbides and nitrides are mutually soluble in each other and, with both carbon and nitrogen present in steels, carbonitrides with compositions depending on the steel composition are formed, as illustrated for niobium steels in Fig. 2.[12] It can be seen that as temperature increases the carbonitride becomes more nitrogen rich, as would be expected from the relative stabilities of the carbide and nitride (Fig. 1), as discussed in detail by Nordberg and Aronsson[13] and more

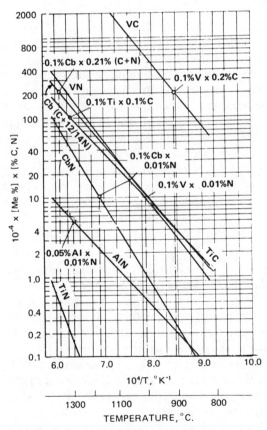

Fig. 1. Solubility products of niobium (columbium), titanium, vanadium and aluminum carbides and nitrides.[11]

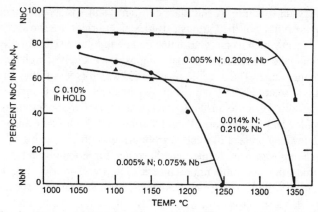

Fig. 2. Change in NbC$_x$N$_y$ composition with temperature and niobium content.[12]

recently by Lakshmanan.[14] In fact, there are considerable discrepancies between the solubility products obtained by different workers, which may be attributed partly to the different techniques used for measurement and partly to differences in steel composition. For example, Koyama et al.[15,16] showed that manganese and chromium increase the solubility of niobium carbide and niobium nitride because they reduce the activity coefficients of carbon and nitrogen in solution, that nickel has little effect and that silicon has the opposite effect.

To illustrate the importance of solubility products, only the relationship obtained by Irvine et al.,[17] which has been found to give good agreement with observed solution temperatures in steels containing about 1.5% Mn and normal nitrogen levels, will be considered. This relationship is expressed as:

$$\log(\text{Nb})\left(C + \frac{12}{14}N\right) = 2.26 - 6770/T \qquad (2)$$

and is illustrated in Fig. 3 for different temperatures in the austenite. The stoichiometric line for Nb(C + 12/14 N) indicates the composition path followed by the austenite as Nb(C + 12/14 N) is taken into solution with increasing temperature or is reprecipitated during cooling. Steels of nonstoichiometric composition follow parallel paths, as illustrated for steels of different (C + 12/14 N) and Nb contents. It can be seen that in a steel with 0.1% (C + 12/14 N) and 0.065% Nb the niobium will just be taken into solution at 1250°C, and that when the Nb content is raised to 0.1% the amount in solution at 1250°C only increases to 0.068%. These two steels would therefore have virtually the same potential for precipitation at lower temperatures after reheating to 1250°C. If the (C + 12/14 N) content is reduced to 0.05 wt %, then, with 0.065% Nb, 1150°C is sufficient to take all the niobium into solution and the potential for repre-

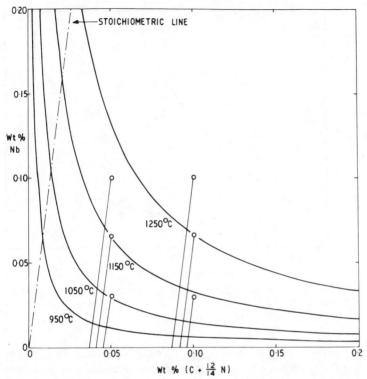

Fig. 3. Solubility limits for Nb(C + $\frac{12}{14}$N) based on the relationship of *Irvine et al.* [17]

cipitation at low temperatures in the austenite is nearly the same as for the steel with the higher (C + 12/14 N) content reheated to 1250°C. The relationship of precipitation potential to the off-stoichiometry of the steel composition has been discussed in detail by Wadsworth et al. [19] and will not be pursued further here. In terms of controlled rolling, the advantage of the steel with the lower (C + 12/14 N) content is that the reheating temperature, and therefore the grain size after reheating and the temperatures of roughing, can be reduced without loss of precipitation potential. This can be of commercial benefit, particularly for thick plates, if production is sufficiently large to justify a reduction in the reheating furnace temperature.

Carbides or nitrides remaining out of solution during reheating can restrain grain growth whenever the volume fraction and particle size meet the critical conditions for pinning of grain boundaries. [20] However, before normal reheating temperatures are reached, coarsening and dissolution of particles usually mean that the critical condition for abnormal grain growth has been reached and mixed grain structures have developed, which on further heating undergo normal growth

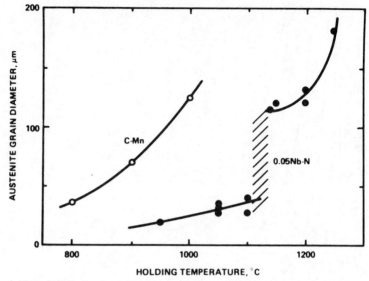

Fig. 4. Effect of reheating temperature on austenite grain size after 30-min hold.[21]

(Fig. 4[21]). The exceptions to this may be steels containing TiN, in which the particle stability leads to grain-coarsening temperatures that may be above normal reheating temperatures.[22–24] It must, however, be emphasized that both particle dissolution and grain coarsening are time-dependent as well as temperature-dependent. In industrial reheating, times are relatively long, but in laboratory experiments they are frequently short so that the starting microstructures before rolling may not be representative of those obtained industrially. In single-pass rolling experiments this starting microstructure, particularly its grain size, has an important effect on recrystallization behavior. In contrast, low volume fractions of precipitates that are out of solution before rolling have little influence on either dynamic or static recrystallization,[24,25] although they may have an important effect in restraining grain growth after recrystallization.

3. Deformation Above the Solubility Limit

For steels of normal microalloy compositions there may be only a small temperature range at the start of roughing which is above the solubility limit. It has been suggested by Yamamoto et al.[18] that the solubility limit is lower in deformed austenite than in recrystallized austenite. However, it is considered that this suggestion arises from the notoriously unreliable nature of volume-fraction measurements made on extraction replicas rather than from any real effect of dislocation structure. The basic solubility curves shown in Fig. 3 are therefore taken as appropriate to reprecipitation in both deformed and recrystallized austenite.

Below the solubility limit there is a driving force for precipitation, and whether it occurs before, during or after deformation depends on the deformation conditions and the kinetics of precipitation, as discussed in the next section. Uncertainty about whether precipitation occurs under particular processing conditions has led to considerable controversy about the roles of microalloy elements in solution or as precipitates in retarding recrystallization.[4-9] In order to resolve this question, deformation must be carried out at temperatures above the solubility limit.

Recently, Luton et al.[26] reported results obtained on a 0.054% Nb steel decarburized and denitrided to 0.0014% C and 0.001% N (by wet hydrogen treatment at 1100°C for 1 week) with an expected maximum solution temperature of 870°C. After reheating to 1150°C and deforming in compression at a strain rate of 10^{-1} or 10^{-2} s^{-1} at 1000, 950 or 900°C, both static recovery and static recrystallization determined from interrupted deformation tests (softening curves) were found to be retarded compared with those in a base C-Mn steel. The time to 50% recrystallization was increased by a factor of 4 at 1000°C and by a factor of 20 at 900°C. It is not clear that these factors can be attributed entirely to niobium in solution, because there were other differences in steel composition and because the original grain sizes, which have been shown to have a major effect on recrystallization kinetics,[27,28] were not reported. Results on a more extensive series of steels decarburized in a similar manner and tested in compression at a strain rate of 10 s^{-1} have been obtained by Yamamoto et al.[18] and are illustrated in Fig. 5. In this figure t_x = 20% and t_x = 50% represent the times to 20 and 50% softening, which by metallographic observations were correlated with the times to the start of recrystallization and to 30% recrystallization. It can be seen that recovery and recrystallization are retarded by a factor which increases approximately linearly with solute content and which is greater for niobium than for titanium, with vanadium having relatively little effect. The retardation factor for niobium was again found to increase with decreasing temperature. In this work, all the steels were reheated to obtain an initial grain size of 140 μm and, although it is not clear that all testing conditions were above the solubility limits, no precipitation was observed, for example, in the 0.097% Nb steel after holding for as long as 2500 s at 900°C after deformation.

These observations establish that the microalloy elements in solution have significant retarding effects, probably related to the distortion they cause in the austenite lattice parameter.[18] At higher temperatures in the austenite, typical of those employed for roughing of commercial microalloy steels, when the temperature may be above the solubility limits, the retarding effects may be relatively small.

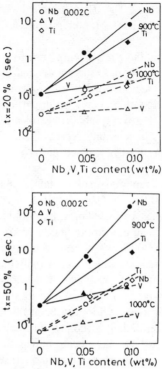

Fig. 5. Effect of niobium, titanium or vanadium in solution on times for 20% softening (~0% recrystallization) and 50% softening (~30% recrystallization).[18]

4. Deformation Below the Solubility Limit

The effects of deformation below the solubility limit have been extensively studied with respect to both precipitation and recrystallization before, during and after a single deformation.

Precipitation

Results from several workers on the time to the start of precipitation of Nb(C, N) are shown in Fig. 6. These curves have been determined by a variety of techniques — chemical or electrolytic extraction and chemical analysis,[25,29] quantitative electron microscopy,[18,34] and analysis of flow-stress curves.[33] Analyses of the steels are given in Table 1 together with the reheating temperatures used and estimates of the solution temperatures made without correction for other elements present. Although there are some discrepancies among the results

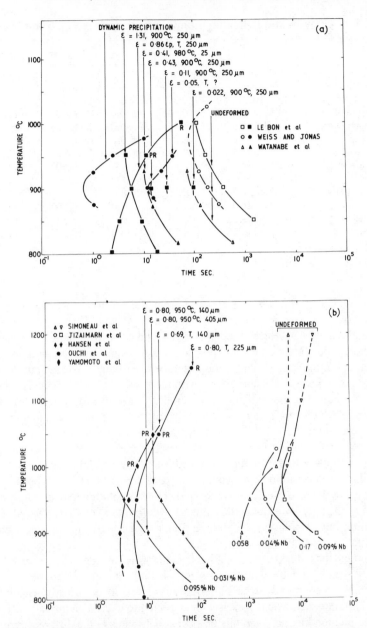

Fig. 6. Precipitation-start (P_s) curves for niobium steels of compositions given in Table 1. "R" and "PR" indicate full and partial recrystallization before precipitation.

Table 1. Analyses and Reheating Conditions of Steels in Fig. 6 and 18

Reference	Composition, %						Reheating temperature, °C	Solution temperature(a), °C
	C	N	Nb	Si	Mn	Al		
LeBon et al.[25] 0.17	0.011	0.040	0.31	1.35	0.017		1250	1264
Weiss and Jonas[33] . . .0.05	0.004	0.035	0.045	0.42	0.057		1100	1084
Watanabe et al.[29] . . .0.063	0.0058	0.084	0.11	1.71	0.024		1260	1230
Simoneau et al.[30] . . .0.07	0.0103	0.04	0.26	0.88	0.027		1350	1149
0.19	0.0072	0.058	0.01	0.65	0.008		1350	1337
Jizaimarn et al.[32]. . . .0.06	· · ·	0.09	0.06	1.26	· · ·		1250	1222
0.06	· · ·	0.17	0.05	1.60	· · ·		1250	1319
Hansen et al.[34] 0.11	0.010	0.031	0.26	1.35	0.023		1250	1170
0.10	0.010	0.095	0.23	1.24	0.014		1250	1321
Ouchi et al.[31] 0.16	0.0054	0.031	0.36	1.41	0.020		1250	1214
Yamamoto et al.[18]. . .0.19	0.0028	0.095	0.25	1.49	0.032		?	1420

(a) Estimated from Eq. 2.

of different workers, which may arise from the different techniques used, a number of important general conclusions can be drawn.

Undeformed Austenite. Precipitation in undeformed austenite is relatively slow but, depending on steel composition, cooling rate and holding time before single-pass rolling experiments, could start before rolling is carried out. Most of the curves indicate a "C" curve form with the "nose" at about 1000°C or above. The exceptions are those of Simoneau et al.,[30] determined by resistivity measurements, which appear to show relatively little time dependence at temperatures all the way up to 1200°C. This is particularly surprising for the lower-niobium steel, for which the estimated solution temperature is 1150°C. However, these results and those of Jizaimarn et al.[32] both show that precipitation is accelerated by higher supersaturations, as would be expected. Note that the higher-niobium steel used by Jizaimarn et al. was reheated below its solution temperature so that, from Fig. 3, only an estimated 0.12% Nb would be in solution. The Nb(C, N) remaining out of solution does not appear to have had a major effect on the precipitation kinetics.

Although the nature of precipitation in undeformed austenite has received little study, it seems probable that it occurs preferentially at grain boundaries, because these have been found to be preferred sites even in deformed austenite[18,34] (Fig. 7). Grain size would therefore be expected to exert a significant influence on precipitation kinetics, and this probably contributes to the differences among the results obtained by Simoneau et al.,[30] LeBon et al.[25] and Watanabe et al.,[29] because the high reheating temperature in the former case would be expected to lead to a very coarse grain size compared with grain sizes of 250 and 25 μm in the other two investigations. Relative shifts in the positions of the curves would

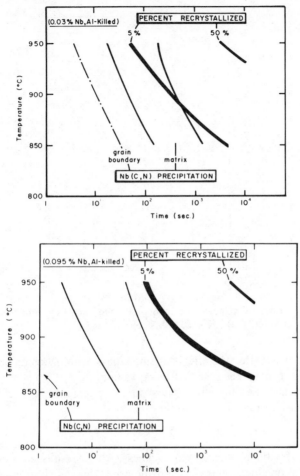

Fig. 7. Recrystallization-precipitation-temperature-time (RPTT) diagrams for niobium steels reheated at 1250°C and hot rolled 50% (ε = 0.80) at 950°C.[34]

also be expected to arise from differences in steel composition. For example, manganese has been shown to have a major effect on the start of dynamic precipitation[36] (Fig. 8). Assuming a similar effect for the curves in Fig. 6 shifts them in the sense which increases the apparent effect of grain size, but it does not eliminate all discrepancies between the results of different workers, even when differences in supersaturation at precipitation temperatures are taken into account. As discussed later, it is also possible that the results of Watanabe et al.[29] were influenced by a prior rolling treatment used to obtain the fine grain size, despite the fact that this treatment was carried out at a temperature above 1095°C.

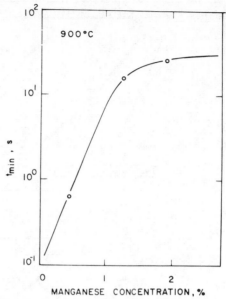

Fig. 8. Effect of manganese content on time to start of dynamic precipitation at 900°C in 0.05C-0.035Nb steels.[36]

Deformed Austenite. The effects of single-pass deformation on precipitation shown in Fig. 6 have been studied using rolling,[18,29,31,34] compression[33] and torsion.[25] In all cases the strain given in the figure has been converted to equivalent tensile strain, i.e., for rolling:

$$\varepsilon = 1.155 \ln 1/(1 - r) \tag{3}$$

where r is the fractional reduction in thickness. All the results show that a major acceleration of precipitation is caused by deformation, and electron microscopic studies of extraction replicas[18,34] and of thin foils[37,38] reveal that strain-induced precipitation in the matrix occurs preferentially on dislocation subgrain boundaries developed during the deformation (e.g., see Fig. 9).

From the data of LeBon et al. in Fig. 6, the sensitivity of precipitation to the amount of prior strain is clear. This is also illustrated by the results of Hoogendorn and Spanraft[40] (Fig. 10). However, Fig. 6 also shows clearly the importance of other conditions of prior straining. It can be seen that precipitation time varies with temperature in different ways for different sets of results. When prior deformation is carried out at a constant temperature before annealing at different temperatures, time for precipitation increases with decreasing temperature[25,29,34] in a manner similar to that of precipitation in undeformed austenite. This type of test provides a constant deformed structure for precipitation if recrystallization does not intervene. Conversely, tests in which defor-

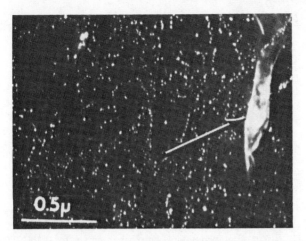

Fig. 9. Dark-field electron micrograph of fine Nb(C, N) particles formed by hot rolling 0.08C-0.02N-0.09Nb steel by 60% reduction ($\varepsilon = 1.06$) in a single pass at 871°C and aging for 1000 s.[39]

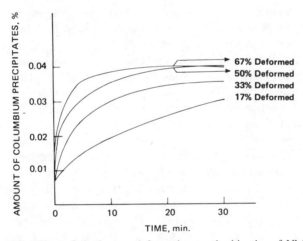

Fig. 10. Effect of single-pass deformation on the kinetics of Nb(C,N) precipitation in 0.06C-0.004N-0.041Nb steel.[40]

mation and holding are carried out at the same temperature tend to show "C" curves with little temperature dependence, except in the work of LeBon et al., in which the strain also increased with decreasing temperature. In this type of test, the dislocation density at the end of deformation will increase with decreasing temperature because of the temperature dependence of flow stress (Fig. 11).[41] In the absence of recrystallization, the effect of deformation temperature is sur-

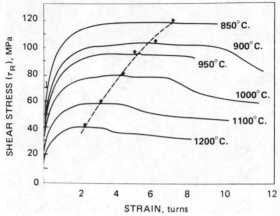

Fig. 11. Effect of temperature on stress-strain curves for 0.16C-0.04Nb steel.[41]

prisingly large compared with the effect of strain if both effects are considered to arise simply from an increase in dislocation density. This leads to the conclusion that deformation temperature *per se* is important, possibly because some "pre-precipitation phenomenon" occurs to a greater extent during deformation at lower temperatures. A "pre-precipitation phenomenon" is suggested because observations by Weiss and Jonas[33] indicate that true dynamic precipitation probably does not occur above a strain rate of about 1 s^{-1}, which is lower than the strain rates in rolling. The curve in Fig. 6 for dynamic precipitation relates to decreasing (low) strain rates with increasing temperature of deformation, which may contribute to the more pronounced "C" shape.

After deformation at the higher temperatures, recrystallization may start before precipitation. This would be expected to have a major effect on precipitation kinetics, as shown schematically in Fig. 12.[34] Such a trend is shown by the results of LeBon et al., but not by those of Yamamoto et al. The observations of Hoogendorn and Spanraft[40] also indicate a surprisingly small effect of recrystallization on precipitation kinetics. The small effect of recrystallization could arise from the refinement of grain size resulting from recrystallization, which, as discussed earlier, would accelerate precipitation in dislocation-free austenite. However, it could again indicate that some "pre-precipitation phenomenon" which is not wiped out by recrystallization may have occurred during or immediately after deformation at well below the solubility temperature.

The term "pre-precipitation phenomenon" does not necessarily imply a clustering or zone-formation stage, although it is interesting to speculate on the nucleation mechanism for Nb(C, N), which is expected to be incoherent with the austenite because of the ~25% difference in lattice spacings. In the present context "pre-precipitation phenomenon" could simply mean precipitation below

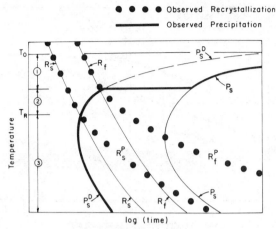

Fig. 12. Schematic recrystallization-precipitation-temperature-time (RPTT) diagram for microalloyed steel.[34]

the detectable limit of the technique used. For example, thin-foil electron microscopy is the most sensitive technique for observing general precipitation (although local precipitation in grain boundaries may be missed), and, using this technique, Davenport et al.[39] detected considerable precipitation on "immediate" quenching of 0.08C-0.02N-0.09Nb-1.25Mn steel after single-pass rolling 60% reduction at 870°C. Extraction replicas probably provide the next most sensitive method, and Hansen et al.,[34] in using this technique (Fig. 7), found that grain-boundary precipitation is detectable after times up to an order of magnitude shorter than those required for matrix precipitation (which are the ones shown in Fig. 6). Electrolytic extraction and analysis of particles may miss particles less than 5 nm in size[39] but should include all other particles, and by use of this technique precipitation (probably at grain boundaries) was detected for times of less than 100 s following rolling at 1200°C after reheating to 1315°C (estimated solution temperature, 1295°C). Loss of secondary hardening would be expected to be sensitive to any prior precipitation in austenite, but, as discussed elsewhere,[35] the data of LeBon et al.[25] show that this technique indicates significantly slower precipitation kinetics than does the chemical extraction method. Finally, detection of dynamic precipitation[33] depends on the state of precipitation which gives detectable retardation of dynamic recrystallization. This is uncertain, but the observations of the lack of effect of aging before deformation[33,37] imply that matrix precipitation on subgrain boundaries must occur. This is consistent with the arguments put forward by Hansen et al.[34] for retardation of static recrystallization. For the above reasons, the curves in Fig. 6 may all tend to overestimate the time to the start of precipitation. This will be considered further when multipass deformation is discussed.

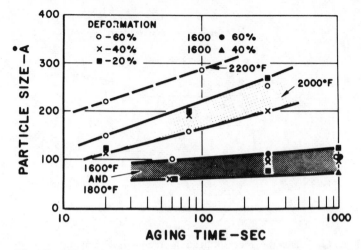

Fig. 13. Effect of aging time on particle sizes in 0.08C-0.02N-0.09Nb steel reheated to 1315°C (2400°F), then rolled and held at 1204, 1093, 982 or 871°C.[39]

All the preceding discussion has related to the start of precipitation, but the effects of variables on the progress of precipitation is consistent with their effects on its start, because 50%-precipitation or precipitation-finish curves are nearly parallel to the start curves (e.g., see Fig. 7). Particle sizes are small at low temperatures but increase in size with increasing temperature and increasing aging time (Fig. 13).[39]

This discussion has also concentrated on precipitation of Nb(C, N), for which most data are available, but precipitation in other microalloy steels is sensitive to similar variables.[18,42–44]

Recrystallization

Recrystallization during hot working may occur either dynamically, during deformation, or statically with time after deformation.

Dynamic Recrystallization. The occurrence of dynamic recrystallization leads to a maximum in the flow stress–strain curve (e.g., see Fig. 11). This occurs at a strain (ε_p) which is always higher than the critical strain (ε_c) for the onset of dynamic recrystallization. The relationship between the two strains is complex, but as a working relationship in steels it can be taken that[28]

$$\varepsilon_c \simeq 0.8\varepsilon_p \qquad (4)$$

As can be seen in Fig. 11, ε_p increases with decreasing temperature. It also increases with increasing strain rate ($\dot{\varepsilon}$) and initial grain size (d_0) and, as discussed previously,[28] for C-Mn steels can be related to the Zener-Hollomon parameter (Z) as

$$\varepsilon_p = A \cdot d_0^{1/2} \cdot Z^{0.15} \tag{5}$$

A similar relationship would be expected for microalloyed steels, at least at temperatures and strain rates higher than those at which dynamic precipitation can occur. The torsion data of Migaud[45] substantiate this and show that the constant A increases with increasing microalloy content. This is also shown by the data of Meyer and Robiller[46] (Fig. 14), from which it can be seen that niobium is more effective than titanium in delaying dynamic recrystallization. This is consistent with their solution effects on static recrystallization (Fig. 5). Figure 14 also shows that at the higher strain rates typical of rolling, it is unlikely that the critical strain will be reached in niobium steels during an industrial plate rolling pass, although it may be reached by the higher reductions of strip rolling passes and is certainly exceeded in some experimental single-pass rolling, e.g., that of Kozasu et al.[47]

It is also apparent from the increase in flow stress with increasing niobium content[45,48] that niobium in solution retards dynamic recovery, which is in agreement with the effect on static recovery[26] reported earlier.

When microalloyed steels are deformed at relatively low strain rates and temperatures where dynamic precipitation can occur, this has a further retarding effect on dynamic recrystallization. This retardation has been used by Jonas and co-workers to examine the effects of composition and other variables on the kinetics of dynamic precipitation.

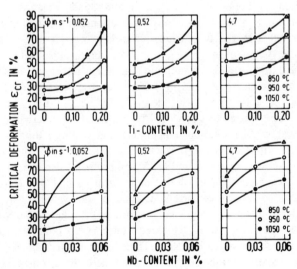

Fig. 14. Effects of niobium and titanium on critical strain for dynamic recrystallization.[46]

Static Recrystallization Kinetics. The effects of deformation variables on the kinetics of static recrystallization have been discussed in detail elsewhere.[27,28] It was shown that for strains less than the critical strain for dynamic recrystallization (ε_c), the time for a constant fraction of recrystallization (t_x) after a single deformation could be described for C-Mn steels by an equation of the form:

$$t_x = B \ \varepsilon^{-n} d_0^p Z^{-q} \ \exp \frac{Q_{rex}}{RT} \qquad (6)$$

where Z is the Zener-Hollomon parameter,

$$Z = \dot{\varepsilon} \ \exp \frac{Q_{def}}{RT_{def}}; \qquad (7)$$

ε is strain; d_0 is original grain size; Q_{rex} and Q_{def} are the activation energies for recrystallization and deformation; T and T_{def} are the absolute temperatures of holding and of deformation; R is the gas constant; and B, n, p and q are constants. For C-Mn steel, n = 4, p = 2 and q $\simeq$ 0. The very strong dependence of recrystallization rate on strain arises partly through the rise in flow stress (σ) with strain, which causes an increase in the driving force by increasing the dislocation density (ρ). However, assuming that $\sigma \propto \rho^{1/2}$ leads to an approximately linear dependence of ρ on ε. The remaining effect of strain is thought to arise because the progressive development of a subgrain structure, the increase in subgrain misorientation, and possibly the development of deformation bands facilitate nucleation. The effect of grain size arises because nucleation occurs preferentially at grain boundaries. For a given grain size, grain-boundary area will also increase with strain, but only slowly. Because flow stress increases with increasing Z, it would be expected that Z would have an effect, e.g., if it is considered that $\sigma \propto Z^{0.1}$ in the hot working range, $\rho \propto Z^{0.2}$, suggesting a value of q $\simeq$ 0.2. However, the extent to which a subgrain structure is developed at a given strain decreases with increasing Z, and in C-Mn steel it appears that the two effects approximately cancel out the influence of Z.

Similarly strong dependences would be expected for microalloyed steels at temperatures above the solution temperature and within a limited temperature range below it, with an increase in microalloy content causing an increase in the constant B, as shown previously in Fig. 5, and probably also increasing the value of Q_{rex}. Use of a relationship of the same form as that of Eq 6 was shown previously[28] to be capable of correlating some of the available data on niobium steels, but a large scatter remained and the form of the relationship for microalloy steels is still open to question. However, there is no doubt that each of the variables discussed above has an effect. Their effect on t_x is reflected by a change in fraction recrystallized if the structure is examined after a constant (short) delay time prior to quenching after deformation. This approach tends to be used by Japanese workers, and typical results are shown in Fig. 15[47] and 16.[49] It should

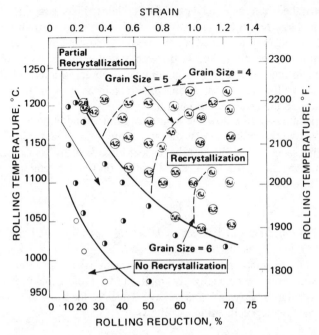

Fig. 15. Austenite recrystallization and recrystallized grain size obtained in 0.03% Nb steel with an initial grain size of ASTM 1 (220 μm) reheated at 1250°C, single-pass rolled at the temperatures shown and iced-brine quenched within 3 s. [47]

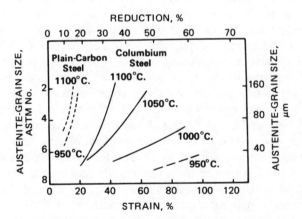

Fig. 16. Effect of initial grain size and strain on critical amount of strain required for recrystallization in 0.12C-0.036Nb steel single-pass rolled at the temperatures shown and quenched within 3 s. [49]

be noted that the strain scales on these figures relate to plane strain (i.e., $\varepsilon/1.155$) and that the large reductions given will result in the critical strain for dynamic recrystallization being exceeded at progressively lower temperatures with increasing strain. Also, in Fig. 16, the different initial grain sizes were obtained by preliminary rolling, and some precipitation was observed after holding for the experimental rolling reduction, so that precipitates as well as grain size could have influenced the critical deformation.

At temperatures well below the solubility limit, in addition to the variables discussed above, precipitation plays an increasingly important role. The effects of precipitates on recrystallization have been discussed by Hansen et al.[34] in terms of the driving force for recrystallization (F_R) and a retarding force (F_P) arising from precipitation:

$$F_R \propto \Delta\rho \tag{8}$$

where $\Delta\rho$, the difference in dislocation density across a recrystallizing boundary, depends on ε and Z, as discussed earlier.

For general precipitation, with the volume fractions available in microalloyed steels, F_P was shown to be negligibly small, but for preferential precipitation on subgrain boundaries, F_P was found to have values of the same order as those of F_R and to vary with precipitation as follows:

$$F_P \propto \frac{f\bar{l}}{r^2} \tag{9}$$

where f is volume fraction, r is precipitate radius, and $\bar{l}$ is mean subgrain size, which decreases as Z increases. Equation 9 may also be considered appropriate for preferential precipitation on grain boundaries before matrix precipitation occurs. In this case $\bar{l}$ must be replaced by $\bar{d}_0$, and it is apparent that, because of the large difference in size between $\bar{d}_0$ and $\bar{l}$, smaller volume fractions or considerably larger particles at grain boundaries could still produce a significant retarding force. Hansen et al. found that particle sizes were from 3 to 4 nm at the end of precipitation at temperatures of 850 to 950°C and only at 950°C did significant coarsening occur after precipitation was complete. In contrast, Chandra et al.[50] observed significant coarsening at temperatures as low as 815°C and also found that it accelerated when recrystallization occurred. Taking these factors into account, and also the effect of temperature on volume fraction and on mean particle size (see Fig. 13), the expected effect of temperature and time on F_P is illustrated schematically in Fig. 17. This figure shows that a maximum in F_P occurs as precipitation proceeds and that F_P decays with coarsening. The increase in the maximum value of F_P with decreasing temperature is shown disproportionately small for clarity in the diagram.

Hansen et al. assume that if $F_R < F_P$ then boundary migration will be completely arrested, but that if $F_R \gg F_P$ then precipitation will not have any significant effect. However, if $F_R > F_P$ and if the magnitudes are comparable, then

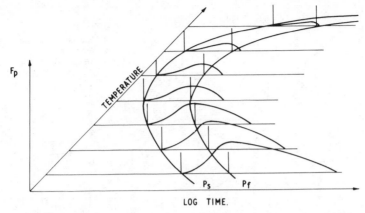

Fig. 17. Schematic representation of the effect of temperature and time on precipitate retarding force (F_p).

boundaries may migrate at a reduced velocity. In terms of retarding recrystallization, this picture may only strictly be true for grain-boundary precipitation, because when particles are precipitated on subgrain boundaries they should also retard the recovery events necessary for nucleation of recrystallization, unless this occurs by a simple grain-boundary bulging mechanism. The retardation of recovery would, however, be expected to depend in a fairly similar manner on the particle parameters, and the concepts of F_R and F_P are extremely useful in understanding the experimental observations.

Figure 18 shows results for the start (5%) of recrystallization determined in some of the investigations for which precipitation kinetics are given in Fig. 6. Figure 18 also shows the curve obtained by Yamamoto et al.[18] for the decarburized steel in which no precipitation occurred, and for comparison a curve is also included for C-Mn steel using value calculated from Eq 6 (with the same grain size but a lower strain, so that ε_c was exceeded). The curve of Yamamoto et al. for their 0.19% C steel deviates progressively from the curve for the decarburized steel. Comparison with Fig. 6 shows that this occurs as the recrystallization-start time becomes increasingly greater than the precipitation-start time (determined by extraction replicas), which would be expected from the increase in F_P and which corresponds closely with the schematic diagram (Fig. 12) for the effect of strain-induced precipitation. In contrast, the results of Hansen et al.[34] (see also Fig. 7) show recrystallization starting well after the start of precipitation. In fact, for the 0.095% Nb steel, precipitation is complete before recrystallization starts, and so the solid-solution retardation is lost and the temperature dependence of recrystallization should be compared with that for C-Mn steel to observe the retarding effect of the strain-induced precipitation. This increases with decreasing temperature probably more because of the increase in

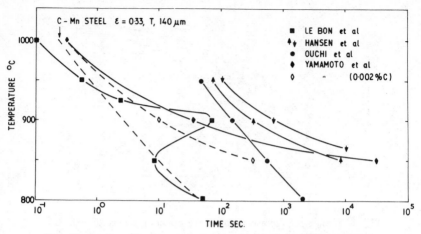

Fig. 18. Recrystallization-start (R_s) curves for niobium steels of compositions given in Table 1.

F_P with the decrease in particle size than because of the increase in volume fraction with the decrease in temperature. The results of Ouchi et al.[31] show a pattern generally similar to those of Hansen et al. but, because of the lower volume fractions of precipitate expected in their steel and the increasing F_R with decreasing temperature, the retardation is less in this case. The results of LeBon et al.[25] show several unusual features. First, retardation appears to begin before the start of precipitation, but, as discussed earlier, this may simply arise because the technique based on loss of secondary hardening does not detect the true start of precipitation. In these experiments the strain was increased with decreasing temperature, giving markedly increasing values of F_R, which could account for the reduced retardation at 850°C. With further decrease in temperature to 800°C, the expected time to the start of recrystallization in the absence of precipitation increases,[35] but the time to the start of precipitation continues to decrease (see Fig. 6), leading to a larger volume fraction of precipitate before recrystallization and hence to enhanced retardation.

Other results in the literature for which precipitation-start times were not measured have a divergence in the shapes of curves similar to that shown in Fig. 18. For example, Lamberigts and Greday[51] and Roberts,[52] working on relatively high-carbon, low-niobium steels with low available volume fractions of carbonitrides, found curves of the same general form as that of LeBon et al., but without the complete inflection. Hoogendorn and Spanraft[40] found, for a lower-carbon steel, a curve generally similar to that reported by Yamamoto et al.[18] but at lower temperatures, whereas Davenport et al.,[39] using a low-carbon, high-niobium steel, found extreme retardation at much higher temperatures.

Niobium content, because of its effects on precipitation kinetics and on the value of F_P through increases in f and possibly decreases in r, would be expected to have a very important effect on the temperature for marked retardation in low-carbon steels. This is clearly demonstrated by the work of Migaud[45] (see Fig. 19). Increasing solute drag with increasing niobium content may also be significant in shifting the R_s curve to longer times with respect to the P_s curve (Fig. 12) so that intersection occurs at a higher temperature. The lower retardation of R_s caused by titanium or particularly by vanadium in solution, as well as differences in precipitation kinetics of TiC or VN, may thus also contribute to the reduced temperature at which recrystallization starts to be severely retarded in steels microalloyed with these elements.

Statically Recrystallized Grain Size. Previous consideration[28] of the effects of variables on recrystallized grain size (d_{rex}) in C-Mn steels led to the relationship

$$d_{rex} = D\varepsilon^{-m}d_0^r Z^{-u} \qquad (10)$$

where m = 1, r = 0.67 and u = 0. Again, the lack of effect of Z is surprising, and in an earlier correlation[27] it had been included. The reasons for the lack of effect of Z were discussed earlier for the recrystallization kinetics. If activation energies for nucleation and growth are similar, the holding temperature should have no effect on recrystallized grain size, and it does not appear in this relationship.

Correlation of fewer data on niobium steels[28] indicated that at temperatures above those for rapid matrix precipitation a similar relationship held, but with m = 0.67. There was some evidence that m could decrease at low strains, and this is shown more clearly by the data of Tanaka et al.,[49] which are compared with the prediction from Eq 10 (labeled "S model") in Fig. 20. In fact, the results

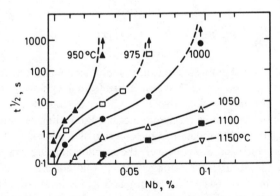

Fig. 19. Effect of niobium content on time for 50% softening in steels containing about 0.06% C and 0.007% N deformed in torsion at a strain rate of 3.6 s⁻¹ to a strain (ε) of 0.44.[45]

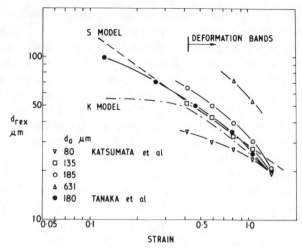

Fig. 20. Effect of strain on recrystallized grain size obtained after single-pass rolling of 0.12C-0.036Nb (*Tanaka et al.*[49]) and AP1 X65 steel (*Katsumata et al.*[53]). The latter data are for rolling at 1050°C.

in Fig. 20 suggest that the simple power relationships in Eq 10 may not be satisfactory because the results of Katsumata et al.[53] indicate an interaction between the effects of strain and grain size. It is, however, also possible that these results are affected by the occurrence of grain-boundary precipitation following the preliminary rolling treatment used to obtain the different initial grain sizes.

An alternative formulation for the effect of grain size and strain has been proposed by Katsumata et al., based on the change of interfacial area/volume (S_v) with strain:

$$S_v = \frac{1000}{d_0}[0.429(1 - r) + 1.571/(1 - r)] + [80r - 24] \qquad (11)$$

where d_0 is the original grain size in μm, r is the fractional rolling reduction, and S_v has the units mm^2/mm^3. The first term arises from the grain elongation and the second term, which only applies when $r > 0.3$, arises from the increasing density of deformation bands with increasing reduction. For rolling reductions of 0.3 or more, they found good correlation between grain size obtained after rolling at a constant temperature and S_v. However, these reductions are much higher than those used in industrial plate rolling, and, if this correlation is extended to reductions less than 0.3 (when no deformation bands form), it leads to a strain dependence much lower than the one observed by Tanaka et al., as shown by the curve labeled "K model" in Fig. 20. A difference between the observations of Tanaka et al. and those of Katsumata et al. is that the former workers found little effect of rolling temperature on recrystallized grain size, whereas the latter found a systematic increase in grain size with increasing rolling temperature. A systematic decrease in grain size with increasing strain rate has also been found by Machida et al.,[54] and effects of both temperature and strain rate have been observed by Cuddy.[21] This led him to propose a relationship between grain size and finishing stress. These results thus indicate that the form of Eq 10 may not be appropriate for some niobium steels and that a real dependence of grain size on Z should be incorporated.

In this consideration of grain size, the possibility of grain growth after recrystallization affecting the observations has been neglected. As shown previously,[27,28] grain-growth rates in niobium steels are much less than those in C-Mn steels. Nevertheless, if small recrystallized grain sizes are obtained by means of large single-pass reductions at high temperature, grain growth could significantly affect the grain size even in short periods of time following the completion of recrystallization. This would lead to a temperature dependence of grain size observed at a constant time after rolling.

Katsumata et al. also observed that, when rolling reductions of less than 10% were given, up to 20 area % of the grains were 500 μm in size or greater. It is not entirely clear whether these grains were present in specimens quenched immediately after rolling or in specimens held at temperature for some time before being quenched. However, these large grains were found at temperatures both above and below the expected solubility temperature, indicating that the presence of particles is not a prerequisite and that coarse grains are probably associated with critical strain recrystallization.

Thus, although the general effects of rolling variables are clearly understood, considerable ambiguity remains concerning the quantitative dependence of grain size on each of them. This could arise because (grain-boundary) precipitation occurred to different extents before rolling in some of the investigations. At lower temperatures, when both matrix and grain-boundary precipitation are known to have started before recrystallization, the results of LeBon et al.[25] show that reducing the deformation temperature has a major effect in reducing the recrystallized grain size,[28] probably through its influence on the state of precipitation. Grains sizes produced by recrystallization under these conditions at finish rolling temperatures are, however, of little practical concern because the recrystallization times far exceed the times available in industrial rolling.

5.   Multipass Deformation

Industrial controlled rolling involves multipass rolling, with temperature decreasing after each pass in both roughing and finishing, but the effects of multipass deformation on precipitation and recrystallization have received relatively little systematic investigation compared with the effects of single deformations.

In C-Mn steels, where there is no precipitation, the effects of multipass deformation on grain size can reasonably be considered by treating each pass in turn on the basis of the relationships between rolling variables and recrystallization kinetics, recrystallized grain size and grain-growth rates determined in single-deformation experiments.[27,28,55] The form of results obtained using this type of modeling is shown in Fig. 21,[55] which relates to a simplified controlled rolling sequence of 15% reduction per pass with 20 s between passes to produce 20-mm plate. Successive recrystallization and grain growth between roughing passes leads to a progressive decrease in grain size, whereas in finishing at the relatively low temperatures for C-Mn steel recrystallization is only partly complete between passes, leading to mixed grain structures, for which the mean is shown in the Fig. 21. The effects of changes in rolling conditions on the evolution of microstructure have been considered in some detail elsewhere,[27,28] and the predictions from modeling have been shown to be in reasonable quantitative agreement with observations.

For microalloyed steels it is convenient to consider the roughing and finishing stages separately because of the differences in the recrystallization and precipitation kinetics in the two temperature ranges.

Roughing

Using the same procedure of modeling as for C-Mn steel, Fig. 21 indicates that a niobium steel grain refinement is more effective, mainly because of the reduction in grain-growth rate. Although the general pattern of grain-size change predicted by modeling[28,59] is in agreement with observations,[56,59] the predictions cannot be considered to be quantitatively accurate because of the uncertainties about the forms of the relationships that should be used in the model and the possible effects of precipitation.

The occurrence of precipitation at high temperatures during multipass rolling is shown clearly by the work of Davenport and DiMicco[60] (Fig. 22). Size distributions of particles precipitated during multipass rolling have also been measured on extraction replicas by Irvine and Baker,[57] who found that the mean particle size increased with increasing pass number, as the volume fraction increased.

At high temperatures, precipitation is expected preferentially at grain boundaries, but even so F_p will be low. When recrystallization and grain refinement take place, many new grain boundaries, which initially are free of precipitates, are formed. During the next pass, growth of existing particles may be accelerated and preferential nucleation sites created for further precipitation at grain boundaries. Because of the short recrystallization times at high temperatures, nucleation must occur extremely rapidly after (or during) deformation. This sequence of events will be repeated so that after several passes the majority of particles will not be on existing grain boundaries. Their contribution to F_p will therefore be very small; also, solid-solution retardation of recrystallization will be lost. Such particles could, however, restrict grain growth after recrystallization and contribute to the marked effect that increasing niobium content has in reducing the grain size obtained at the end of a multipass roughing sequence.[59]

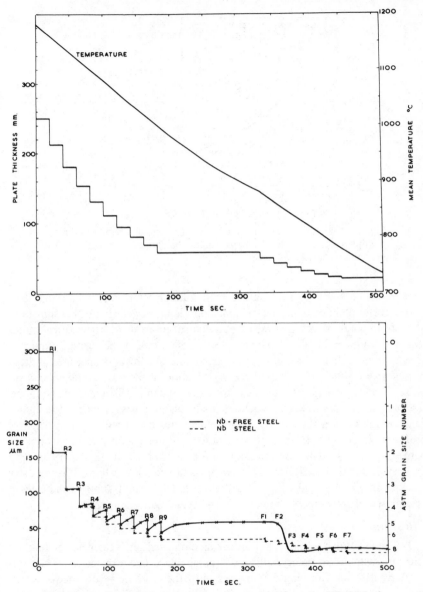

Fig. 21. Predicted evolution of microstructure during a simplified schedule of controlled rolling. [55]

During industrial rolling, small pass reductions are frequently used and very coarse grains can be formed during roughing. This type of grain-coarsening behavior has been found in simulative tests carried out by Cuddy [21,58] (Fig. 23). This shows that coarse grains can be produced by small reductions either at

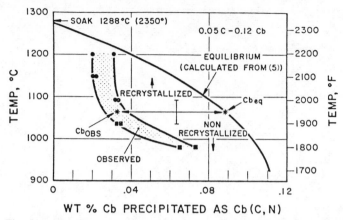

Fig. 22. Variation in the amount (wt%) of niobium precipitated during multipass rolling of a low-carbon, niobium bainitic steel. Rolled by 20% reduction per pass.[60]

temperatures above or below the expected solution temperatures, and that they also occur in plain carbon steel, which supports the conclusion of Katsumata et al.[53] from single-pass experiments that coarse grains arise from critical strain recrystallization rather than from particle effects. Although the average grain size is reduced by further passes after grain coarsening, the lower tempertures of these passes may result in only parital recrystallization of the coarse grains between passes, leading to a mixed structure at the end of roughing. To avoid this problem, it appears necessary to keep reductions in all roughing passes above 10% and to finish roughing at a sufficiently high temperature so that strain-induced matrix precipitation does not retard recrystallization sufficiently for it to be incomplete between passes. Under these conditions it is possible to obtain final recrystallized grain sizes as small as 20 μm, but this appears to be a lower limit. It is not clear whether this is imposed by grain growth after recrystallization or indicates a breakdown of relations of the form of Eq 10 at small grain sizes.

Finishing

The aim of finish rolling is to produce an elongated, dislocated austenite grain structure without any recrystallization taking place between passes. The temperature at which finishing commences must therefore be below that at which strain-induced precipitation can produce a sufficient volume fraction of small particles for F_p to be greater than F_R before recrystallization starts. The relatively low pass reductions of industrial rolling compared with those of experimental single-pass rolling lead to lower values of F_R and also increase recrystallization time more than they increase precipitation time, e.g., from the data of LeBon et al[25] (Fig. 6), reducing the strain from 0.43 to 0.11 only increases the

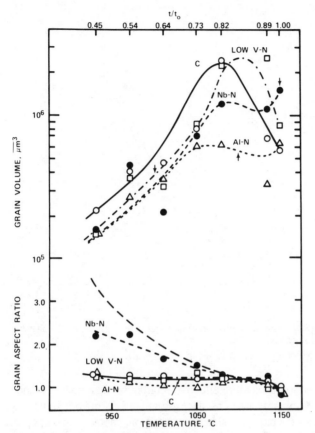

Fig. 23. Changes in austenite grain volume and grain aspect ratio in a plain carbon steel and microalloyed steels during a 12-pass hot deformation schedule comprising a total reduction of 55%.[58]

precipitation-start time by about a factor of 2, whereas, from Eq 6, the recrystallization start time is increased by a factor of about 200. It might therefore be expected that finishing could start at temperatures higher than those suggested by Fig. 18 for a given steel composition. However, this effect of strain per pass is partly counteracted by the reduced supersaturation resulting from precipitation during roughing.

As finishing proceeds and the strains from several passes are accumulated, recrystallization times are reduced. If data from single-pass experiments are used to predict recrystallization times,[28] it is found that recrystallization could occur before the end of finishing in low-niobium steel, but this is not observed in practice. Thus, the accumulated strain from a series of small passes at decreasing

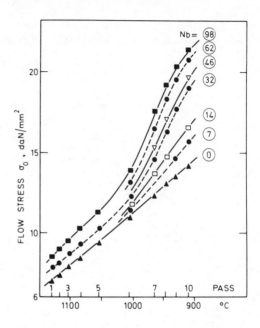

Fig. 24. Hot strength of niobium steels in a simulated 10-pass hot strip rolling schedule: $\bar{\varepsilon} = 0.174$ (0.8 rev); $\dot{\bar{\varepsilon}} = 3.62$ (1000 rev/min). Values of Nb content (circled) are in wt % × 10^3. Ref 45.

temperatures produces more effective retardation than does a single large reduction. This presumably arises because growth of existing precipitates is accelerated during successive passes, and because additional nucleation takes place after each pass as a result of the decreasing solubility. This does not appear to have been studied. The presence of the particles may also influence the dislocation distribution, and reduced heterogeneity would also tend to retard recrystallization. During normal controlled rolling, precipitation is incomplete so that a supersaturation remains, which leads to precipitation in the ferrite during cooling, with resulting precipitation hardening.

A consequence of the lack of recrystallization during finishing is that the flow stress and hence the rolling load rise rapidly in successive passes (Fig. 24). It is not clear whether this is entirely due to the retained strain hardening or whether there is an additional contribution from precipitation strengthening of the austenite.

6. Concluding Remarks

This paper has concentrated on niobium steels, for which most data are available, but the principles governing recrystallization, precipitation and their interactions with each other and with rolling variables are similar in other microalloyed steels.

It is now established that microalloy elements in solution, particularly niobium, have a retarding effect on recrystallization, but that for the severe retardation of recrystallization required in finish rolling of plate, extensive matrix precipitation on subgrain boundaries must occur. The precise role of grain-boundary precipitates, which are formed earlier or at higher temperatures than matrix precipitates, is still unclear because the majority of investigations have been carried out on coarse-grain austenite subjected to a relatively large single deformation. It is argued that in multipass roughing operations the cycles of recrystallization and precipitation will tend to minimize any retarding effects of such particles and that they do not appear to be directly responsible for the coarse grains that can be formed during roughing with small reductions.

Increasing the microalloy content and reducing the interstitial content, particularly carbon in relation to nitrogen, accelerates precipitation and provides a potentially larger volume fraction of precipitates so that the temperatures for the end of roughing and for the beginning of finishing are increased.

Although the relationships between rolling variables and recrystallization and precipitation have been reasonably quantified for single deformations and subsequent holding, there are still uncertainties about the precise forms of the algebraic relationships, particularly for the lower rolling reductions typical of industrial rolling. Industrial rolling with successive passes at lower temperatures also leads to more complex interactions between the variables. These require further investigation if a fully quantitative description of controlled rolling, capable of being used in a predictive sense by computer modeling, is to be obtained.

References

1. *Metallurgical Developments in Carbon Steels,* ISI Special Report No. 81, Iron and Steel Institute, London, 1963.
2. *Symposium on Low-Alloy High-Strength Steels,* The Metallurg Companies, Nuremburg, BRD, 1970.
3. *Processing and Properties of Low Carbon Steels,* edited by J. M. Gray, AIME, New York, 1973.
4. *The Hot Deformation of Austenite,* edited by J. Ballance, TMS-AIME, New York, 1977.
5. *Micro-alloying 75,* edited by M. Korchynsky, Union Carbide Corp., New York, 1977.
6. "Recrystallization in the Development of Microstructure," *Met. Sci.,* Vol 13, 1979.
7. *Hot Working and Forming Processes,* edited by C. M. Sellars and G. J. Davies, Metals Society, London, 1980.
8. *International Conference on Steel Rolling,* The Iron and Steel Institute of Japan, Tokyo, 1980.
9. *Thermomechanical Processing of Micro-alloyed Austenites,* edited by P. J. Wray and A. J. DeArdo, AIME, Warrendale, PA, 1982.

10. A. J. DeArdo, J. M. Gray and L. Meyer, *Niobium 81,* San Francisco, 1981.
11. F. O. M. Obehauser, F. E. Listhuber and F. Wallner, Ref 5, p 665.
12. Hoesch-Estel, Dortmund, unpublished results quoted in Ref 10.
13. H. Nordberg and B. Aronsson, *J. Iron Steel Inst.,* Vol 206, 1968, p 1263.
14. V. K. Lakshmanan, M. Met. Eng. Thesis, McMaster University, Canada, 1977.
15. S. Koyama, T. Ishii and K. Narita, *J. Jap. Inst. of Met.,* Vol 35, 1971, p 698.
16. S. Koyama, T. Ishii and K. Narita, *ibid,* p 1090.
17. K. J. Irvine, F. B. Pickering and T. Gladman, *J. Iron Steel Inst.,* Vol 205, 1967, p 161.
18. S. Yamamoto, C. Ouchi and T. Osaka, in Ref 9.
19. J. Wadsworth, J. H. Woodhead and S. R. Keown, *Met. Sci.,* Vol 10, 1976, p 362.
20. T. Gladman, *Proc. Roy. Soc.,* Vol 294A, 1966, p 298.
21. L. J. Cuddy, *Met. Trans.,* Vol 12A, 1981, p 1313.
22. T. J. George and J. J. Irani, *J. Australian Inst. Met.,* Vol 13, 1968, p 94.
23. T. J. George, G. Bashford and J. K. MacDonald, *ibid,* Vol 16, 1971, p 36.
24. L. A. Leduc and C. M. Sellars, in Ref 9.
25. A. LeBon, J. Rofes-Vernis and C. Rossard, *Mem. Sci. Rev. Met.,* Vol 11A, 1975, p 411.
26. M. J. Luton, R. Dorval and R. A. Petkovic, *Met. Trans.,* Vol 11A, 1980, p 411.
27. C. M. Sellars and J. A. Whiteman, Ref 6, p 187.
28. C. M. Sellars, Ref 7, p 3.
29. H. Watanabe, Y. E. Smith and R. D. Phelke, Ref 4, p 140.
30. G. Simoneau, G. Begin and A. H. Marquis, *Met. Sci.,* Vol 12, 1978, p 381.
31. C. Ouchi, T. Sanpei, T. Okita and I. Kazasu, Ref 4, p 316.
32. J. Jizaimarn et al., *Tetsu-to-Hagane,* Vol 60, 1974, p 177.
33. I. Weiss and J. J. Jonas, *Met. Trans.,* Vol 10A, 1979, p 831.
34. S. S. Hansen, J. B. Vandersande and M. Cohen, *ibid,* Vol 11A, 1980, p 387.
35. C. M. Sellars, *Recrystallization and Grain Growth of Multiphase and Particle Containing Materials,* Riso National Laboratory, Roskilde, Denmark, 1980, p 291.
36. M. G. Akben, I. Weiss and J. J. Jonas, *Acta Met.,* Vol 29, 1981, p 111.
37. T. Chandra, I. Weiss and J. J. Jonas, *Met. Sci.,* Vol 16, 1982, p 97.
38. E. L. Brown and A. J. DeArdo, Ref 4, p 250.
39. A. T. Davenport, R. E. Miner and R. A. Kot, Ref 4, p 186.
40. T. M. Hoogendorn and M. H. Spanraft, Ref 5, p 75.
41. A. LeBon and L. N. De Saint-Martin, Ref 5, p 90.
42. B. Bacroix, M. G. Akben and J. J. Jonas, in Ref 9.
43. M. G. Akben, B. Bacroix and J. J. Jonas, *Acta Met.,* Vol. 31, 1983, p 161.
44. T. Chandra, M. G. Akben and J. J. Jonas, *Sixth Int. Conf. on the Strength of Metals and Alloys,* Melbourne, 1982.
45. B. Migaud, Ref 7, p 67.
46. L. Meyer and G. Robiller, as Ref 35, p 311.
47. I. Kozasu, C. Ouchi, T. Sampei and T. Okita, Ref 5, p 120.
48. J. R. Everett, A. Gittins, G. Glover and M. Toyanu, Ref 7, p 16.
49. T. Tanaka, N. Tabata, T. Hatomura and C. Shiga, Ref 5, p. 107.
50. T. Chandra, I. Weiss and J. J. Jonas, *Canadian Met. Quart.,* Vol 20, 1981, p 421.
51. M. Lamberigts and T. Greday, Ref 4, p 286.
52. W. Roberts, *Scandinavian J. Metall.* Vol 9, 1980, p 13.
53. M. Katsumata, M. Machida and H. Kaji, in Ref 9.
54. M. Machida, M. Katsumata and H. Kaji, Ref 8, p 1249.
55. C. M. Sellars and J. A. Whiteman, presented at *Product Technology Conference on Controlled Processing of HSLA Steels,* York, 1976 (unpublished).

56. H. Sekine and T. Maruyama, *Microstructure and Design of Alloys,* 3rd Int. Conf. on Strength of Metals and Alloys, Inst. Metal/Iron Steel Inst., London, Vol 1, 1973, p 85.
57. J. Irvine and T. N. Baker, Ref 6, p 228.
58. L. J. Cuddy, J. J. Bauwin and J. C. Raley, *Met. Trans.,* Vol 11A, 1980, p 381.
59. J. M. Chilton and M. J. Roberts, *ibid,* p 1711.
60. A. T. Davenport and D. R. DiMicco, Ref 8, p 1237.

The Role of Thermomechanical Processing in Tailoring the Properties of Aluminum and Titanium Alloys

J. C. WILLIAMS
Carnegie-Mellon University
Pittsburgh, PA 15213

E. A. STARKE, JR.
University of Virginia
Charlottesville, VA 22903

Abstract

The effects of thermomechanical processing (TMP) on the microstructures and properties of aluminum and titanium alloys are described. The microstructural variables which are affected by TMP include grain size, degree of recrystallization and grain aspect ratio. In addition, TMP can be used to impart specific, controlled amounts of preferred orientation or texture. Both the microstructural variables and texture which are controlled by TMP have strong effects on the properties of high-strength Al and Ti alloys. The relationship among microstructure, texture and properties for these alloys is described. In particular, strength, ductility, fracture toughness, stiffness (modulus), fatigue strength, fatigue-crack growth, stress-corrosion cracking and other properties are discussed. These variations in properties are rationalized to the extent possible using the current understanding of deformation and fracture in Al and Ti alloys.

Introduction

The properties of structural materials depend on alloy composition and on microstructure. Traditionally, use of compositional control has been practiced to provide alloys which have reproducible strength levels after a given heat treatment cycle. However, the recent increase in emphasis on damage-tolerant design has led to the realization that microstructure must also be controlled on a scale which is not significantly affected by heat treatment if enhanced fracture-related properties are to be achieved in a reproducible fashion. An effective way of manipulating microstructure on such a scale is through combined thermal and mechanical working. These combined processes allow a broader range of microstructures to be obtained than is possible through thermal treatment alone. Such manipulation is now known as thermomechanical processing (TMP) and is becoming an integral part of modern processing practice for high-performance materials such as high-strength Al and Ti alloys.

More recently, TMP has been used to control the type and extent of preferred crystallographic orientation (texture) which is present in high-strength, high-performance materials. Production and control of texture have not been as thoroughly integrated into commercial practice as has TMP for control of microstructure. Nevertheless, texture can have significant effects on properties and is thus worthy of discussion.

In this article, we review the roles of TMP in altering and controlling microstructure and texture in high-strength Al and Ti alloys. For convenience we have elected to separate the discussions of these two classes of alloys, for which the motivations for and responses to TMP are quite different. The Al alloys consist of face-centered cubic matrixes containing solid-solution additions, precipitates and constituent phases whereas the commercially important high-strength Ti alloys consist of mixtures of two matrix phases — hexagonal close packed α and body-centered cubic β. Thus there is little common basis for comparing these two classes of alloys and they are more appropriately discussed separately, as will be done in subsequent sections. There are, however, some common concepts employed in processing both classes of alloys. These will be briefly summarized before the separate discussions of Al and Ti alloys. The final section will contain some speculation about the course of further developments in TMP, and some comments on the relationship between TMP and some new, alternate processing methods will be offered for the reader's consideration.

Processing

Mechanical working has been historically used as the primary means of changing the sizes and shapes of materials while transforming the cast structure of an ingot into what is generically referred to as a wrought product. The two principal methods of working are forging and rolling, although other means, such as extrusion and drawing, are sometimes used. As shown schematically in Fig. 1, forging and rolling result in significantly different flow patterns, the former being axisymmetric and the latter being unidirectional, although cross rolling is sometimes used. These operations typically are conducted at relatively high temperatures because the lower flow stress of the material permits larger section reductions to be achieved, thereby lowering power requirements and processing

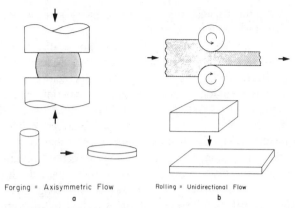

Forging = Axisymmetric Flow   Rolling = Unidirectional Flow
             a   b

Fig. 1. Schematic illustrations of two common working operations used during TMP which give rise to very different flow patterns in the workpiece.

times. Literally all working operations are conducted under conditions of falling temperature, because this is the easiest way of conducting the working operation.

However, as TMP has become a more common means of controlling microstructure, additional requirements related to the working operation have become commonplace, and these requirements place additional restrictions on the working temperature range and the amount of work done at temperature. There are a variety of characteristics of the final product which are sensitive to the details of working history. These include:

- Grain size
- Degree of recrystallization
- Grain aspect ratio
- Texture.

Grain size and degree of recrystallization depend on the deformation strain introduced during working and on the temperature at which the working operation is conducted. During elevated-temperature working, the competition between energy storage within the material and dynamic recovery is complex, but the final grain size is strongly dependent on the recrystallization response of the material either during working or during subsequent reheating. This recrystallization process is driven by the residual stored energy which remains after working. Thus, the more extensive the recovery, the greater the amount of work which must be subsequently introduced in order to retain sufficient stored energy to ensure recrystallization. If a strong tendency for recrystallization and a small recrystallized grain size are desired, then higher strain rates and lower working temperatures, which minimize concurrent recovery during working, are desirable. If, on the other hand, a recrystallized structure is not desired, then working at a lower strain rate, a lower total strain and a somewhat higher temperature is desired because this permits recovery and the development of extensive substructure without recrystallization.

Grain aspect ratio is determined both by the amount of directional flow and by the distribution of second-phase particles which restrict lateral growth of grains during recrystallization. That is, large grain aspect ratios are most readily achieved by extensive unidirectional working of materials which contain insoluble second phases. These phases become aligned in the working direction and act as grain-growth inhibitors. The mean free path between inclusions is greater in the working direction than in directions perpendicular to it. Thus, the recrystallized grains are elongated in the working direction, because they grow faster in this direction. This is discussed in considerably more detail in the section on aluminum alloys. It is sufficient here to say that an elongated grain structure imparts anisotropic behavior to the material and that this directionality must be accounted for in multiaxial loading situations. Moreover, while grain aspect ratio and texture can be related, these two factors are derived from different sources, so that no direct relationship between them exists. It must be emphasized that the two are quite distinct sources of anisotropy and thus should be dealt with separately.

Texture refers to the extent to which the lattices in adjacent grains are crystallographically aligned. Two types of textures exist in wrought products, and these must be distinguished when specifying or discussing texture. They are the deformation texture, which results from alignment of lattice planes during deformation, and the recrystallization texture, which is present after recrystallization. The two are usually related, because the recrystallization texture is derived from a product which contains a deformation texture. The type and intensity of the texture depends on the amount of deformation and, often, on the deformation temperature. The specific textures and their effects on properties will be discussed in subsequent sections.

Aluminum Alloys

Microstructural features of importance for property control of aluminum alloys include: (*a*) coherency and distribution of strengthening precipitates; (*b*) degree of recrystallization; (*c*) grain size and shape; (*d*) crystallographic texture and (*e*) size and distribution of intermetallic particles including the disperoids (present by design) and constituent phases (which result from iron and silicon impurities). Our ability to understand the effects of these features on the properties of aluminum alloys has led to efforts to upgrade product performance by modifying conventional processing.

Conventional Processing

Conventional processing normally consists of mechanical working operations designed to effect shape changes efficiently and economically and heat treatments

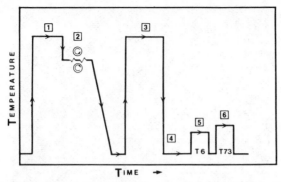

Fig. 2. Schematic representation of conventional processing sequence used for age-hardenable aluminum alloy plate. Steps 1 to 6 are described in the text.

for structure control to achieve desired physical, mechanical and chemical properties. A typical sequence for an age-hardenable aluminum alloy is shown schematically in Fig. 2. Beginning with the as-cast ingot, conventional processing includes a homogenization treatment (step 1) to reduce segregation, remove the low-melting-point phases and thus improve workability. This thermal treatment also serves to precipitate dispersoid-forming elements such as chromium, manganese and zirconium, so that they may perform their role of grain control during processing. Step 2 is hot working for ingot breakdown and shape change to the appropriate product form. Because aluminum and its alloys have high stacking-fault energies, sufficient dynamic recovery normally occurs during hot deformation, i.e., at temperatures above $0.5T_m$, to give rise to stable polygonized substructures (Fig. 3). The subgrains remain equiaxed during hot working although the grains elongate in the direction of flow and do not recrystallize (Fig. 4). Some of the alloying additions are in solution during hot working, but the dispersoid-forming transition elements have already precipitated and may become strung out in the working direction. A fine distribution of dispersoids delays or prevents static recrystallization and aids in retaining the elongated or "pancake"-shape grains during subsequent processing. The wrought product is heat treated to develop the desired microstructure, i.e., either annealed for stress relief or solutionized (step 3), quenched and aged (steps 4, 5 and 6).

Thermomechanical Processing

Metallurgists have long recognized that the sequence of thermal and mechanical treatments greatly affects final microstructure and properties, and so various termomechanical processing procedures have been developed for microstructure and property control. These procedures are normally divided into two classifications for aluminum alloys: intermediate thermomechanical treatments (IT-MT's), which are used to develop a fine recrystallized grain structure and the

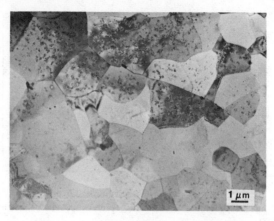

Fig. 3. TEM showing equiaxed substructure typical of those produced during hot working of aluminum alloys.

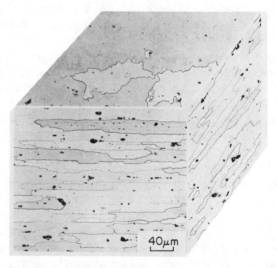

Fig. 4. Optical micrograph of alloy 7075 showing elongated grains typical of hot rolled structures. *Courtesy of J. A. Wert.*

desired texture; and final thermomechanical treatments (FTMT's), which are used to develop a dislocation structure that modifies precipitation behavior during subsequent aging treatments.

Recrystallization Mechanisms. ITMT procedures require an understanding of recrystallization mechanisms. Recrystallization depends on the deformed state of the material[1] and can occur continuously in a homogeneously deformed structure by subgrain growth and coalescence or discontinuously by strain-induced grain-boundary migration. Discontinuous recrystallization occurs by nucleation in regions of strong orientation gradients such as deformation bands, kink bands, etc. Schematic representations[2] of these mechanisms in single-phase alloys are shown in Fig. 5.

Regardless of the mechanism, recrystallization may be either stimulated or retarded by the presence of particles, depending on particle size, spacing and coherency. Small particles (normally less than 0.4 μm in diameter) retard continuous recrystallization by pinning subgrain boundaries and preventing subgrain coalescence,[3-6] as illustrated in Fig. 6. They retard discontinuous recrystallization by pinning migrating high-angle grain boundaries—a process simply described by Zener. Zener[7] suggested that a grain would cease to grow when its radius, R, was approximately equal to the ratio of the particle radius, r, divided by the particle volume fraction, f, i.e.:

$$R_{crit} = \frac{3\beta}{4}\left(\frac{r}{f}\right) \tag{1}$$

where β is a constant approximately equal to one.[8] Equation 1 suggests that small, closely spaced particles have the greatest effect in retarding recrystallization. Coherent particles are more effective than incoherent particles in inhibiting boundary migration, because the precipitate/matrix interface must either change from coherent to semicoherent or incoherent as the boundary passes

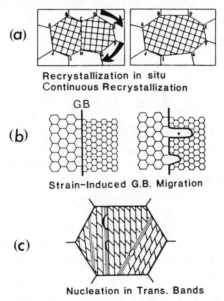

(a)

Recrystallization in situ
Continuous Recrystallization

(b)

G.B.

Strain-Induced G.B. Migration

(c)

Nucleation in Trans. Bands

Fig. 5. Schematic representations of recrystallization mechanisms. *After Cotterill and Mould.*[2]

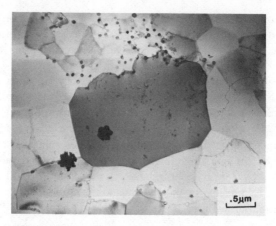

Fig. 6. TEM showing a subgrain boundary pinned by small Al₆Fe dispersoids. *Courtesy of Henry E. Chia.*

the precipitate[9] or, alternatively, dissolve and reprecipitate after the boundary passes through its local.[10] Both processes require considerable energy.

Large particles (diameters normally greater than 1 μm) create strain concentrations, i.e., deformation zones, at particle/matrix interfaces during deformation, and these zones can subsequently act as nucleation sites for recrystallization. The width of the deformation zone, λ, can be related to the particle diameter, d, and true strain, ε, by[11,12]

$$\lambda = Ad\varepsilon^{(n/n+1)} \tag{2}$$

where

$$A = \frac{K\sqrt{2}}{8\tau_{cr}} \left[\frac{4\tau_{cr}^{(n+1)}}{\sqrt{2}nK} \right]^{n/n+1} \tag{3}$$

and τ_{cr} is the critical resolved shear stress required to move dislocation loops away from the particle. The other parameters are from Ludwick's strain-hardening relationship, $\sigma = K\varepsilon^n$. Nes[8] suggests that a new grain has to exceed a critical diameter δ_{crit} before it will expand into the matrix:

$$\delta_{crit} = 4\gamma/E \tag{4}$$

where γ is the grain-boundary energy and E is the stored deformation energy in the matrix. Nes substitutes the deformation-zone size in Eq 2 for the critical diameter of Eq 4 to obtain the critical particle diameter for recrystallization after a true strain of ε:

$$d_{crit} = \frac{4\gamma}{E} \left[2A\varepsilon^{(n/n+1)} + 1 \right]^{-1} \tag{5}$$

McQueen[13] has recently compiled all known experimental data concerning the effect of particle size and spacing on the recrystallization behavior of alloys. His analysis (see Fig. 7) clearly shows the importance of particle spacing on both inhibition and acceleration of recrystallization. Particle-size distribution also has an important effect. When a bimodal particle-size distribution is present, the small particles may pin the migrating boundaries of the recrystallized nuclei associated with large particles, and the driving force for recrystallization, E, in Eq 5 must be reduced by the boundary/particle interaction energy[8]

$$\frac{3\beta}{4}\left(\frac{r}{f}\right)\gamma \qquad (6)$$

Because most commercial alloys contain both large constituent particles and small disperoids, their combined effects must normally be considered in development of TMP procedures for grain-structure control.

ITMT for Control of Grain Structure. ITMT's involve a series of heat treatment and deformation steps directed toward producing a fine grain size by discontinuous precipitation. The process has been applied to a wide variety of age-hardenable aluminum alloys. A representative ITMT is shown schematically in Fig. 8. After initial homogenization (step 1) and hot working (step 2), an appropriate overaging treatment (step 3) is used to precipitate a high density of particles approximately 1 μm in diameter. This is followed by mechanical deformation (step 4) at temperatures lower than conventional hot working temperatures, to minimize dynamic recovery, and to introduce a high degree of strain hardening. The deformation zones that form around the large particles in step 4 serve as nucleation sites for recrystallization during a subsequent solutionizing

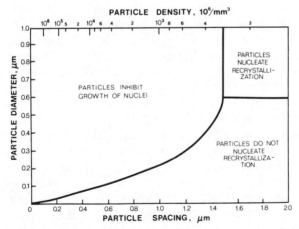

Fig. 7. Schematic representation of the relationship among particle size, particle spacing and recrystallization. *Courtesy of H. J. McQueen.*[13]

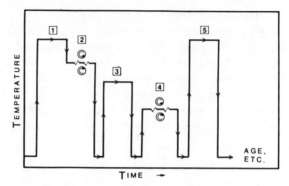

Fig. 8. Schematic illustration of typical ITMT processing for grain refinement. Steps 1 to 5 are described in the text.

treatment (step 5). This is illustrated in the sequence of micrographs shown in Fig. 9 (courtesy of John A. Wert[14]). Figure 9 (top) shows the deformation associated with a large particle in 7075 after step 4, and Fig. 9 (center) shows the beginning of recrystallization (step 5) associated with the deformation zone. Figure 9 (center) clearly shows the pinning effect that the small particles have on the migrating grain boundary. Figure 9 (bottom) shows the fine grain structures after completion of the recrystallization process.

The density of particles greater than 0.75 μm in diameter has been observed to correlate well with grain density.[15,16] However, Wert et al.[16] observed a coarse-particle density of about ten times the recrystallized grain density in a commercial high-strength alloy, suggesting that not all particles were effective in nucleating recrystallization. This may have been due to a high density of small particles inhibiting the growth of some of the nuclei, as described previously.

Deformation and Recrystallization Textures. Texture, i.e., the orientation distribution of crystallites in a material, depends on the deformation characteristics and recrystallization mechanisms.[8,17–19] The grain orientations of cold worked face-centered cubic (fcc) metals and alloys usually spread from $\{110\}\langle1\bar{1}2\rangle$ to $\{112\}\langle11\bar{1}\rangle$ and are generally characterized as either copper-type $\{123\}\langle\bar{4}\bar{1}2\rangle$ (Fig. 10a) or brass-type $\{110\}\langle1\bar{1}2\rangle$ (Fig. 10b). Aluminum alloys normally have the copper-type deformation texture.[17–19] The recrystallization textures of polycrystalline fcc materials usually fall into one of three categories: cube texture $\{001\}\langle100\rangle$ (Fig. 10c); retained deformation, i.e., R texture (Fig. 10d); or random texture (Fig. 10e).[20,21] Cube textures are believed to develop by the oriented-growth mechanism from recrystallization nuclei formed in transition (deformation) bands.[22,23] The R texture is believed to result from continuous recrystallization and/or strain-induced grain-boundary migration. A random texture may develop from a weak deformation texture, or from recrystallization nuclei formed in the deformation zones associated with coarse

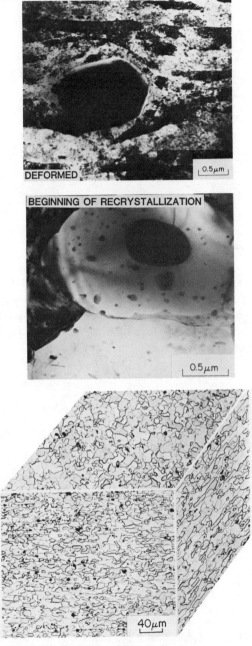

Fig. 9. (Top) TEM showing the deformation structure associated with a large particle in ITMT-processed 7075. (Center) TEM after recrystallization in the deformation zone. (Bottom) optical micrograph of the recrystallized structure. *Courtesy of John A. Wert.*

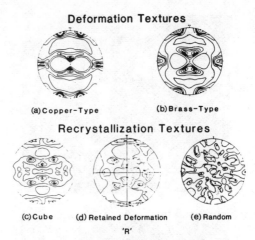

Fig. 10. Representative textures in fcc metals and alloys: (a and b) deformation textures (c, d and e) recrystallization textures.

precipitates located randomly throughout the matrix.[15,20,21,24] However, when the coarse precipitates are associated with grain boundaries, they accelerate strain-induced grain-boundary migration and the R texture; and when associated with transition bands they enhance the formation of the cube texture.[8] A schematic representation[8] of texture formation is given in Fig. 11.

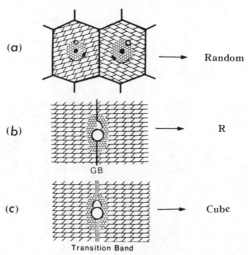

Fig. 11. Schematic representation of formation of recrystallization textures in polycrystalline materials. *After Nes.*[8]

The recrystallization rate normally decreases in the following sequence: (1) particle-stimulated nucleation, (2) nucleation at transition bands or grain boundaries and (3) continuous recrystallization. However, when particles are located at transition bands, these regions are the most favorable sites for recrystallization due to the superposition of the deformation zones of the particles and the transition bands. Consequently, the formation of a cube texture is normally preferred to an R or random texture based simply on kinetic considerations.

ITMT for Texture Control. The ITMT procedure described in the previous section for producing a small, equiaxed grain structure normally results in a random texture. This may be due to a high density of randomly distributed coarse particles, resulting from step 3 in Fig. 8, and the presence of small particles which slow down the growth rate of the preferred nuclei, i.e., those having cube and R orientations. Consequently, the nuclei with other orientations are allowed to develop and grow and are present at the completion of recrystallization. There is no relationship between grain shape and crystallographic texture. The grain shape is determined by the distribution particles that inhibit grain growth. An ITMT procedure for producing an R texture has been experimentally developed for an X7091 P/M alloy[25] (Fig. 12, top). This procedure also involves grain-size control. Hot working above the solvus temperature provides a homogeneous, dynamically recovered structure with a sharp R texture. Introduction of a small amount of cold deformation to this structure stimulates strain-induced grain-boundary migration during subsequent recrystallization, and the R texture is preserved (Fig. 12, bottom).

FTMT for Precipitate Control. FTMT's are used for age-hardening alloys and involve a mechanical working step after solution heat treatment and prior to warm aging (precipitation step). In some cases the mechanical working step exerts an accelerating influence on the precipitation reaction by increasing the vacancy concentration, which enhances the diffusion necessary for precipitation. The primary role of deformation, however, is to increase the dislocation density and thereby the number of nucleating sites for heterogeneous precipitation. The effectiveness of the process is related to the interfacial strains of the precipitates, being highly effective for precipitates having large interfacial strains, e.g., those in Al-Cu alloys, since the total strain present is minimized when the interfaces are associated with the dislocations. The TEM's in Fig. 13 show the effect of deformation on an Al-Cu alloy.[26] In some cases, e.g., in high-temperature aging, the deformation structure may be unstable and anneal out during excursion to the aging temperature. Therefore, low-temperature aging should precede warm aging, or the deformation step should follow a preliminary aging treatment so that it is stabilized by the strengthening precipitates. This type of FTMT is especially effective for 7xxx alloys, which do not respond to working prior to aging because of the low interfacial strains associated with precipitates that occur in these

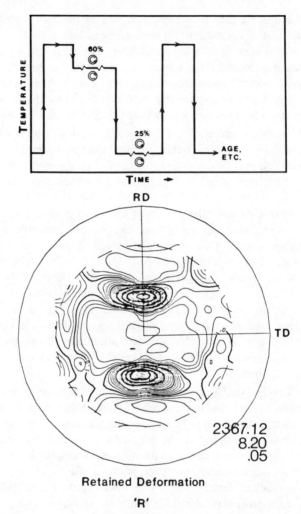

Retained Deformation

'R'

Fig. 12. (Left) Schematic of ITMT procedure used to produce retained deformation textures in 7xxx alloys. (Right) {111} pole figure of recrystallized plate. *After Victor Kuo.*

systems. A component of substructure strengthening results when working is carried out after an aging step,[27] as shown schematically in Fig. 14.

Relationship Between TMP-Affected Properties and Microstructure

Conventional processing of commercial aluminum alloys typically results in highly elongated grains with resulting directional mechanical properties. Because grain boundaries are almost continuous in the TL plane, most high-strength

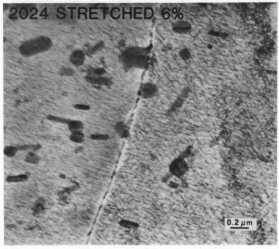

Fig. 13. TEM's showing the effect of plastic deformation on the precipitate distribution in 2024 (top) undeformed, aged 12 h at 190°C (375°F) and (bottom) stretched 6%, aged 12 h at 190°C (375°F).

aluminum alloys have lower ductility, lower toughness and lower stress-corrosion resistance in the short transverse direction. The directionality of grain structure is very significant for corrosion processes, such as exfoliation corrosion, that involve intercrystalline attack.[28]

Early work on ITMT was concerned with producing a fine, equiaxed, recrystallized grain structure directed toward eliminating the anisotropy present in

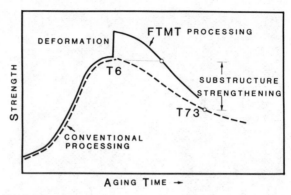

Fig. 14. Schematic comparison of aging treatments of FTMT and conventional processing for Al-Zn-Mg alloys. *After Paton and Sommer.*[27]

thick plate and in improving both mechanical properties and corrosion resistance.[29-31] More recently,[14,32] there has been considerable interest in the superplastic properties of ITMT-processed materials, because superplastic forming frequently results in significant reductions in fabrication costs.

Superplasticity. Fowler et al.[33] provided evidence that reductions in grain size can cause improvements in the superplastic properties of high-strength aluminum alloys. This is especially true when the refinement is accompanied by a change from elongated to equiaxed grains,[32] and arises from the nature of the flow-stress/strain-rate relationship for fine-grain materials[34] and the mechanisms of superplastic deformation.[14] Superplastic forming requires temperatures of at least $0.4T_m$ and low strain rates, usually in the vicinity of 10^{-3} s^{-1}. It is generally believed that grain-boundary sliding and diffusional creep are the primary deformation mechanisms during superplastic flow. In any event, the basic microstructural requirement is small, equiaxed, stable grains, and the amount of superplasticity usually increases with decreasing grain size, as discussed by Sherby and Wadsworth.[35] ITMT processing produces the small equiaxed grains in many aluminum alloys, and grain-size stability is usually maintained by dispersoid-forming alloying additions. Figure 15 shows a comparison of the tensile ductility of conventionally processed and ITMT-processed 7475.[32] The fine grain structure (d ~ 10 μm) produced by the process represented schematically in Fig. 8 has an elongation of 525% compared with 120% for the conventionally processed material (d ~ 60 μm).

Tensile Properties.

Strength. The strengthening mechanisms most often employed for aluminum alloys include those associated with precipitates, dispersoids, and grain- and subgrain-boundary effects. Texture strengthening has not been used extensively for aluminum alloys due to the large number of easily activated slip systems. However, texture can significantly affect formability, grain-size effects and

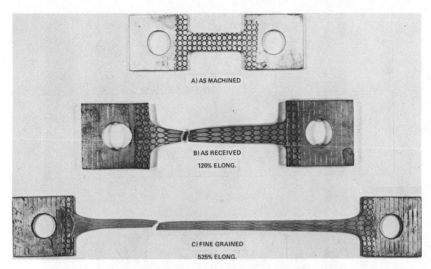

Fig. 15. Comparison of tensile ductility for conventionally processed and ITMT-processed 7475. (a) As-machined. (b) CP, normal ductility. (c) ITMT, superplastic behavior. *Courtesy of Paton, Hamilton and Wert.*

strength in certain product forms. Although each strengthening mechanism may be used independently, they are most often used in combination, and collectively the contributions of the various mechanisms are somewhat less than would be anticipated on the basis of their individual effects.

Effective precipitate strengthening requires a uniform distribution of precipitates which are either sheared or looped and bypassed by dislocations. Figure 16 illustrates how FTMT (2% stretching prior to aging) refines the precipitate structure in an Al-Cu-Li-Mg alloy.[36] The FTMT processing results in a strength of 555 MPa and a reduction in area of 29%, compared with 475 MPa and 27% for the unstretched material.

Grain and subgrain boundaries generally inhibit dislocation motion, and their presence is a source of strength. The effect of grain size (d) on the yield stress is normally described by the Hall-Petch relationship[37,38]

$$\tau_y = \tau_0 + kd^{-1/2} \tag{7}$$

where τ_0 and k are constants. A similar relationship holds for strengthening by subgrains or cells;[39] however, for substructure strengthening the constant k is usually somewhat smaller than for grains and the exponent of d normally has a value of one. The value of τ_0 depends on the various matrix-hardening mechanisms, and a large τ_0 reduces the grain-size contribution to strength. A yield stress/grain size relationship does exist for age-hardening aluminum alloys when the precipitates are coherent and are sheared by dislocations, because the slip length is then determined by the grain size. A sharp texture may decrease the

Fig. 16. TEM's showing effect of FTMT on precipitate distribution in Al-3Cu-1.5Li-1Mg-0.2Zr alloy. (Top) SHT, aged 6 h at 190°C. (Bottom) SHT, stretch 2%, 6 h at 190°C. (*Ref 36*)

effectiveness of grain boundaries as slip barriers due to the alignment of slip systems in adjacent grains and hence the ease of propagation of slip.[40,41] When the precipitates are incoherent, as in the overaged condition, dislocations bypass and loop the precipitates, slip length is determined by the interparticle spacing, and strength is independent of grain size.

The principles described above can be illustrated by the following examples. Carter et al.[42] processed alloy 7475 so as to obtain both a fine grain structure (d = 18 μm) and a coarse grain structure (d = 80 μm). Each product had a random texture, and tensile properties were obtained for both underaged (shearable precipitates) and overaged (nonshearable precipitates) conditions (Table 1). Reducing the grain size by a factor of ~4 resulted in an increase in yield strength of 54 MPa for the underaged condition but an increase of only 10 MPa for the overaged condition. Similar results have been obtained by other workers.[43]

Table 1. Effect of Grain Size on Yield Strength of Alloy 7475

Condition	Yield strength, MPa
18-μm Grain Size	
Underaged......................	505
Overaged......................	466
80-μm Grain Size	
Underaged......................	451
Overaged......................	445

The influence of texture on the mechanical and physical properties of materials has been reviewed recently by Penelle.[44] Sharp textures which result in slip systems of low resolved shear stress also result in high yield strengths. The following example taken from the work of Palmer et al.[45] illustrates this point. A set of extrusions of an Al-3Li-2Cu-0.2Zr alloy was prepared so as to have different aspect (width/thickness) ratios. Pole figures of round extrusions indicate that a majority of [111] poles are contained in a narrow angular region of 10° centered about the extrusion axis (Fig. 17). With increasing aspect ratio, the intensity of [111] poles near the extrusion axis is decreased and the angular distribution is widened. Yield-strength values, for the same aging condition, were found to range from 521 MPa for round axisymmetric extrusions to 414 MPa for sheet bar extrusions with an aspect ratio of 8:1. The higher yield strengths can be explained by the larger number of (111) planes oriented nearly perpendicular to the tensile axis and therefore unfavorably oriented for slip.

Ductility. Refinement of grain size frequently improves ductility in aluminum alloys[46] since it enhances homogeneous deformation. Low ductility in high-strength aluminum alloys has often been associated with coherent strengthening precipitates and precipitate-free zones (PFZ's) adjacent to grain boundaries.[47,48] The fracture mechanisms may be summarized as follows: For aging conditions that do not produce PFZ's, deformation occurs within the grains and the coherent or partially coherent precipitates are sheared by moving dislocations. This results, successively, in a local decrease in resistance to further dislocation motion, coarse planar slip, strain localization, high stress concentrations across grain boundaries, and premature crack nucleation. For aging conditions that produce PFZ's, plastic deformation is localized in these soft regions, leading to high stress concentrations at grain-boundary triple junctions and premature fracture. These mechanisms are schematically illustrated in Fig. 18. Because the magnitude of the stress concentration, τ^*, depends on the slip length,[46,48] TMP offers ways of improving ductility through control of grain structure. Obviously,

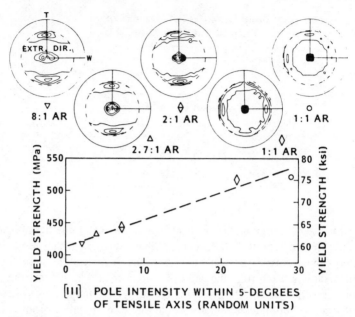

Fig. 17. {111} pole figures and yield strength versus [111] pole intensity for various extrusions of an Al-3Li-2Cu-0.2Zr alloy. *From Palmer et al.*[45]

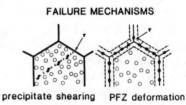

Fig. 18. Schematic representations of tensile-deformation and crack-nucleation mechanisms of age-hardened aluminum alloys.

reducing the grain size reduces the slip length and therefore τ^*, but sometimes it is also necessary to control grain shape, degree of recrystallization, and texture. A few examples are presented below.

An Al-Cu-Li alloy (2020) having very attractive strength, modulus, density and corrosion properties was developed in the 1950's. However, the low ductility of this alloy led to termination of its production in 1969. Alloy 2020 was strengthened by coherent and partially coherent precipitates, and had PFZ's along high-angle grain boundaries.[49] In addition, large recrystallized grains, which formed during primary processing due to the presence of coarse constituent phases, enhanced the strain-localization effects.[50] TMP was used to develop fine recrystallized and partially recrystallized grain structures (TMP-I) and un-recrystallized grain structures (TMP-II).[50] The effects of these procedures on the strength-ductility relationship are shown in Fig. 19. Using a yield strength of 520 MPa as an example, the as-received 2020 having large grains has a strain to failure of 5.4%; the TMP-I processed to produce a much smaller grain structure has a strain to failure of 8.2% (a 43% improvement); and the TMP-II processed to produce unrecrystallized grains has a strain to failure of 11.6% (a 115% improvement). The improvement brought about by TMP-I is associated with a reduction in slip length and therefore in τ^*. However, the failure mechanism was still associated with cracks nucleating at grain boundaries and then propagating intergranularly along PFZ's. The improvement associated with TMP-II is due to the presence of a sharp deformation texture. Stress concentrations at grain boundaries are relieved by shear in neighboring grains, because the low degree of misorientation reduces the effectiveness of the grain boundary as a slip barrier and guarantees nearly parallel slip systems. Therefore, the adverse effect of the PFZ is removed, and transgranular shear fracture, with higher ductility, results.[56]

However, if PFZ's form along subgrain boundaries, unrecrystallized grains may not be desirable. As mentioned previously, subgrains formed during high-temperature processing are normally equiaxed. PFZ's adjacent to the subgrain boundaries will be favorably oriented for slip, and intersubgranular fracture with associated low ductility can result.[51] Processing procedures which produce elongated recrystallized grains can improve ductility in both the longitudinal and

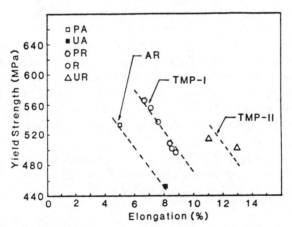

Fig. 19. Yield strength versus elongation for alloy 2020 (data from tensile tests). AR indicates as-received (i.e., no TMP) and includes peak-aged (PA) and underaged (UA) data. TMP-I indicates processed to produce partially recrystallized (PR) and recrystallized (R) structures. TMP-II indicates processed to produce unrecrystallized (UR) structures. TMP-I and TMP-II in PA condition.

transverse directions since the PFZ's associated with the high-angle grain boundaries will not be favorably oriented for slip.[52] A sharp texture, which minimizes stress concentrations at grain boundaries associated with strain localization in the matrix, is also desirable. Such a structure can significantly improve ductility, as noted in Table 2. Figure 20 presents optical SEM[52] pairs of the grain structures and fracture surfaces of the unrecrystallized structure (intersubgranular; Fig. 20a and b) and of the elongated, recrystallized structure (transgranular; Fig. 20c and d).

Fatigue Properties. In general, for high-strength aluminum alloys, a fine grain structure produced by ITMT improves fatigue-crack-initiation resistance but reduces fatigue-crack-propagation resistance when compared with a typical pancake-shape, partially recrystallized, hot worked structure.[53] This effect is more pronounced when the strengthening precipitates are shearable and the grain size determines the slip length. Figure 21(a) compares the LCF curves of ITMT and CP 7xxx alloys. The ITMT material shows a significant increase in reversals to initiation for all strain amplitudes. The ITMT fine grain structure homogenizes the deformation, and the decrease in strain localization improves the resistance to fatigue-crack initiation.[54] The convergence of the curves at high strain amplitudes results from homogenization of deformation by high strains. Unfortunately, homogeneous deformation increases crack-propagation rates because it results in a single straight-running crack during subcritical crack growth. The planar slip and inhomogeneous deformation of CP material enhance crack branching, increase the total crack path, and lower the effective stress intensity at the tip of the crack, all of which lower crack-growth rates.[55] Figure 21(b) compares the FCP curves of the same materials used for Fig. 20(a) and shows the detrimental effect that a fine grain structure has on the FCPR.

Combined effects of grain size, deformation mode and environment on FCP behavior (tensile data given in Table 1) are shown in Fig. 22 for ITMT-7475. A detailed explanation of these data is beyond the scope of this paper and will be presented elsewhere.[42] However, the following general conclusions can be drawn from these results. There are significant effects of both aging treatment and grain size on the FCGR measured in vacuum (Fig. 22a). Decreasing the grain

Table 2. Effect of Microstructure on Tensile Properties of a Peak-Aged Al-2.7Li-2.3Cu-0.2Cd-0.2Zr Alloy

Microstructure	Yield strength, MPa	Strain to fracture, %
Unrecrystallized	442	3.2
Partially recrystallized	420	6.5
Recrystallized (elongated grains)......	435	10.1

Fig. 20. Microstructure and tensile fracture of Al-3Li-2Cu-0.2Zr TMP processed to produce (a and b) unrecrystallized structure and (c and d) recrystallized structure. *Courtesy of F. S. Lin.*

size by ITMT's and overaging, both of which homogenize deformation and decrease the reversibility of slip, increase FCGR's. Although the same trends are observed in air (Fig. 22b), the magnitude of the effect is considerably reduced due to environment-enhanced growth.

Titanium Alloys

Pure titanium has two allotropic forms, the low-temperature, hexagonal close-packed (hcp) α-phase and the elevated-temperature, body-centered cubic (bcc) β-phase. Various alloying elements tend to preferentially stabilize one or the other of these forms. As a result, Ti alloys are generally classified according to their equilibrium constitution, which varies depending on the type and concentration of alloying elements. According to this classification scheme, Ti alloys are

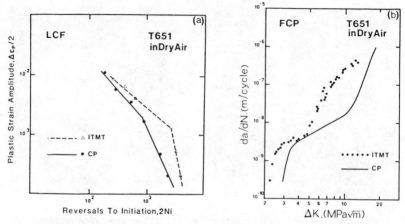

Fig. 21. Effects of ITMT on (a) FCI and (b) FCP of 7*xxx* aluminum alloys. *Courtesy of H. Chang.*

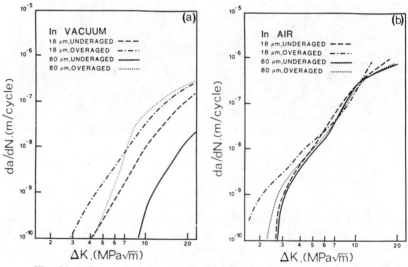

Fig. 22. Effects of grain size and aging treatment on the FCGR of ITMT alloy 7475. (a) Vacuum tests. (b) Laboratory air tests. *From Carter et al.*[42]

referred to as α-phase alloys, α + β alloys or β-phase alloys. Although this system is useful, it is not rigorously correct, because most α-phase alloys contain small amounts of β phase and nearly all β-phase alloys are metastable and thus are prone to α formation. The α + β alloys find the largest usage, and although they contain a mixture of the α and β phases, the α phase is typically the major constituent. The distribution and morphology of these phases depends on process-ing history and can be manipulated extensively both by heat treatment and by TMP. The morphology of the α and β phases has a pronounced effect on the properties of these alloys, as will be discussed later. However, since the α phase is hcp, it is quite anisotropic, both elastically and plastically. Thus, texture also has a significant effect on the properties of α and $\alpha + \beta$ alloys.

One of the vexing problems associated with structure-property studies of Ti alloys is the difficulty encountered in studying microstructure or texture effects *per se*. This is because many of the TMP schedules which can be used to alter the microstructure also simultaneously alter the texture, and vice versa. As a result, unambiguous interpretation of much of the structure-property work which has been reported in the literature is not possible, because these two variables (microstructure and texture) are interrelated to such an extent that separating their individual roles in affecting properties is hard if not impossible. Nevertheless, a major portion of this section will be devoted to a discussion of these factors in $\alpha + \beta$ alloys, because these alloys are more important from a techno-logical standpoint.

Effects of TMP on Microstructure and Texture

To understand the effects of TMP on microstructure it is first necessary to summarize the microstructural variations which can be produced by thermal treatment alone. If this is viewed as the baseline range of attainable micro-structures, then variations which occur outside this range during TMP are reason-ably described as those which can only be obtained by TMP. It is not the intent of this paper to provide an in-depth review of the phase transformations which occur in Ti alloys. The reader interested in a detailed treatment of this topic should refer to any of several earlier reviews.[56-60]

The basis for microstructural manipulation during heat treatment of Ti alloys centers around the $\beta \rightarrow \alpha$ transformation which occurs in these alloys during cooling. This transformation can occur by nucleation and growth (N & G) or it can occur martensitically, depending on the alloy composition and the cooling rate. The martensitic product is usually hcp and is designated α'. There also is an orthorhombic martensite, designated α'', which forms in alloys which contain higher concentrations of refractory elements such as Mo, Ta or Nb. However, because literally all TMP is conducted above the M_s temperature for either α' or α'', little of the remaining discussion will be devoted to these martensitic phases.

The effect of cooling rate on the constitution of an $\alpha + \beta$ alloy is shown schematically in the continuous cooling transformation (CCT) diagram shown in Fig. 23. In general, the C-curve for the $\beta \rightarrow \alpha + \beta$ N & G transformation moves to the right for alloys with increasing β-stabilizer concentrations. Alloys which contain enough β-stabilizing elements to depress the M_s temperature below room temperature can be rapidly cooled to retain the metastable β phase. As the alloy content is increased further, the N & G α curve shifts farther to the right and

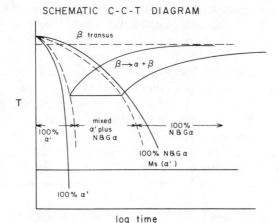

SCHEMATIC C-C-T DIAGRAM

Fig. 23. Schematic CCT diagram showing the effect of cooling rate on the constitution of an α + β Ti alloy after continuous cooling.

the β phase can be retained even in large sections during air cooling. The β phase is still metastable, however, and is usually transformed to an α + β mixture during a subsequent isothermal aging treatment.

The resulting microstructures which correspond to 100% α′ and 100% N & G α* are shown in Fig. 24(a) and (b). These structures are fairly typical of those which would be found in a range of α + β alloy compositions in which the β → α + β transformation temperature (the β transus) occurs between ~900 and ~1050°C. The α′ structure consists of an acicular hcp martensite which is supersaturated in β-stabilizing elements such as V or Mo. The N & G structure consists of colonies of laths of α-phase which are separated by thinner laths of β phase. Both the α′ and the N & G α have the Burgers orientation relation with respect to the parent β phase. The Burgers relation is $(0001)_\alpha \| \{011\}_\beta$; $\langle 11\bar{2}0 \rangle_\alpha \| \langle 111 \rangle_\beta$. There are 12 crystallographic variants of this relation. The martensite structure is always finer, and the individual plates are more randomly oriented than in the N & G structure. It is this variation in scale and randomness that becomes important later when recrystallization response, grain size and texture are considered.

An additional feature which can occur in most alloys that have been cooled from above the β transus is grain-boundary α (GB α). Grain-boundary α is heterogeneously nucleated at prior grain boundaries. Examples of GB α are shown in Fig. 25. The thickness and continuity of this layer depend on cooling

---

*As used here, 100% N & G α means that all alpha present has formed by a N & G reaction. Depending on alloy concentration, there may be (and usually is) some β phase present also. Likewise, reference to 100% α′ ignores the possibility of small amounts of retained β.

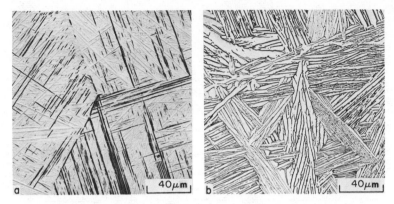

Fig. 24. Light micrographs showing the difference in morphology and scale of the martensitic transformation product (a) and the nucleation and growth product (b) in Ti-6Al-4V quenched and air cooled, respectively, from above the β transus.

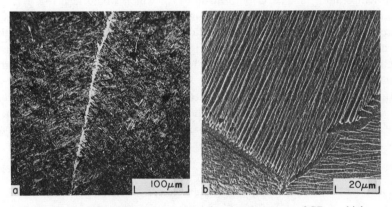

Fig. 25. Light and SEM micrographs showing the nature of GB α which forms in Ti-6Al-4V during cooling from above the β transus.

rate and alloy composition; however, the kinetics of nucleation and growth of GB α formation tend to be less sensitive to alloy composition than those for intragranular α formation. As a result, alloys which are fairly rich in β-stabilizing additions, e.g., Ti-6Al-2Sn-4Zr-6Mo, exhibit a stronger tendency for grain-boundary α formation than do leaner alloys such as Ti-6Al-4V. The relative kinetics of GB α and intergranular α formation are schematically represented in the CCT diagram, Fig. 26. This diagram is similar to that in Fig. 23 except that it also contains the start curve for GB α formation. The time range, Δt, also shown on this diagram, represents the range of differences in cooling times to the

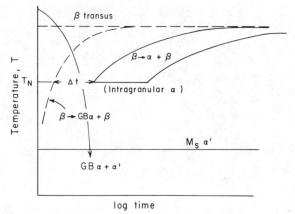

Fig. 26. Schematic CCT diagram similar to that in Fig. 23 but also showing the dashed C curve for GB α formation.

nose temperature for N & G α formation, T_N, which would lead to GB α formation without accompanying intragranular α formation. This range tends to increase with increasing β-stabilizing alloy content since the curve for formation of intragranular N & G α shifts to the right, whereas that for the GB α remains relatively fixed. Thus, GB α formation becomes more prevalent in more highly β-stabilized alloys.

If mechanical working is superimposed on cooling so that the N & G α phase is mechanically deformed as it forms, then subsequent globularization of the α phase is possible due to recrystallization. A typical structure in which the α phase has been changed from acicular to equiaxed by TMP is shown in Fig. 27.

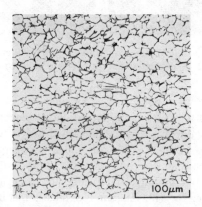

Fig. 27. Light micrograph of α + β–worked Ti-6Al-4V, showing the equiaxed α which forms due to recrystallization during and after α + β working.

Depending on the details of the working operation (T, ε, ε̇), α-phase aspect ratios (length:thickness) ranging from that of the plates in the colony structure (~30) to spherical (~1) can be achieved. This provides a considerable degree of flexibility in manipulating the microstructure using TMP as a tool. There are two ways of representing this type of TMP schedule. The first is to qualitatively indicate the working which goes on during cooling, as done schematically in Fig. 28(a). The second (Fig. 28b) is to actually plot the three-parameter temperature (T) – time (t) – strain (ε) profile. Of course, the slope of the ε-t trace is the instantaneous strain rate (ε̇). Two such profiles are shown in Fig. 28(b). The cooling rates, represented by the slope of the trace of these profiles on the T-t plane, are identical, but the amount of deformation in path "a" is less than that in path "b." The strain rate is also higher for path "b" since the cooling rates are identical and thus the elapsed time for introducing greater amounts of strain is the same. As a result, there also is less extensive recovery in material which has been processed along "b" in addition to the greater amount of deformation associated with "b." Thus, the recrystallization response of "b" will be much more rapid and complete than that of "a."

The recrystallization response of hot worked plate in alloys such as Ti-6Al-4V is frequently very nonuniform, even when plates from the same lot of material are compared. This reflects the difficulty that is encountered when reproduction of a particular TMP cycle (ε, t, T path) is attempted under production processing conditions. The nonuniformity of recrystallization becomes especially obvious if the standard mill-annealed condition is considered. In Ti-6Al-4V, this condition is achieved by annealing hot rolled plate at ~700°C (1300°F) for times ranging from 30 min to several hours. This tends to leave a well-recovered, partially recrystallized structure, as illustrated in Fig. 29(a). Part, but not all, of this

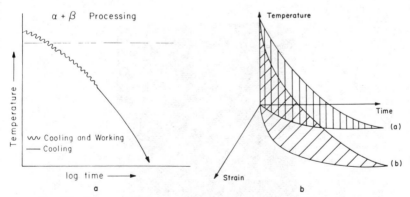

Fig. 28. Schematic representations of TMP cycles which include α + β working. Schematic (a) makes working relative to β transus easier to visualize whereas schematic (b) provides more detail regarding the temperature-time-strain combinations.

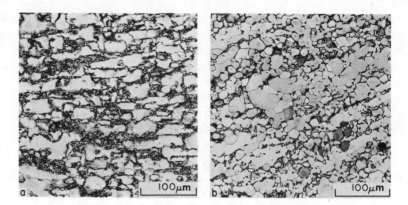

Fig. 29. Light micrographs showing (a) a typical α + β–worked and incompletely recrystallized microstructure of Ti-6Al-4V and (b) the effect of reannealing on this microstructure.

variability can be overcome by using a more effective annealing cycle, but even then the α morphology and degree of recrystallization are variable. This point can be illustrated using the material shown in Fig. 29(a). If this material is reannealed at ~925°C (1700°F) for 4 h, and then slowly cooled (at ~100°F/hr), it recrystallizes completely (Fig. 29b), although the α grain size is still not very uniform.

In general, forging is an easier way of imparting a reproducible TMP cycle to a material. Forging also is a more efficient method of introducing large amounts of stored energy into the material. Thus, forgings usually recrystallize more uniformly. Figure 30 shows the microstructure of a well-annealed, completely

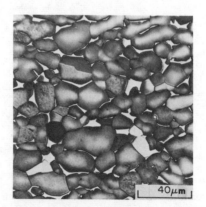

Fig. 30. Light micrograph showing the structure of a Ti-6Al-4V forging which has been recrystallization annealed (see Table 3).

recrystallized forging of Ti-6Al-4V. Microstructures such as these are relatively difficult to produce in hot rolled plate.

Recently, there has been some progress made in understanding the use of TMP in refining α grain size in alloys such as Ti-6Al-4V.[61,62] The central idea in this work is that the recrystallized primary α grain size is directly related to the dimensions of the α-phase regions which are subjected to deformation and re-crystallization. Thus, in Ti-6Al-4V, for example, rapid cooling from above the β transus to refine the α plate size, followed by reheating in the α + β region and forging or rolling with fairly heavy deformations, permits recrystallized α grains, 1 to 2 μm in diameter, to be achieved during annealing at ~800°C. This approach is a very useful, although reasonably arduous, method for producing very fine α grain sizes. There also is a section-size limitation associated with this technique, because refinement of the scale of the β → α + β microstructure requires high cooling rates, and this places obvious restrictions on section size.

TMP also is an effective means of controlling grain-boundary α formation. An alloy such as Ti-6Al-4V can be worked during continuous cooling from above the β transus to 25 to 50°C below the transus without significantly affecting the microstructure which results from the β → α + β transformation. This working does lead to continuous recrystallization of the β phase, however, and it would appear that this ongoing process tends to eliminate the β grain boundaries as primary sites for heterogeneous nucleation of α phase. It is also possible that continuous working down through the transus introduces sufficient alternate heterogeneous nucleation sites in the form of dislocations that the role of β grain boundaries as sole heterogeneous nucleation sites is considerably reduced. In any case, working during continuous cooling through the β transus is a very effective method of eliminating GB α.

In metastable β alloys TMP affects not only the microstructure but also the kinetics of decomposition of the metastable β phase during aging.[63-65] The increased dislocation density which remains after working these alloys leads to extensive heterogeneous nucleation of the equilibrium α-phase. This is demon-strated in Fig. 31, in which the variation in density of α precipitates between a recrystallized and an unrecrystallized region in a Ti-11.5Mo-4.5Sn-6Zr alloy (β-III) can be directly compared. In addition to N & G of α phase during aging, many metastable β-phase alloys form an intermediate decomposition product known as the ω phase (cf Ref 57). The presence of heterogeneous nucleation sites such as dislocations lead to direct α-phase formation, thereby causing this inter-mediate reaction product to be bypassed in the precipitation sequence. An ex-ample of this is given in Fig. 32, which shows the heterogeneously nucleated α particles at dislocations in a matrix which contains ω phase. If the dislocation density is increased by TMP then the diffusion fields which surround the α precipitates nucleated at these dislocations are more closely spaced and begin to overlap. When this occurs, ω-phase formation is completely suppressed. Thus,

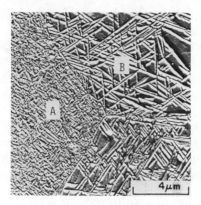

Fig. 31. Replica TEM micrograph showing the effect of nonuniform recrystallization on the size and distribution of α precipitates in Ti-11.5Mo-4.5Sn-6Zr (β-III). The higher density of precipitates at "A" than at "B" is the direct result of differences in the extent of recrystallization.

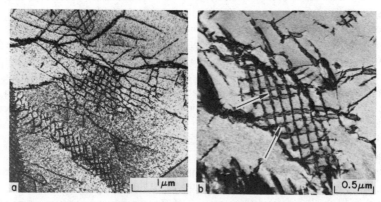

Fig. 32. Thin-foil TEM micrographs showing the heterogeneous nucleation of α-phase precipitates on dislocation substructure in β-III. The precipitates are the small laths, two of which are marked by arrows in (b).

there is a minimum strain required which can be introduced by TMP that will actually change the precipitation sequence and eliminate ω-phase formation.

TMP is also effective in altering the α-phase morphology in β alloys just as it is in α + β alloys.[66,67] Figure 33 shows the effect of TMP on α morphology in the metastable β alloy Ti-10V-2Fe-3Al. The structure of this alloy after cooling from above the β transus without TMP is shown in Fig. 34. Comparison of these two figures also shows the effectiveness of TMP in eliminating GB α in alloys such as this. As will be discussed later, this factor is a major consideration in control of the ductility and toughness of this class of alloys.

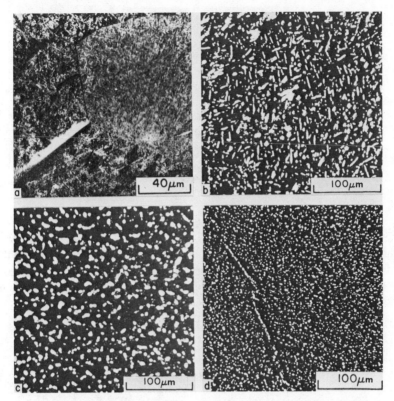

Fig. 33. Light micrographs showing the effect of varying TMP schedule on the distribution and morphology of primary α in Ti-10V-2Fe-3Al. All structures have been aged to produce strengthening due to α-phase formation. (a) β processed and rapidly cooled to minimize GB α. (b) α + β processed with light (~15%) α + β work. (c) α + β processed with heavy (~65%) α + β work. (d) α + β processed with moderate (~40%) α + β work (note remnant GB α).

The development of preferred orientation or texture, especially in the α phase, can lead to considerable anisotropy in the properties of Ti alloys. The role of TMP in the development of texture in Ti alloys has been the subject of a number of studies.[61,68-72] These textures are most conveniently represented by pole figures which describe the statistical distribution of grain orientations using a stereographic plot to depict the spatial distribution of the poles of a particular crystallographic plane. In hexagonal metals the texture is conveniently represented by the pole figure which shows the distribution of the poles of the basal plane, (0001). However, a complete description of the texture requires determination of a second pole figure also, because of the rotational symmetry of the basal pole. In Ti alloys the prism (10$\bar{1}$0) pole is usually used for this purpose, because it is

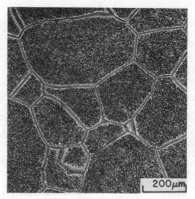

Fig. 34. Light micrograph showing heavy GB α layer formed in Ti-10V-2Fe-3Al as a result of slow cooling from above the β transus.

orthogonal to (0001) and because it is an active slip plane in α-Ti and is thus intrinsically useful also. Schematic examples of (0001) and (10$\bar{1}$0) pole figures are shown in Fig. 35. In general, two distinct basal textures are commonly observed. These are illustrated in Fig. 36 for the case of directional working. The

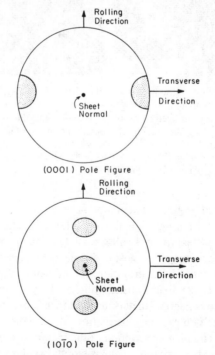

Fig. 35. Schematic representation of commonly observed basal- and prism-plane pole figures for α-Ti.

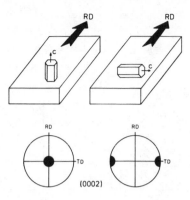

Fig. 36. Schematic representation of crystallite orientation and corresponding basal-plane pole figures for basal and transverse textures in α-Ti. *Courtesy of Dr. M. Peters.*

texture in which the basal poles are aligned with the sheet normal (Fig. 36, left) is referred to as a symmetrical or basal texture. The other texture, in which the basal poles lie in the plane of the sheet and are aligned with the transverse direction (Fig. 36, right), is referred to as a basal transverse or simple transverse texture. When axisymmetric flow is used instead of unidirectional working, these textures become rotationally symmetric about the principal strain axis. For example, the basal transverse texture is spread into a circularly symmetric distribution of basal poles around the perimeter of the pole figure. Many other combinations of these two basic textures can be obtained, depending on the details of the type of working operation (e.g., rolling versus forging) and the TMP schedule (T-t-ε profile). In addition, there are two other possibilities: no texture, and a highly symmetrical texture due to the β → α transformation. Figure 37 summarizes all of these possibilities together with the type of working and the working temperature range which lead to each texture (relative to the β transus).

The extent of development of the preferred orientation or the texture intensity depends on the amount of work. The type of texture depends on working method and working temperature. In principle, these variables can all be controlled to produce a material with a type and degree of anisotropy which vary from random to extensive. However, in current commercial practice, control over the necessary process variables is sufficiently difficult that materials with controlled textures are seldom produced on a commercial scale.

Effects of Microstructure and Texture on Properties

The properties of Ti alloys are strongly dependent on microstructure and texture. As demonstrated in the previous section, these two factors are, in turn, directly related to the TMP history of the material. Thus the use of TMP to

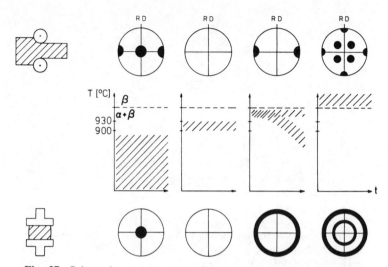

Fig. 37. Schematic representation of the relationship among working method (rolling at top, axisymmetric at bottom), working temperature and texture for Ti-6Al-4V.

manipulate microstructure and texture in order to alter, control or optimize properties is, indeed, a very useful concept. In fact, the use of TMP in controlling microstructure is already practiced extensively. In the remainder of this section, the effects of microstructure and texture on properties will be reviewed. Where possible, the explicit role of TMP will be identified even though such identification is somewhat redundant in view of the previous discussion of the relation between TMP history and microstructure and texture. This section will be organized according to individual groups of properties. The property groups of central focus will be tensile properties (strength and ductility), fracture toughness and fatigue behavior (crack growth and life). Other properties such as modulus, stress-corrosion cracking susceptibility and creep strength, will be briefly mentioned at the end. Each of these groups of properties will be discussed below. In each discussion, $\alpha + \beta$ and β-phase alloys will be considered separately.

Tensile Properties. Microstructure has significant effects on both strength and ductility of Ti alloys. For many applications, these properties represent both the initial and the primary consideration in selecting TMP and heat treatment conditions. All Ti alloys show reductions in ductility with increasing strength in much the same way that is well-known for other Fe-, Ni- and Al-base structural alloys. However, unlike these alloys, the primary α morphology also is a microstructural variable which can be controlled by TMP and which has a significant effect on ductility. The extent to which primary α morphology affects ductility depends strongly on the strength level. However, at all strength levels, this factor is important enough so that separate curves representing strength-ductility trends are often plotted for materials with equiaxed and elongated α-phase morphologies, as shown in Fig. 38. Since most high-strength alloys (yield strength ≳1100 MPa) tend to be metastable β alloys and most moderate-strength alloys (yield strength ≲1100 MPa) tend to be α + β alloys, it is convenient to consider these two classes of alloys separately.

Alpha-Plus-Beta Alloys: Because of the central importance of strength in sizing components, a minimum strength level is often fixed by the design requirements. Once the strength level is fixed, there still remain opportunities for optimizing the fracture resistance of the component by selecting a structure which has superior tensile ductility or fracture toughness. This selection process is complicated by a somewhat unexpected inverse dependence of ductility and toughness on microstructure for a given strength level. As a general rule, the tensile ductility of α + β – processed material is from 1.5 to 2 times that of β-processed material of comparable or somewhat lower strength. As will be discussed further in a subsequent section, the fracture toughness K_{IC} exhibits the reverse trend. These unusual trends in properties have led to considerable confusion in material-selection criteria, especially during the period when fracture mechanics properties such as K_{IC} were just gaining recognition and acceptance. At present, K_{IC} is usually given precedence over ductility for heavy-section applications. Nevertheless, tensile ductility is still important in applications where the material will be subjected to room-temperature deformation such as forming or straightening operations. Tensile ductility also can be a reasonable indicator of low-cycle fatigue behavior if empirical relationships such as the Coffin-Manson law[73] are used.

In addition to the uncertainties in relating microstructure, strength and ductility, another consideration pertains: the effect of microstructure on the strength of Ti alloys is not well understood in a fundamental way. Other factors besides microstructure which affect strength include solid-solution strengthening, especially by interstitial oxygen, and texture strengthening.[74–78] Notwithstanding these latter two factors, the role of microstructure in influencing yield strength of

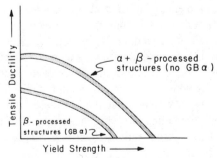

Fig. 38. Schematic strength-ductility trend curves for α + β and β-processed structures.

α + β Ti alloys is somewhat more pronounced than would be predicted on theoretical grounds. Consider the example of α + β worked and annealed Ti-6Al-4V, the microstructure shown in Fig. 30. In this condition, this alloy consists of ~85 vol % α present as equiaxed grains ~15 μm in diameter and small regions of β phase located largely at grain-boundary triple points. The strength of a monolithic specimen of α phase would be ~660 MPa,* whereas the strength of the β phase would be ~725 MPa.* Nevertheless, the relatively coarse mixture of the two phases described above will invariably exhibit yield strengths in excess of 860 MPa.* The detailed reasons for the higher-than-anticipated yield stress are not clear, but it has been pointed out elsewhere that this effect is general to dispersions of two deformable phases.[74] In this connection, it has been suggested[74] that the increase in yield stress reflects the difficulty with which slip can be transferred across interphase boundaries; in the case of α + β Ti alloys, transfer of slip across bcc/hcp boundaries between the β and α phases would be the pertinent consideration. This matter is made additionally complex by the existence of a third constituent at most α/β interfaces — a constituent known as the interface phase. This subject clearly merits further investigation using oriented bicrystals or heavily textured polycrystals.

There are a number of well-established trends relating microstructure and tensile properties which, although not well understood in a fundamental sense, provide useful guidelines for selection of TMP and heat treatment conditions. As a general rule, the yield stress of α + β alloys is highest when the microstructure consists of a mixture of primary α and either a fine α + β mixture resulting from N & G or martensitic α'. In alloys such as Ti-6Al-4V, quenching from above the β transus can also lead to high strengths. The microstructures which contain α' are produced by a solution treatment followed by water quenching and an aging treatment; this sequence is often called STA, for "solution treat and age." A typical heat treatment cycle used to produce these structures is shown schematically in Fig. 39. In alloys such as Ti-6Al-2Sn-4Zr-2Mo and Ti-6Al-4V, the solution treatment is followed by water quenching, which transforms most of the β phase to α' martensite. In more heavily β-stabilized alloys such as Ti-6Al-2Sn-4Zr-6Mo, solution treatment is followed by air cooling, which retains much of the β phase in a metastable state. Both of these treatments are followed by an aging reaction, although subsequent aging of the as-quenched α' or the metastable β phase leads to different metallurgical reactions. These reactions are tempering of α' martensite and N & G decomposition of the metastable β phase, as described in detail elsewhere.[56,59] There are a number of standard and frequently used heat treatment cycles which are applied to these alloys. These cycles are summarized in Table 3, including the pertinent designation for each cycle. These designations will be used subsequently to identify particular microstructural con-

---

*These strengths all depend on oxygen content but have been selected to be self-consistent for the purposes of this example.

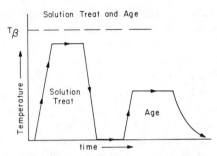

Fig. 39. Schematic heat treatment cycle for the solution treat and age sequence used in Ti-6Al-4V and other α + β Ti alloys.

Table 3. Summary of Heat Treatments for α + β Ti Alloys

Heat treatment designation	Heat treatment cycle	Microstructure
Duplex anneal (DA)	Solution treat at 50-75°C below T_β(a), air cool and age for 2-8 h at 540-675°C	Primary α, plus Widmanstätten α + β regions
Solution treat and age (STA)	Solution treat at ~40°C below T_β, water quench(b) and age for 2-8 h at 535-675°C	Primary α, plus tempered α′ or a β + α mixture
Beta anneal (BA)	Solution treat at ~15°C above T_β, air cool and stabilize at 650-760°C for 2 h	Widmanstätten α + β colony microstructure
Beta quench (BQ)	Solution treat at ~15°C above T_β, water quench and temper at 650-760°C for 2 h	Tempered α′
Recrystallization anneal (RA)	925°C for 4 h, cool at 50°C/h to 760°C, air cool	Equiaxed α with β at grain-boundary triple points
Mill anneal	α + β hot work + anneal at 705°C for 30 min to several hours and air cool	Incompletely recrystallized α with a small volume fraction of small β particles

(a) T_β is the β-transus temperature for the particular alloy in question. (b) In more heavily β-stabilized alloys such asTi-6Al-2Sn-4Zr-6Mo or Ti-6Al-6V-2Sn, solution treatment is followed by air cooling. Subsequent aging causes precipitation of α phase to form an α + β mixture.

ditions. Strength depends significantly on the scale and distribution of these aging products. Thus, aging temperature and time at temperature are effective means of controlling strength in a manner which is almost independent of the microstructure as it is affected by TMP.

In this connection, considerable effort has been expended in studying the effects of microstructure and processing history on the properties of α + β Ti alloys. Of all the available information, there are several systematic studies which have been especially helpful in establishing property trends.[79-81] The Sparks and Long[79] work probably represents the most extensive study of the properties of forgings made from the same heat of material with the same billet microstructure. Furthermore, because Sparks and Long, and also Chesnutt et al.,[80] investigated forgings, minimum amounts of texture were present to complicate their interpretation of the effects of microstructure on properties.

Tensile properties of three α + β alloys which have widely different heat treatment responses are shown in Table 4. From this table it can be seen that heat

Table 4. Tensile Properties of Three α + β Ti Alloys

Condition	Yield strength, MPa	Tensile strength, MPa	Elonga-tion, %	Reduction in area, %	Ref
Ti-6Al-4V					
α + β forge + recrystallization anneal (RA)(a)	711	876	12.4	36	80
α + β forge + mill anneal (minimum values)	828	897	10.0	25	82
α + β forge + STA (age 4 h at 594°C)	876	938	15.2	34	80
α + β forge + STOA (age 24 h at 594°C)	904	973	15.5	47	80
β forge + AC {BA} (+ 705°C/2 h/AC)	773	856	11.2	23	80
β forge + WQ {BQ} (+ 705°C/2 h/AC)	863	932	5.9	6	80
α + β forge DA (870°C/2 h/AC + 705°C/2 h/AC)	856	911	15.3	47	80
Ti-6Al-2Sn-4Zr-6Mo					
10-20% primary α + STA(b)	1118	1214	13	37	79
+ STOA(c)	1021	1090	16	42	79
40-50% primary α + STA(b)	1152	1242	14	42	79
+ STOA(c)	1070	1145	14	41	79
β forge + STA	1049	1201	6.5	13	79
Ti-6Al-2Sn-4Zr-2Mo					
α + β forged + STA(d)	1028	938	16	35	83
Ti-6-2-4-2+0.25 wt % Si					
α + β forge + 965°C/1 h/AC + 595°C/8 h/AC	1007	1063	13	34	83
α + β forge + 965°C/1 h/OO + 595°C/8 h/AC	1090	1187	12	20	83
β forge + 965°C/1 h/AC + 595°C/8 h/AC	952	1035	11	25	83
β forge + 965°C/1 h/AC + 595°C/8 h/AC	1042	1145	7	15	83

(a) RA = 925°C/4 h/cool at 50°C/h to 760°C/air cool. (b) STA = 885°C/1 h/air cool + 595°C/8 h/air cool. (c) STOA = 885°C/1 h/air cool + 705°C/1 h/air cool. (d) STA = 985°C(T_β − 15°C)/air cool + 595°C/8 h/air cool.

treatment has the greatest effect on strength, whereas forging practice (TMP) has the greatest effect on tensile ductility. For example, the effect of forging practice is reflected by the properties of Ti-6Al-4V in the BA and STOA conditions. The tensile ductility of the BA material is lower, as is its strength. To examine the effect of heat treatment on strength, compare the properties of Ti-6Al-4V in the BQ and BA conditions or in the recrystallization annealed (RA) and STA or STOA conditions. It also should be noted that the strength in the STOA condition, shown in Table 4, is higher than that in the STA condition. This result is not well understood, but the data in Table 4 were obtained from lower-oxygen material than is typically used in the STA or STOA condition.[80] The aging response of the lower-oxygen material may be quite different, and this could account for the above result. In general, strength appears to be controlled by the fineness of the microstructure. Thus, the refinement in microstructure which occurs between the BA and BQ conditions can qualitatively account for the higher strength of material in the BQ condition. Oxygen content also has large effects on strength and other properties, as will be discussed later.

Comparable but larger effects of processing and heat treatment on tensile properties can be seen in the data for Ti-6Al-2Sn-4Zr-6Mo also shown in Table 4. This alloy contains a higher concentration of β-stabilizing elements and thus is more responsive to heat treatment than are alloys such as Ti-6Al-4V. Because of the relatively strong heat treatment response of this alloy, it is seldom used in an annealed condition. Thus, if the annealed strength is considered to be a baseline level, the effects of heat treatment on strength are much larger than those shown here, but these larger effects are not relevant to commercial practice. For example, the difference in strength between the STA and STOA conditions reflects the effect of heat treatment, but the scale of these micostructures is similar with respect to that affected by TMP. The differences in ductility between β-forged STA material and $\alpha + \beta$ forged 40 to 50% primary α STA material reflect the effect of TMP. Thus, it is clear that variations in microstructure which occur on a scale greater than 1 μm dominate the fracture process and the ductility of the material, whereas variations in microstructural features which occur on a scale smaller than 1 μm have larger effects on strength.

Trends similar to those mentioned above, especially for Ti-6Al-4V, exist in the data for Ti-6Al-2Sn-4Zr-2Mo except that the heat treatment response is not as great. Thus, while the effect of processing on ductility is comparable to that observed in the two alloys already described, this alloy exhibits more limited increases in strength as the result of heat treatment. This is typical of alloys which contain low concentrations of β-stabilizing elements.

Most of the data for Ti-6Al-2Sn-4Zr-2Mo shown in Table 4 pertain to the version of the alloy which contains 0.25 wt % Si,* although baseline data for Si-free material is also included. Silicon is added to the alloy to impart creep

---

*Current practice usually limits Si content to 0.1 wt % or less.

resistance, as will be discussed later. In addition to improving creep resistance, Si also improves the response to heat treatment, probably because it is a relatively potent β stabilizer. Concomitant with the increased heat treatment response in the Si-bearing alloys is an increased tendency for metallurgical instability during elevated-temperature service. This factor must also be considered.

The extent of heat treatment response exhibited by the alloys discussed here increases with increasing β-stabilizer content, as can be seen from the data in Table 4. This trend is quite general among α + β Ti alloys, because alloys with higher concentrations of β-stabilizing elements have larger volume fractions of β phase when heated to a given temperature in the α + β – phase field. These β-phase regions remain as discrete microstructural constituents at room temperature even though they may decompose into an α + β mixture. Moreover, this enriched β phase tends to have slower N & G kinetics, which typically results in a finer dispersion of α-phase precipitates after aging. When these β-phase decomposition reactions occur on a finer scale, increased strength after heat treatment is observed. Thus, alloys with increased volume fractions of β phase have the potential for higher strength when these β-phase regions are strengthened by a decomposition reaction. As a result, the more heavily β-stabilized alloys tend to exhibit greater responses to heat treatment.

As mentioned previously, increases in oxygen content significantly increase the strengths of α + β Ti alloys, because interstitial oxygen has a potent solid-solution strengthening effect on the α phase. This effect is most clearly documented in commercial-purity Ti, for which an increase in oxygen content from 0.1 to 0.3 wt % can change the yield stress from <240 MPa to >500 MPa.[84] Significant increases in strength also are observed in the higher-strength alloys. Tensile data for Ti-6Al-4V with constant microstructure and different oxygen contents is shown in Table 5.[85] From this table it can be seen that strength increases significantly in the higher-oxygen material. The variations in tensile

Table 5. Effect of Oxygen Content on Tensile Properties of Ti-6Al-4V

Condition	Oxygen content	Yield strength, MPa	Tensile strength, MPa	Elonga-tion, %	Reduc-tion in area, %	Ref
BA	Low(a) 773		856	11	23	80
	Std(b) 883		994	11	20	79
STOA	Low 904		973	16	47	80
	Std 945		1042	14	33	19
RA	Low 711		876	12	36	80
	Std(c) 1056		1076	13	...	...
BQ	Low 863		932	6	6	80
	Std 980		1076	10	20	79

(a) 0.12 wt % oxygen. (b) 0.19 wt % oxygen. (c) 0.20 wt % oxygen (plate).

ductility are minor and in some cases seem to favor the higher-oxygen material, but this may reflect other factors such as grain size or texture. As will be shown later, an increase in oxygen content always results in lower fracture toughness.

Another factor which affects both strength and ductility is preferred orientation or texture. The effect of texture is to give an oriented polycrystalline material rather than a random one. Because the α phase is hexagonal close packed, it is plastically anisotropic. Thus, the presence of texture introduces variability and directionality of properties. Because of texture, comparisons of the properties of alloys with different microstructures must be approached cautiously to ensure that part of the observed property variations are not due to the presence of textures of varying type and/or intensity. As mentioned earlier, several common α-phase textures are observed in α + β Ti alloys, but the two most common ones are the basal transverse and symmetrical basal textures. The effects of texture on the properties of Ti alloys have been reviewed by Larson and Zarkades.[68] This review presents an account of texture development and of the variations in properties in textured Ti alloys. For the purposes of examining the extent of texture effects, the Larson-Zarkades article is quite useful. However, for readers interested in mechanisms of texture strengthening or in mechanistic interpretation of the role of texture in fracture or fatigue, this article contains significant deficiencies. For example, it is well established that $\bar{c} + \bar{a}$ slip is an important deformation mode in the α phase,[75] and yet, in the Larson-Zarkades article, it tends to be ignored and the importance of this nonbasal slip mode is generally underestimated.

The effects of texture on the tensile properties of one alloy (Ti-6Al-4V), in one microstructural condition, will be shown in order to demonstrate the magnitude of texture effects. The material from which this data was obtained had a basal transverse texture with 14 times the random number of basal-plane poles aligned along the transverse direction.[86] The data in Table 6 show that the textured material is much stronger when loaded parallel to the basal poles. Also shown in Table 6 is the higher modulus in this direction. As mentioned earlier, these property variations result from the elastic and plastic anisotropy of the hexagonal α phase, which is the majority phase in Ti-6Al-4V. Finally, it should be noted that the tensile ductility in the transverse direction is equal to or slightly greater than

Table 6.   Effect of Basal Texture on Tensile Properties of Ti-6Al-4V

Test direction	Yield strength, MPa	Tensile strength, MPa	Elonga- tion, %	Reuc- tion in area, %	E × 10^6
Longitudinal(a)	883	918	13	32	16.0
Transverse(b).	1063	1132	16	33	21.2

(a) Normal to high density of basal poles. (b) Parallel to high density of basal poles.

that in the longitudinal direction. Thus, it appears that texture strengthening can be achieved with essentially no penalty in tensile ductility.

Beta-Phase Alloys: Considerably less effort has been devoted to the study of the effects of TMP on the tensile properties of β-phase alloys, although recently there has been renewed interest in an alloy which has yield-strength levels around 1200 MPa along with acceptable tensile ductility and toughness. This is an ambitious goal, the achievement of which may require the use of TMP with tighter controls than are currently typical of commercial practice. In the remainder of this section, the results of TMP on the strength and ductility of β-phase alloys will be reviewed. The results will be taken from several studies in which this point has been examined.[87-91]

The effects of TMP on the strength and ductility of β-phase alloys tend to follow trends similar to those shown earlier for α + β alloys except that the strength levels are typically considerably higher. Three major factors which are affected by TMP appear to be important: primary α morphology, primary α volume fraction, and grain-boundary α. The strengths of these alloys are controlled by the scale of the α-phase precipitates in the matrix, and this can be controlled independent of the primary α morphology and volume fraction to a significant extent. As a result, data on yield strength versus tensile fracture strain can be plotted for each of several primary α volume fractions, as shown for alloy Ti-10V-2Fe-3Al in Fig. 40.[88] These data show that the alloy in the most ductile condition at any of the strength levels studied is that which contains a small (~0.1) volume fraction of primary α. This condition represents a compromise in the sense that alloys containing no primary α unavoidably have GB α, whereas, at higher volume fractions of primary α, strain localization tends to occur between the primary α particles. Both GB α and strain localization lead to premature fracture initiation, and thus the alloy which does not exhibit either of these has better ductility.

The role of texture in β alloys is not well understood, but it should be noted tha β-phase alloys produced in sheet or plate form exhibit a characteristic anisotropy in yield strength between the longitudinal and transverse orientations, with the strength being greater in the transverse direction. The ductility in the transverse direction is lower. These effects are illustrated by the typical data presented in Table 7.[89,90] These variations in strength and ductility have been discussed in terms of texture. Although the effects of texture are potentially less pronounced for bcc materials such as the β phase, it seems certain that texture plays a role in the directionality observed in these alloys.

Fracture Toughness. Fracture toughness is a measure of the resistance of a material to catastrophic propagation of a pre-existing sharp crack. Because many applications for moderate-to-high-strength Ti alloys involve heavy sections which are fracture-critical, resistance to catastrophic fracture is an important consideration in selection of an alloy and a heat treatment condition. Further-

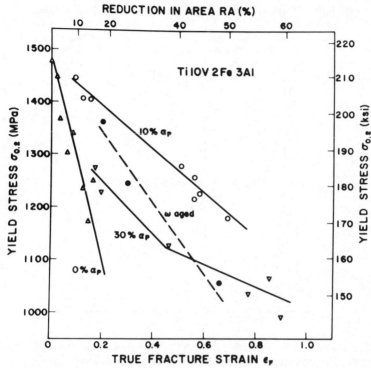

Fig. 40. Strength-ductility trend curves for Ti-10V-2Fe-3Al containing various amounts of primary α (cf. Ref 67).

more, at a constant strength level, the maximum size of pre-existing flaw that will not propagate unstably is proportional to the square of the fracture toughness/ yield stress ratio. Thus, as toughness increases at constant strength, the critical flaw size also increases dramatically. As a result, metallurgical factors which can be controlled to improve toughness offer the potential for improved structural integrity and efficiency.

As was done in the previous section for tensile properties, the effects of microstructure and texture on fracture toughness will be discussed separately for α + β and β-phase alloys. However, it is worth mentioning that most of the data presented for both types of alloys will be on plane-strain fracture toughness. This is because most fracture toughness testing has focused on test conditions which result in plane strain, since these conditions are pertinent to heavy sections. As a result, the data presented here will be largely K_{IC} or plane-strain fracture toughness data. For the purpose of comparison, K_{IC} is the most useful toughness parameter to consider, because it is independent of thickness and also represents a lower-bound value for toughness.

Table 7. Typical Tensile Data for Selected β-Phase Alloys

Orientation	Yield strength, MPa	Tensile strength, MPa	Elongation, %	Reduction in area, %	Ref
Alloy 4 (Ti-8Mo-2.5Al-4.5Cr)					
L	1432	1435	1.0	1.6	90
T	1465	1633	2.0	3.2	90
L	1229	1229	1.0	1.6	90
T	1665	1732	1.0	2.8	90
L	1179	1254	3.5	2.8	90
T	. . .	. . .	. . .	. . .	. . .
Alloy 334 (Ti-10Mo-2.5Al-6Cr)					
L	1260	1330	7	15	89
T	1275	1340	2	4	89
L	1150	1185	4	14	89
T	1225	1260	2	6	89
L	1095	1105	13	36	89
T	1180	1235	5	9	89
L	890	940	23	52	89
T	905	930	15	44	89

Moreover, the plane-stress fracture toughness (K_c) of Ti alloys rolled into sheet for thin-section applications is generally excellent and is virtually never a design-limiting factor in structures which are built up from sheet. As a result, other factors, such as strength, ductility, buckling resistance and creep resistance, usually dominate design decisions for thin-section components. Thus, all subsequent discussions of the effect of microstructure on toughness will pertain to variations in K_{IC}.

Alpha-Plus-Beta Alloys: As is the case with other structural materials, $\alpha + \beta$ Ti alloys typically have lower plane-strain fracture toughness (K_{IC}) at higher strength levels. In these alloys, however, unlike in aluminum alloys and high-strength steels, the processing and heat treatment combinations which lead to variations in strength can cause much larger attendant variations in microstructure. These variations have been discussed above, and so it is sufficient to point out that changes in morphology, size and distribution of microstructural constituents which occur on a scale comparable to the crack-tip plastic-zone size at the onset of unstable fracture can be expected to influence the crack path and fracture toughness. Microstructural features or constituents which are much smaller or larger have little effect on toughness. Thus, in $\alpha + \beta$ Ti alloys, gross microstructural variations such as those between equiaxed $\alpha + \beta$ and BA materials might be expected to affect toughness independently of strength. This is indeed observed; moreover, as a general rule, BA materials have significantly higher fracture toughness than do materials of comparable strength which have equiaxed $\alpha + \beta$ microstructures. These toughness variations have been shown to cause an accompanying change in fracture topography,[59,92] from a highly irregular fracture in BA material to transgranular fracture in $\alpha + \beta$ – processed material. This change in fracture will be discussed further, below.

Other material variables such as oxygen content and texture also affect fracture toughness of Ti alloys. Oxygen has two effects: it solid-solution strengthens the α phase, and it tends to promote planar slip and strain localization in the α phase. As a result, toughness decreases relatively rapidly as oxygen is increased, and thus material containing 0.13 wt% oxygen or less is usually used for toughness-critical applications. Because toughness naturally decreases with increasing strength, higher-strength alloys such as Ti-6Al-2Sn-4Zr-6Mo sometimes are specified with even lower oxygen contents, e.g., 0.10 wt %, because the strengthening contribution of oxygen is not needed and it is desirable to avoid strain localization. In lower-strength alloys, especially those used in the annealed condition, unnecessarily low oxygen levels can lead to superfluous toughness at a significant strength penalty. Texture also affects fracture toughness, and thus unexpected variations in toughness at constant strength, oxygen content and microstructure may be traceable to texture variations. This is especially true in flat-rolled products such as plate and sheet, which typically exhibit textures more intense than those observed in forgings.

A representative sampling of data which show the effect of microstructure on K_{IC} is presented in Table 8. Here again, data for three $\alpha + \beta$ alloys have been included because these alloys are representative of the range of strengths typically observed in $\alpha + \beta$ alloys.

The data for Ti-6Al-4V show that this alloy generally exhibits attractive properties, although the toughness of the standard oxygen mill-annealed (MA) plate can be quite low. The toughness of the mill-annealed material can be significantly improved with little loss of strength by reannealing, as shown by the increased toughness of the plate in the RA condition. These data for the MA and RA conditions were obtained by reannealing the same plate, and thus the changes in properties accurately reflect the effect of heat treatment at essentially constant texture. The highest toughness in plate material is typically achieved by β annealing, as also shown in Table 8.

The properties of the Ti-6Al-4V forgings in the BQ and the DA (50% primary α) conditions shown in Table 8 permit comparison of the effects of acicular and equiaxed microstructures on toughness and tensile ductility at essentially constant strength. These data show that the $\alpha + \beta$ – forged + DA materials have strength-toughness and strength-ductility combinations which are superior to those of either BA or BQ material. This result is at variance with the data for plate. Although the reasons are not entirely clear, it is possible that the inevitable presence of texture in $\alpha + \beta$ – processed plate affects the apparent toughness-

Table 8. Fracture-Toughness Variations in α+β Ti Alloys

Condition	Yield strength, MPa	Tensile strength, MPa	Elongation, %	$K_{IC}(K_Q)$, MPa$\sqrt{m}$
Ti-6Al-4V				
Plate:				
α + β roll + mill anneal(a)	1096	1171	14	32
α + β roll + recrystallization anneal (RA)(a)	1054	1144	13	51
α + β roll + duplex anneal (954°C/1.5 h/AC + 760°C/1 h/AC).....	882	971	14	59
BA (α + β roll + 1038°C/20 min/AC + 732°C/2 h/AC)	875	951	11	87
α + β roll + recrystallization anneal (RA)(b)	785	882	...	94
Forging:				
α + β forge + RA	710	875	12	(121)
α + β forge (15% primary α) + DA.......	813	889	12	(123)
α + β forge (50% primary α) + DA.......	854	909	15	(111)
β forge (BA).........................	772	854	11	(119)
β forge + quench (BQ).................	861	930	6	101
α + β forge + STA....................	875	937	15	(92)
α + β forge + STOA	903	971	16	84
Ti-6Al-2Sn-4Zr-6Mo				
Forging:				
α + β forge + STA(c) (10% primary α)	1116	1213	13	34
α + β forge + STA(c) (50% primary α)....	1150	1240	14	26
α + β forge + ann(d) (50% primary α)	1061	1130	13	26
β forge + STA(c)......................	1047	1199	7	57
Ti-6Al-2Sn-4Zr-2Mo				
Forging:				
α + β forge–β ST + age	903	...	12.5	81
β forge–α + β ST + age	896	...	11.0	84

(a) Standard-oxygen (~0.20 wt %) material. (b) Lower-oxygen (~0.13 wt %) material. (c) 885°C/1 h/AC + 593°C/8 h/AC. (d) 704°C/1 h/AC.

microstructure comparisons for this product form. In this connection, it is important to note that the forgings from which the data in Table 8 were obtained were virtually untextured.[80] Thus, if the variability between plate and forgings is due to texture, then it would appear that any texture may be deleterious to toughness.

As mentioned earlier, fractographic analysis[59] has shown that the fracture surfaces of K_{IC} specimens are much rougher and more irregular for acicular than for equiaxed microstructures. The irregularity of the fracture surface and the density of secondary cracks are higher in the BA than in the BQ condition. In the former condition it appears that prior β grain boundaries and Widmanstätten colony boundaries are the primary sites for crack deviation and secondary crack formation. Such tortuous, irregular fracture paths tend to increase toughness. This correlation between irregular crack path and high toughness is especially evident in the higher-strength alloys such as Ti-6Al-2Sn-4Zr-6Mo. The relationship between fracture topography and toughness has been discussed in detail elsewhere.[59,92]

The inverse trends in tensile ductility and fracture toughness that are exhibited by equiaxed and acicular microstructures are important to notice. This is characteristic of all α + β alloys but is especially pronounced in higher-strength alloys such as Ti-6Al-2Sn-4Zr-6Mo. The data in Table 8 show that β-processed forgings of Ti-6Al-2Sn-4Zr-6Mo have decidedly superior strength-toughness combinations. Moreover, in the α + β – processed condition this alloy can exhibit very low toughness values. For example, at a yield-strength level of 1150 MPa, a toughness of 26 MPa√m corresponds to a critical flaw size of ~0.5 mm. The high toughness of the β-forged material correlated well with the change in fracture topography from a flat transgranular fracture in α + β – forged material to an irregular intergranular fracture in β-forged material, as discussed above. This alloy is currently used almost exclusively as a forging alloy, and, as a result, no data are available for commercially produced plate.

The data in Table 8 for Ti-6Al-2Sn-4Zr-2Mo also show that this alloy is generally quite tough. Thus the improvement in toughness that can be achieved by β forging is less significant than that attainable in Ti-6Al-2Sn-4Zr-6Mo. Fortunately, this condition also has other attractive properties, as will be discussed later.

The effect of texture on toughness has been studied to a limited degree for Ti-6Al-4V and Ti-6Al-2Sn-4Zr-6Mo.[86] These studies have shown that materials exhibiting relatively intense transverse basal textures have greater fracture resistance when the crack-propagation direction is perpendicular to the majority of the basal planes, i.e., the transverse direction. The one exception to this appears to be ELI (extra low interstitial) Ti-6Al-4V, for which the opposite appears to be true. Fracture-resistance data obtained from R curves of materials with strong basal transverse textures are shown in Table 9. Note that the R curves for the ELI material differ from the others inasmuch as the incremental crack growth, Δa, which preceded unstable fracture was smaller by a factor of 5 in the LT specimens compared with the TL specimens. This may account for the apparent reversal in the effect of texture on K_R in this material. One possible explanation for this apparent reversal is based on the relative amounts of twinning which occur in the crack-tip plastic zone. Both oxygen content and texture affect the extent of twinning. Because twinning occurs in response to c-axis loading, the extent of twinning in material of basal transverse texture should be considerably greater in

Table 9. Toughness Directionality in Textured Material

Alloy (grade)	TL orientation	LT orientation
Ti-6Al-4V (ELI) (Ref 86)	123(b)	113(b)
Ti-6Al-4V (Ref 86)	113	124
Ti-6Al-2Sn-4Zr-6Mo (Ref 86)..	106	124
Ti-6Al-4V Plate (Ref 93)......	46	80

(a) Maximum K prior to onset of unstable fracture. (b) $\triangle a$ corresponding to K_R^{max} was 5 times smaller for the LT orientation than for the TL orientation.

TL specimens than in LT specimens. Extensive twinning would make the material more compliant, which is consistent with the R curves for low-oxygen material. Increasing oxygen dramatically reduces the incidence of twinning.[94] Thus, in the low-oxygen material, crack-tip plasticity may be enhanced by twinning, with an attendant increase in toughness.

The use of R curves to determine the toughness of the textured material was necessary because the material was fabricated in sheet thickness. Thus the LT and TL values of K_R must be examined on a relative basis, although the absolute relationships between these values and the K_{IC} values which would be expected from thicker plate are not obvious. One advantage of measuring the effects of texture in sheet is that the texture is much more uniform in the thickness direction. In thick plate, a surface-to-center texture gradient is nearly always present, because during continuous rolling the amount of work near the surface increases as the working temperature decreases. Because the final texture depends on both amount of work and working temperature, it is understandable that surface-to-center texture gradients are developed in plate during continuous rolling.

Some K_{IC} data on textured plate[93] have been reported also. These toughness data agree with the variation in crack-propagation resistance, K_R, shown earlier, as can be seen in Table 9. This provides further support for the incidence of lower toughness values when the crack propagates along the basal plane. The exception to this trend may be low-oxygen material, in which twinning may lead to increased plasticity and reductions in compliance.

Beta-Phase Alloys: The effect of microstructure on toughness has not been studied as extensively for β-phase alloys as for α + β alloys, but several limited studies have been conducted in conjunction with alloy-development programs.[89,90] These studies have consistently demonstrated that the toughness of these alloys is considerably higher after TMP treatments in the α + β field which lead to equiaxed primary α and no grain-boundary α. In sharp contrast to those of low- and medium-strength α + β alloys, microstructures which lead to tortuous fracture paths usually do not exhibit correspondingly higher toughness values.[89,90] This has been qualitatively explained on the basis of the strength differences between the grain-boundary α and the interiors of the prior β grains.[90] When the grain interiors are relatively weak, the crack-tip plastic zone is distributed between the grain-boundary α and the grain interiors. However, as the α-phase precipitates are refined by lower aging temperatures, as in the case of β alloys, the grain interiors are strengthened considerably. Under such circumstances the cracks are essentially constrained to follow the grain-boundary α layer. The net result of this is an artificial contraction of the plastic zone, a reduction in critical crack-opening displacement for crack extension, and an attendant drop in toughness. A schematic illustration of this relationship is shown in Fig. 41. This model seems to be consistent with the available data, which show that grain-boundary α is especially detrimental to toughness in higher-strength alloys. A general schematic which summarizes strength-toughness trends for equiaxed and acicular structures is presented in Fig. 42.

In summary, it is clear that TMP becomes an essential element of the processing of β-phase alloys if GB α is to be eliminated. Moreover, it is reasonably clear

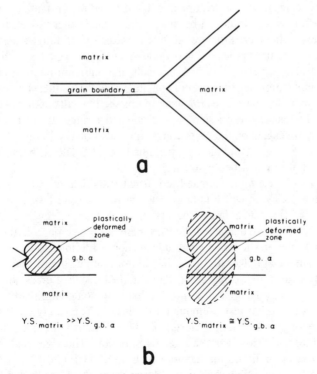

Fig. 41. Schematic drawing showing the relationship among GB α, plastic-zone size and matrix strength.

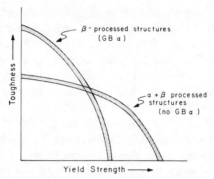

Fig. 42. Schematic plot of strength-toughness trends in Ti alloys with α + β and β-processed structures.

that GB α must be eliminated if good toughness is to be obtained at strength levels above ~1200 MPa.

Fatigue Behavior. The wide variations in microstructure attainable in α + β Ti alloys by controlled heat treatment and TMP lead to substantial variations in fatigue performance. Moreover, because of the large changes in tensile ductility with variations in microstructure, different measures of fatigue performance, e.g., notched versus smooth and low-cycle versus high-cycle tests, show different microstructural dependences. As a result, the optimum microstructure for a particular application depends not only on the particular loading situation but also on the design philosophy. The remainder of this section will attempt to review the effects of microstructure on smooth and notched fatigue life in the high- and low-cycle regimes and on the fatigue-crack-propagation (FCP) rate. The preponderant fraction of the available data pertains to α + β alloys, but a few observations on β alloys will be included.

Alpha-Plus-Beta Alloys: Fatigue life is usually divided into two regimes; low-cycle fatigue (LCF), in which failure occurs in 10^4 cycles or less, and high-cycle fatigue (HCF), in which failure occurs in more than 10^4 cycles. Coffin and Manson[73] have shown that at ~10^4 cycles there is a transition from the life being dominated by the plastic strain amplitude at shorter lifetimes to the elastic strain (stress) amplitude at longer lifetimes. In addition, fatigue life of smooth test bars is generally considered to represent the combined result of crack initiation and propagation. The conventional wisdom is that initiation comprises 85 to 95% of total life in smooth bar tests, although Tomkins[95] has recently questioned this. In the case of notched fatigue tests, the foreshortened lifetimes are usually considered to result from very early crack initiation. Thus, notched fatigue life depends more critically on the factors which affect FCP rate.

With the foregoing factors in mind, the available fatigue-life data for specimens with controlled microstructures will be examined. Smooth-bar life will be examined first, notched bars will be considered next, and the section will close with a discussion of factors which affect FCP rate.

There now is compelling evidence regarding Ti-6Al-4V which indicates that superior high-cycle ($N_f = 10^7$ cycles) smooth-bar fatigue strength is obtained when the slip length is small.[96-100] Small slip lengths can be obtained by using some of the special TMP methods for obtaining very fine-grain equiaxed material or by quenching from the β-phase field to produce fine, acicular α'. There is general agreement that the Widmanstätten or colony $\alpha + \beta$ microstructure has decidedly poorer fatigue strength. In the coarser, equiaxed microstructure the fatigue strength is significantly lower, but it is still better than that in the colony microstructure.[80] STA material has good fatigue strength, but not as good as that of the fine-grain equiaxed material or the beta-quenched material.[80,97,98] These trends in microstructure–fatigue strength dependency follow from the general body of available data on this subject. However, it is difficult to find a single set of data which includes all microstructures in the above ranking. There are several reasons for the scarcity of self-consistent data. These include variations in texture, variations in test method (axial versus bending), variations in test conditions (load ratio, frequency) and finally, but possibly most important, variations in surface-preparation method. In this connection, Leverant et al.[101] have shown that any machining method which leaves a compressive surface stress results in increased fatigue strength. Presumably, this is due to an extended preinitiation portion of the over-all specimen life.

A broader range of microstructures have been examined by Sparks and Long,[79] who have presented extensive tensile and smooth-bar fatigue data for 23 different conditions of Ti-6Al-4V. These data are summarized in Table 10. Using their data, regression analysis has been performed to see if a correlation exists between 10^7-cycle fatigue strength and yield or tensile strength. In both cases the coefficient of correlation was smaller than 0.1, indicating that essentially no correlation exists. This tends to point out an important difference between Ti alloys and steels — namely, that microstructure and strength affect fatigue life. However, since microstructure and strength can vary simultaneously, it is not surprising that the 10^7-cycle fatigue strength did not correlate well with strength alone: that is, in any one specimen, the effects of microstructure and strength can be offsetting factors so that no change in fatigue performance might be observed even when strength is increased.

The low-cycle fatigue life of Ti-6Al-4V appears to be most attractive in the STA condition,[102] although systematic studies of the effect of microstructure on LCF life are not readily available in the open literature. The data of Lucas also suggest that LCF life is best in STA material. One of the difficulties of using Lucas' data[97] for LCF is that his tests were constant-stress (HCF-type) tests. Such tests have only limited use in the LCF regime, where plastic strain amplitude dominates life. In contrast to the HCF case, the effect of primary α grain size

Table 10. Fatigue and Tensile Data for Various Microstructural Conditions

Condition	Yield strength, MPa	Tensile strength, MPa	Elonga- tion, %	Reduc- tion in area, %	Stress (smooth) at 10^7 cycles, MPa	Stress (notched) at 10^7 cycles, MPa
Ti-6Al-4V						
10% equiaxed primary α + ann(a)	971	1068	14	35	537	214
40% equiaxed primary α + ann	930	1013	15	41	579	255
10% equiaxed primary α + STOA(b)..........	978	1061	15	41	489	220
10% equiaxed primary α + ann(c)	958	1040	14	37	606	262
50% elongated primary α + ann(c)	923	1020	13	32	620	227
β forge + ann	882	992	11	20	565	220
β forge/water quench + ann..........	951	1054	10	21	606	186
β forge + STOA	978	1075	10	20	586	220
10% equiaxed primary α + ann(d)	882	985	13	33	620	214
Ti-6Al-2Sn-4Zr-6Mo						
10% equiaxed primary α + ann(e)	1020	1109	15	37	620	289
10% equiaxed primary α + STA(f)............	1116	1213	13	37	620	248
50% equiaxed primary α + ann	1061	1130	13	34	620	282
50% equiaxed primary α + STA..............	1151	1240	14	42	675	262
50% equiaxed primary α + STOA(g)..........	1068	1144	14	41	620	262
50% elongated primary α + STA..............	1096	1206	10	23	751	276
20% elongated primary α + STA..............	1109	1206	11	26	620	282
β forge + STA..........	1047	1199	7	13	675	262

(a) ann = 705°C/2 h/AC. (b) STOA = 955°C/1 h/WQ + 705°C/2 h/AC. (c) Water quenched off forging press. (d) Low-oxygen material. (e) ann = 705°C/1 h/AC. (f) STA = 885°C/1 h/AC + 595°C/8 h/AC. (g) STOA = 885°C/1 h/AC + 705°C/1 h/AC.

on LCF life has not been systematically studied. In a different study, Gilmore and Imam[103] showed that specimens solution treated at 900°C have excellent LCF life. The applicability of these results is subject to question, however, because Gilmore and Imam performed their testing in torsion. Since torsion suppresses crack opening, the torsion-initiated cracks propagate in mode II. Thus, comparison of the life data from this type of test with that from an axial or tension-tension test (mode I) is very difficult. Nevertheless, Gilmore and Imam's work tends to corroborate the earlier result that LCF life is optimized in STA material which contains primary α in an α' matrix.

In connection with the general low-cycle-fatigue behavior of α + β alloys, the cyclic stress-strain behavior of Ti – 6Al – 4V during fully reversed loading recently has been shown to vary with microstructure.[80] Seven microstructures were examined: RA, two DA conditions, BA, BQ, STA and STOA. Of these, three showed pronounced cyclic work softening whereas the others showed cyclic σ-ε curves which crossed the monotonic curves. The three conditions which resulted in softening were BQ, STA and STOA. This is not surprising because the α' in these structures contains a high initial dislocation density which tends to "shake down" during cyclic straining. The effects of these variations in cyclic stress-strain behavior of LCF life merit systematic investigation. Nevertheless, the softening behavior is generally undesirable in applications which are load-limited (as opposed to displacement-limited), because the plastic strain amplitude increases during service under constant load amplitude.

It also is important to mention that the microstructural conditions mentioned above as having good LCF crack-initiation resistance also have fracture-toughness values which vary from 65 to 95 MPa√m and tensile strengths ranging from 990 to 1015 MPa. Thus, selection of a material for attractive LCF performance may be done at the expense of toughness or tensile strength, or both. Again, this underscores the importance of identifying the performance-limiting property in design.

Finally, Bowen[104] has examined the effect of texture on the smooth-bar HCF behavior of Ti-6Al-4V bar stock. He tested specimens cut from a bar with a transverse basal texture and a microstructure in which the primary α grains were elongated in the rolling direction. Bowen's data (Table 11) show that the fatigue strength is significantly lower when the loading axis coincides with the direction of maximum basal-pole density (in this case, the transverse direction). These data also show that the fatigue strength is greater in the short transverse than in the longitudinal direction. Bowen has attributed this to the elongated grain structure

Table 11.   Effect of Texture on Fatigue Strength of Ti-6Al-4V[104]

Specimen orientation	Yield strength, MPa	Tensile strength, MPa	E $\times 10^3$ MPa	Stress at 10^7 cycles, MPa
Longitudinal	813	889	110.9	482
Transverse	909	958	124.7	413
Short transverse . . .	868	951	110.9	551

which makes the short transverse grain size effectively smaller in comparison with the longitudinal direction. The dependence of the yield-stress values on orientation, also shown in Table 11, is consistent with this interpretation, because the rotational symmetry of the texture about the bar axis excludes strength differences in the longitudinal and short transverse directions due to texture. Thus, the observed strength differences probably result from grain-size effects.

Other studies of the influence of texture on smooth-bar fatigue life of Ti-6Al-4V also show a significant effect.[61,86] These studies have shown that the type of texture and the test environment both affect fatigue performance, as shown by the S-N curves presented in Fig. 43 to 45. The curves in Fig. 43 show that the fatigue strength is greater when the stress axis coincides with the direction of a high density of basal poles. The data in Fig. 44(a) show a reverse trend, although comparable data obtained from tests run in vacuum agree with the trend shown in Fig. 43. Figure 44(b) shows that testing in an aqueous 3½% NaCl solution reduces fatigue strength when the stress axis is along $[0001]_\alpha$. The data in Fig. 45 show that a mixed texture lowers the high-stress end of the S-N curve preferentially. Perhaps most surprising, however, is the environmental sensitivity which can be seen by comparing Fig. 43 and 44(a).

The amount of fatigue data available for Ti-6Al-2Sn-4Zr-6Mo is small compared to that for Ti-6Al-4V. Nevertheless, HCF data exists for the 11 conditions studied by Sparks and Long. No systematic trends are evident for these conditions. Good HCF strength at 10^7 cycles was obtained in material with both equiaxed and elongated primary α. The best fatigue performance in this alloy was obtained in the STA condition, which corresponds to solution treating for 1 h at 885°C and air cooling, followed by aging for 8 h at 595°C and air cooling. This is in contrast to Ti-6Al-4V, where the highest fatigue strengths were obtained by annealing for 2 h at 705°C after forging. It should be noted that the air-cooled microstructures in Ti-6Al-2Sn-4Zr-6Mo do not have Widmanstätten $\alpha + \beta$ colo-

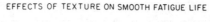

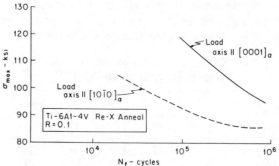

Fig. 43. S-N curves showing the effect of texture on the fatigue strength of Ti-6Al-4V.[86]

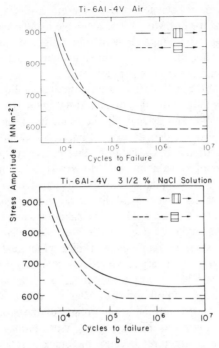

Fig. 44. S-N curves for $\alpha + \beta$ – processed Ti-6Al-4V, showing the effects of texture and environment on fatigue strength.[61] (a) Tested in air. (b) Tested in 3½% NaCl.

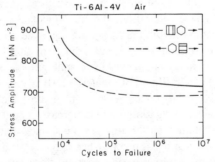

Fig. 45. S-N curves showing the effect of more complex texture on fatigue strength of Ti-6Al-4V. Compare Fig. 44(a).[61]

nies such as those present in Ti-6Al-4V. As a result, BA or β-forged material does not have an increased slip length in Ti-6Al-2Sn-4Zr-6Mo in comparison with Ti-6Al-4V. Thus, there is no significant lowering of the HCF strength in this alloy when elongated α is present.

Only limited data have been reported on the effect of texture on the fatigue behavior of Ti-6Al-2Sn-4Zr-6Mo.[86] The available data pertain to $\alpha + \beta$ rolled and annealed material. These data show that the fatigue strength at 10^6 cycles is ~ 105 MPa greater when the stress axis is perpendicular to the basal-pole texture direction ($\langle 0001 \rangle_\alpha$) than when it is parallel to it.

Some discussion of the smooth-bar fatigue-life data summarized above seems appropriate. The low fatigue strength of Ti-6Al-4V in the BA condition has been examined by Stubbington and Bowen,[99] who found that stage I cracks initiate at α/β interfaces. Because the density of α/β interfaces is very high in the Widmanstätten structure of the BA condition, this microstructure should be susceptible to initiation of stage I cracks. Moreover, as mentioned earlier, the interface phase at $\alpha + \beta$ boundaries retards slip transfer across the α/β interfaces, and thus a local strain incompatibility exists in the region of the interface. Finally, since all α plates in a single colony have the same orientation, stage I cracks can propagate for larger distances in a constant maximum shear stress orientation. In contrast, fine-grain equiaxed microstructures contain many α/α boundaries but few α/β boundaries. Moreover, smaller grain sizes lead to reduced strains which accumulate at grain boundaries by reducing the slip length and, therefore, the dislocation pile-up size. Also, since α/α boundaries are more prevalent in this microstructure, the occurrence of the strain incompatibility mentioned earlier is minimized. Finally, stage I crack propagation is more difficult in equiaxed $\alpha + \beta$ structures because the crack is reoriented each time it crosses an α/α grain boundary. Obviously, the extent to which texture is present becomes a central consideration in this matter. Thus, there are a number of factors which help explain the superior HCF behavior of fine-grain equiaxed microstructures relative to Widmanstätten BA microstructures. From the standpoints of HCF life and stage I crack initiation, the fine acicular α' in BQ material has several features in common with the fine-grain equiaxed microstructures. These include almost no α/β interfaces, small grains (in this case, the individual martensite plates), and frequent changes in crystallographic orientation. It is thus not surprising that the HCF life of BQ material also is very good. Furthermore, because BQ material typically contains almost no texture, the even more frequent changes in stage I crack propagation orientation may account for the BQ material being typically better than even the fine-grain equiaxed material.

In contrast to smooth-bar tests, notched-bar fatigue tests are generally considered to result in early crack initiation. Furthermore, because of the stress state due to the notch, the extent of stage I crack growth is typically quite limited. Thus, notched life is dominated by the rate at which stage II cracks propagate. In this regard, Eylon and Pierce[105] have shown that the notched life of BA material is superior to that of RA, MA, DA or STA material. However, these authors are somewhat vague in their explanation of this effect, leaving two possible causes:

large numbers of cycles to crack initiation and longer stage I crack lengths.* They further show that microstructure has a significant effect on the extent of stage I crack propagation prior to the transition to stage II cracking. Even though these authors studied notched specimens, their findings must be compared with the earlier discussion of stage I cracking in smooth specimens. In this connection it is appealing to relate good notched fatigue strength to lower fatigue-crack-propagation rates. However, other available notched fatigue data, especially those of Sparks and Long,[79] make this appear to be an oversimplification. For example, their data shows that the BA and BQ conditions are among the better conditions but that a DA condition has the best notched fatigue strength. Thus, the factors which control notched fatigue strength appear to be more complex than suggested above or by others.[105]

In an attempt to produce an empirical correlation, regression analysis similar to that mentioned earlier in connection with smooth fatigue of Ti-6Al-4V was also conducted on notched fatigue data for 23 conditions of Ti-6Al-4V and 11 conditions of Ti-6Al-2Sn-4Zr-6Mo. There was essentially no correlation between 10^7-cycle notched fatigue strength and either tensile or yield strength. However, the variations in notched fatigue strength are also smaller, so that adoption of minimum values of this property in design leads to only modest penalties in structural efficiency. The effect of texture on notched fatigue is neither well characterized nor well understood. However, limited test data[86] show that the trends in Ti-6Al-4V follow those for smooth-bar fatigue — that is, the fatigue strength is lower when the stress axis is parallel to the basal texture direction. In Ti-6Al-2Sn-4Zr-6Mo the presence of basal texture tends to reverse the trends between smooth and notched strength. In this latter case, the notched fatigue strength is higher when the loading direction is parallel to the basal texture direction whereas the smooth-bar fatigue strength is lower for the same orientation. No obvious explanation is available for this result.

Fatigue-crack-propagation rates have become more heavily emphasized in recent years because of the growing interest in damage-tolerant design. As a result, significant amounts of data are now available which show that microstructural variations can alter the FCP rates at constant cyclic stress intensity (ΔK) by more than an order of magnitude.[59,80,98,100,106–112] It is well known that a log-log plot of FCP rate versus ΔK for all materials has the characteristic shape shown in Fig. 46.[113] In this figure it can be seen that the curve bends downward at low growth rates and eventually becomes parallel to the ordinate. When the curve becomes parallel to the ordinate, the ΔK intercept represents a threshold value (ΔK_{th}) below which no crack growth occurs. The crack-growth rate at ΔK_{th} is

---

*Eylon and Pierce call this initial crack the initiated crack, but examination of their Fig. 14(b) strongly suggests that this crack is a stage I crack.

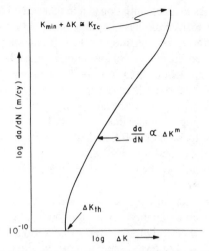

Fig. 46. Schematic plot showing characteristic shape of fatigue-crack-growth rate (da/dN) versus cyclic-stress-intensity (ΔK) curves.

typically $\sim 10^{-10}$ m/cycle. Referring again to Fig. 46, it can be seen that at higher growth rates there is a linear portion of the curve. This linear portion was represented as ΔK^m by Paris and Erdogan[114] and is now frequently referred to as the Paris law regime of fatigue-crack growth. Most structural materials show variations in near-threshold FCP rate and in ΔK_{th} but fewer show significant variations in FCP rate in the Paris law regime.[113]* In contrast, Ti alloys show significant variations in FCP rate over the entire range. An example is shown in the FCP curves of Fig. 47.[80] At the highest crack-growth rates shown in Fig. 46, the FCP-rate curve bends upward. This is controlled by fracture toughness. However, since crack-growth rates are uncontrollably rapid in this latter regime, it is of little interest and will not be discussed further here. Moreover, since the majority of the lifetime of a crack component is spent in the low-FCP-rate regime, factors which control FCP at rates less than $\sim 10^{-6}$ m/cycle are probably most important. These factors include microstructure and texture.

Effects of microstructure are most frequently studied because this variable is relatively easy to control. However, caution must be exercised not to let effects of texture cloud the interpretation of the influence of microstructure on FCP. Among the commonly used microstructures described in Table 3, the FCP rate is typically lowest for BQ and BA and highest for Ma and STA or STOA. The FCP rates for the other microstructures fall between these extremes. Furthermore, the FCP rate for BQ is consistently lower than for BA at very low rates. Recent work[98] has shown that the FCP-rate curves for these two microstructures cross

---

*One exception is some recent work on fully pearlitic plain carbon steel.[115]

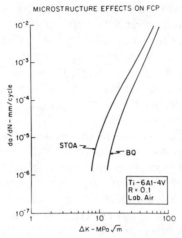

Fig. 47. Actual da/dN-ΔK curves for two microstructures of Ti-6Al-4V.[80]

at $\sim 10^{-7}$ m/cycle. Above this rate the BA condition is superior. In addition, the BQ material has a ΔK_{th} of ~ 9 MPa$\sqrt{m}$, whereas the BA-annealed material has a ΔK_{th} of ~ 6 MPa$\sqrt{m}$. Such variations are consistent with the higher fatigue strength of BQ material, which was verified in the same study[98] — that is, higher fatigue strength and higher ΔK_{th} are self-consistent inasmuch as they both indicate a low rate of growth of short (low-ΔK) cracks.

In an attempt to provide an improved understanding of the effect of microstructure on FCP rate, detailed electron fractography[59,80,98,104] has been conducted on FCP specimens with different microstructures. These fractographic studies have shown that, depending on ΔK level, FCP is accompanied by formation of either of two distinctly different fracture features. The dimensions of these features depend on microstructure but typically range from 5 to 150 μm.

At low ΔK levels (low FCP rates) the crack growth occurs by a faceted growth mode, whereas at higher ΔK levels crack propagation is accompanied by formation of ductile fatigue striations. Furthermore, the details of the topography of the facets on a submicron scale vary with microstructure. For those microstructures which exhibit low FCP rates (BQ and BA), the fracture surfaces contain facets which have extensive secondary cracking. In contrast, the facets on the fracture surfaces of material in the MA or the RA condition are smooth and exhibit no evidence of secondary cracking. It has been reasoned[80] that the secondary cracking reduces the local effective stress intensity and thereby leads to lower FCP rates.

At higher FCP rates, where ductile fatigue striations are observed, the crack-growth rate is consistently lower in those microstructures which have more tortuous crack-propagation paths. In order to effectively reduce FCP rate, the wavelength of the tortuosity must be hundreds of times that of the crack-growth rate per cycle. Thus, in structures such as DA or STA the individual martensite or Widmanstätten α plates are too small to reorient the crack, probably because these microstructural units fail in a single load cycle and thus cannot divert the crack. Fractographic examination bears this out, because FCP specimens with these structures exhibit a high density of tear ridges in the α + β or α' regions and striations in the primary α regions.

The variations in fracture topography with microstructure have been studied in greater detail[116] and are discussed elsewhere.[117] It is now quite clear that the rough, faceted fatigue-fracture surfaces do not reclose as perfectly as the smoother ones do during unloading. As a result, there is a residual compressive stress left at the tip of the faceted fracture which must be overcome during reloading. This effect has been termed "roughness-induced closure" and is discussed in detail elsewhere.[115,116,117] Closure measurements[116] which verify this effect have shown that accounting for the roughness-induced closure variations as a function of microstructure permits the effect of microstructure on FCP rate to be rationalized. Further work is needed to provide a detailed understanding of the relation between microstructure and fracture-surface roughness. However, it should be mentioned here that microstructures which give rise to faceted irregular fatigue fractures typically exhibit superior FCP properties. The well-known effect of load ratio (R) on FCP ratio (Fig. 48) is also consistent with these closure effects. Thus, the role of closure seems at least to be consistent with the bulk of available experimental data.

The above discussion raises questions regarding the low FCP rates in BQ material in the near-threshold regime. It appears that the α' plates in BQ material

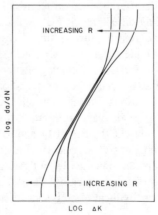

Fig. 48. Schematic plot showing the effect of load ratio (R) on da/dN-ΔK curves.

are significantly larger than those in STA material. Similar trends pertain to the Widmanstätten α plate sizes in BA and DA materials. The reasons for these microstructural-scale variations have not been analyzed in detail but it is suggested that the size of these transformation products (Widmanstätten α and α' plates) scales with the mean free path of β phase. Thus, large β regions will have significantly larger transformation products. This is certainly true in BA and BQ materials, where the α' or Widmanstätten α plates are much larger because the specimen is fully β at the onset of the transformation. It is also noteworthy that the larger α' plate size in the BQ condition may not be large enough to divert the crack at higher ΔK values. Thus, the crossing of the FCP-rate curves for BA and BQ material may be accounted for on this basis.

The effect of microstructure on the FCP rate of Ti-6Al-2Sn-4Zr-6Mo is smaller,[80] with the maximum variation being a factor of 4 or 5. As for Ti-6Al-4V, the lowest FCP rates correspond to β-forged material, whereas the highest FCP rates correspond to α + β – forged material. Fractographic examination of these specimens showed a significantly higher density of secondary cracking in the β-forged condition. This observation is consistent with the earlier discussion of the effect of secondary cracking on FCP rate. The smaller differences in FCP rate between microstructures in Ti-6Al-2Sn-4Zr-6Mo may be due to the considerably finer microstructural scale which typifies all conditions of this alloy. The effect of increased alloy content in reducing the microstructural scale of α + β alloys has been discussed elsewhere.[59]

Separation of the effects of texture and microstructure on FCP rate can be elusive. This is due in part to the increased intensity of texture which is typically found in equiaxed microstructures in comparison with acicular microstructures. The reason for this is related to the randomizing effect of the nucleation and growth or martensitic β-phase decomposition reactions. Since there are 12 variants of the α/β orientation relation, the β-phase decomposition products tend to exhibit a relatively random texture which reflects a nearly random population of these 12 variants.

Because examination of microstructural effects on FCP rate should be conducted under conditions of constant texture, comparison of FCP rates in RA or MA material with those in BA or BQ material is almost certainly complicated by variations in texture. Such complications may account for the apparent variability in FCP rates in the same microstructure. This is particularly true for α + β microstructures containing equiaxed primary α, which is most prone to texture. Further, since texture also affects modulus and yield strength, variations in texture lead to changes in other factors such as crack-opening displacement (COD) and plastic-zone size (r_p). Thus, variations in texture complicate the design of FCP experiments which permit fair assessment of microstructural effects.

In view of the above complications, Bowen[69] has attempted to assess the intrinsic effect of texture on FCP in Ti-6Al-4V. In order to minimize microstructural effects he has used α + β – worked material with a single micro-structure and with a strong transverse basal texture. He has compared the FCP rates for all six possible orientations in a directionally worked product. He has found that the FCP rate in the Paris law regime exhibits different exponents, m, for different crack-growth directions. Where the FCP rate is proportional to $(\Delta K)^m$ the corresponding m values are shown in Table 12. From these experiments Bowen has concluded that texture affects the symmetry of the plastic flow in the crack-tip region. When flow is symmetrical, m values are low; when flow is nonsymmetrical, m values are high. He also has shown that the crack plane is normal to the stress axis when the local orientation permits symmetrical flow in the crack-tip region. When the flow is nonsymmetrical, the crack plane varies considerably, leading to a tortuous crack path.

More recently, another study[86] of the effect of texture on FCP rate has been reported. This study shows that, for specimens of Ti-6Al-4V, the FCP rate was approximately 5 to 6 times higher when the loading direction was parallel to ⟨0001⟩ so that the crack was growing along the basal plane, compared with a specimen rotated 90° so that the crack-growth direction was parallel to ⟨0001⟩. A similar but much smaller effect was observed also for textured Ti-6Al-2Sn-4Zr-6Mo, but in this alloy the FCP rates differed by a factor of about 2 to 3. In the same study, textured ELI (extra low interstitial) Ti-6Al-4V was also tested. In this material there were essentially no differences in FCP rate between the two orientations. One possible explanation for this is the significant increase in propensity for twinning in the ELI material compared with normal interstitial material. Twinning alters the flow characteristics in the crack-tip plastic zone, which might cause the FCP rate to slow down in the orientation where cracking occurs along ⟨0001⟩$_\alpha$. These data also show that texture does have an intrinsic effect on FCP rate, but they do not provide information regarding the relative FCP rates in textured and random material. It does not necessarily follow that the lowest FCP rate in textured material will be lower than the corresponding rate in

Table 12. Fatigue-Crack-Growth Exponents for Textured Ti-6Al-4V[69]

Test-piece orientation(a)	m value	Test-piece orientation(a)	m value
LS	2.7	ST	3.6
LT	4.1	TS	2.5
SL	2.4	TL	3.1

(a) L = longitudinal direction (also the rolling direction), T = transverse direction (also the width direction), S = short transverse direction (also the thickness direction). The first letter designates the loading direction, and the second letter designates the crack-propagation direction.

random material. This may be particularly true at low FCP rates where the faceted mode of crack growth occurs. Because the facet plane is near $(0001)_\alpha$, random material should exhibit more frequent reorientation of the crack front. The situation for FCP in textured material when $(0001)_\alpha$ is parallel to the loading axis is not clear, because combined FCP-rate measurement and fractographic examination have not been carried out for these circumstances. It is clear, however, that further investigation of both FCP rates and attendant fracture modes in materials of varying controlled texture and microstructure is warranted.

Beta-Phase Alloys: There has been considerably less work done on FCP in β alloys, although several studies have provided some interesting results.[66,118,119] In general, the scale of the microstructure is small enough so that subtle variations in microstructure, e.g., α-phase precipitate size or volume fraction, can affect the strength while leaving the FCP rate essentially unchanged. However, when the stability of the β phase is changed markedly or when ω phase is present, the deformation character can vary from twinning to wavy slip to planar slip. In such cases the FCP rates do vary considerably, as shown in Fig. 49. The three microstructures represented by the curves in Fig. 49 exhibit wavy slip, planar slip and reversible twinning, respectively.[118] The variations in FCP rate can be qualitatively understood on the basis of slip reversibility — that is, the ω-aged condition has considerably more reversible slip in the crack-tip plastic zone than does the α-aged condition, whereas the β – solution treated condition tends to exhibit reversible twinning. Similar trends have been observed in Ti-10V-2Fe-3Al[66] and in binary Ti-V alloys.[119] In general, the more completely reversible the crack-tip deformation, the lower the amount of residual fatigue damage that is deposited

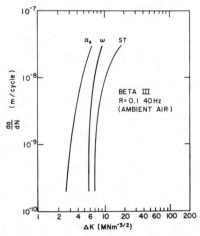

Fig. 49. FCP curves for Ti-11.5Mo-4.5Sn-6Zr (β-III) in three microstructural conditions, showing a large effect of microstructure on da/dN.[118]

in the plastic zone per cycle and the lower the average crack extension per cycle. This viewpoint does not take roughness-induced closure into account, and this must be reconciled with the compelling evidence for closure effects mentioned earlier for α + β alloys. However, increased slip reversibility usually occurs in alloys which exhibit planar slip. These alloys also tend to exhibit greater fracture-surface roughness due to the presence of more extensive facets. Thus, slip reversibility and increased fracture roughness may be interrelated (at least these two factors tend to vary in a fashion which is qualitatively consistent with the FCP data). Closure and fracture-roughness measurements in these alloys would be very useful in helping to quantify these factors.

Other Properties. There are several other properties of Ti alloys which can be important under certain circumstances. These include modulus, creep strength and stress-corrosion resistance. Among these, stress-corrosion resistance has been the most extensively studied, because stress corrosion of some Ti alloys has been the source of major concern for some applications (e.g., during the U.S. Supersonic Transport Program). While each of these properties can be very important in particular applications, they generally are treated as subsidiary properties relative to the properties already discussed in some detail. Thus, only a brief summary of the factors which affect these subsidiary properties will be included here.

The effect of texture on modulus was mentioned earlier and has been reported in a number of studies.[68,71,72,86] The origin of this texture-dependent variation in modulus is elastic anisotropy of the α phase, due to the variation in elastic constants along directions parallel and perpendicular to the c axis. These variations amount to ~35% in pure titanium, with the stiffer direction being along ⟨0001⟩. As Al is added as a solid-solution strengthening agent, all of the α-phase elastic constants increase, but the increases are more rapid normal to ⟨0001⟩ than along it.[71] Thus, addition of Al to α-Ti increases the modulus but reduces the anisotropy and thus the maximum variation in modulus that can be achieved through texture. Nevertheless, modulus values in excess of 138 GPa have been achieved in heavily textured Ti-6Al-4V[86] and in Ti-8Al-1Mo-1V.[120] Thus, texture can be a useful means of increasing the stiffness of Ti alloys.

The phenomenology of the susceptibility of Ti alloys to stress-corrosion cracking is generally well understood and has been described in detail elsewhere.[121,122] One of the central issues in susceptibility to aqueous stress-corrosion cracking (SCC) is the cleavagelike cracking which occurs near the basal plane of the α phase. The occurrence of this mode of cracking appears to be normal-stress-controlled as opposed to plastic-strain-controlled. Thus, textured alloys tend to be considerably more susceptible to SCC when loaded in the stiffer direction, i.e., along ⟨0001⟩, than when loaded normal to this direction. Fager and Spurr have studied this in some detail.[123] This occurrence places some severe limitations on the use of texture strengthening, since the high-strength, high-molulus direction is also the most susceptible to SCC. Susceptibility to SCC also depends on composition (particularly aluminum and oxygen contents) and on microstructure. These factors have been discussed in detail elsewhere.[121,122] Here we will summarize by saying that increases in aluminum and oxygen tend to increase susceptibility to SCC. For example, alloys with Al contents of 5 wt % or less tend to be immune to SCC and whereas alloys with ~6 wt % Al tend to be susceptible at oxygen contents $\geqq 0.1$ wt % and immune at oxygen levels below this. Slip length also affects susceptibility, and thus alloys with equiaxed microstructures tend to be more susceptible than the same alloys in the β-processed condition. The dependence of SCC susceptibility on microstructure also depends on the alloy in question. The reader is referred elsewhere for details.[122]

Other environments, such as N_2O_4, methanolic solutions, chlorinated hydrocarbons and others, also cause susceptibility to SCC. Again, a detailed account can be found elsewhere.[122] Elevated-temperature exposure to NaCl also causes SCC, but this is referred to as "hot salt" SCC and has been clearly shown to be a form of hydrogen embrittlement by Gray.[124] Fortunately, there is a relatively narrow temperature range over which hot salt SCC occurs. Thus, structures which operate either above or below this range appear to be relatively immune.

Finally, the creep resistance of Ti alloys deserves brief mention. Ti alloys are commonly used at temperatures up to 1000°F in gas-turbine engine applications. At these temperatures, creep frequently is the limiting factor. As a result, several α + β alloys are used nearly exclusively at elevated temperatures because they have excellent creep resistance and good, or at least acceptable, metallurgical stability during long-time, elevated-temperature service. The most common of these alloys is Ti-6Al-2Sn-4Zr-2Mo(+Si). Because essentially all of the Ti-6Al-2Sn-4Zr-2Mo alloy produced in the U.S. now contains between 0.05 and 0.10 wt % Si, the designation Ti-6Al-2Sn-4Zr-2Mo implies the Si-bearing version of this alloy. The role of Si will be further discussed below. In the open literature, creep data for Ti alloys are very limited. If, in addition, the available data are restricted to include only those which pertain to controlled microstructure, the amount of data becomes almost vanishingly small. Nevertheless, the available data show that Ti-6Al-4V has acceptable creep resistance at temperatures up to and including 400°C. Furthermore, the available data show that the BA condition is the most creep resistant and that there is essentially no difference between the MA and STA conditions. At temperatures above 400°C, Ti-6Al-2Sn-4Zr-2Mo is decidedly superior. Even in the range from 350 to 400°C, Ti-6Al-2Sn-4Zr-2Mo in the conventionally α + β –forged condition is considerably better than Ti-6Al-4V, and it is better still in the β-forged condition. Because Ti-6Al-2Sn-4Zr-2Mo is significantly better than Ti-6Al-4V for creep-limited applications, most of the available creep data pertain to this alloy. The remainder of this discussion will summarize the relationships among microstructure, composition and creep behavior for Ti-6Al-2Sn-4Zr-2Mo.

The creep behavior of Ti-6Al-2Sn-4Zr-2Mo is strongly dependent on micro-structure.[125,126] In general, β-processed material has better creep resistance than α + β – processed material. However, there are numerous combinations of heat treatment and processing each of which lead to differences in creep rate. For example, α + β solution treatment following β forging appears to consistently improve creep strength in comparison with β forging and β solution treating, α + β forging and β solution treating or α + β forging and α + β solution treating. Furthermore, forging right at or slightly above the β transus leads to improved creep strength compared with forging well above the transus. The origins of these subtle differences must lie in small microstructural variations, although, at present, detailed microstructural characterization has not been performed on each of these processing conditions. Some possible causes of these differences will be discussed later in connection with creep mechanisms.

The effect of chemical composition on creep rate also has been evaluated.[127–129] These studies have shown that Si significantly increases the creep strength of α + β Ti alloys of the Ti-6Al-2Sn-4Zr-2Mo type. In an alloy with a base composition of Ti-6Al-2Sn-2Zr-1Mo, the creep rate fell continuously, over a wide range of applied stress, with increasing Si content up to 0.5 wt %. Increasing the Mo content from 0 to 2.0 wt % also reduced the creep rate. However, when both the Si and Mo contents were varied simultaneously the optimum creep strength was observed at 1% Mo and 0.5% Si. Thus, complex interactional effects exist with respect to the various alloying elements. The same study also showed that increasing the Fe content is deleterious to creep resistance. This also has been found empirically for Ti-6Al-2Sn-4Zr-2Mo. As a result, the high-creep-strength grades of Ti-6Al-2Sn-4Zr-2Mo have specfied limits on Fe content. These maximum Fe contents vary with specification but typically lie in the range 0.05 to 0.1 wt %. In any evaluation of compositional effects it is imperative that microstructure be held constant. This requirement places limitations on the range over which the composition can be varied, because changes in Fe, Mo or Si content also change the microstructure. Among those features which can vary with composition are the equilibrium volume fraction of β phase and, possibly more important, the transformation kinetics leading to simultaneous variation of Widmanstätten α plate and packet sizes.

The mechanism of creep in Ti alloys has not been the subject of extensive investigation, at least by comparison with the Ni-base superalloys. Moreover, the maximum operating temperature for Ti is low on a homologous scale, being less than $0.4T_m$. Thus, the pertinence of the mechanisms of creep which have been extensively investigated for service temperatures greater than $0.6T_m$ is not obvious in the case of α + β Ti alloys.

Investigations of creep mechanisms[127,128,130] have shown that the principal contribution to creep strain is the thermally activated motion of dislocations (dislocation creep). In addition, boundary sliding provides a small contribution to creep strain.[127] The boundaries which show the greatest tendency for sliding are the prior β boundaries. However, the rate-controlling creep mechanism is dislocation creep.

It has been shown in Ti alloys that Si forms atmospheres around moving dislocations in the α phase during creep. These atmospheres retard the motion of the dislocations and, as a result, reduce the creep rate. Thus, the beneficial effect of Si additions on creep rate is due to reductions in the mobility of the dislocations which give rise to the creep strain.

The better creep strength in Widmanstätten microstructures than in equiaxed microstructures is less well understood. However, as described earlier, the Widmanstätten α + β microstructures consist of parallel plates of α phase separated by thin regions of β phase. In addition, each α/β interface contains a layer of interface phase. During creep by a dislocation mechanism, the dislocations must pass from one Widmanstätten α plate to the next in order to move over significant distances. In Widmanstätten microstructures this requires that the dislocations pass through the β-phase regions. Because the β phase has a different crystal structure, this requires significant adjustments in the detailed nature of the displacement field surrounding the dislocation core. These readjustments also impede the motion of the dislocations, and this also reduces creep rate. By comparison, equiaxed α + β microstructures have mostly α/α grain boundaries. The transfer of dislocations across these boundaries does not require such extensive adjustments and, therefore, should be significantly easier.

Additions of Mo improve creep resistance by increasing the β-phase volume fraction. This ensures that there is β phase between all the Widmanstätten α plates. In addition, it has been shown that Mo additions raise the athermal component of flow stress.[128] This apparently is due to solid-solution strengthening of the β phase. Elevated-temperature solid-solution strengthening by low-diffusivity solutes is greater. Thus, Mo is much more beneficial than Fe.

It also appears that there is an optimum β-phase volume fraction that is necessary for obtaining the greatest creep resistance. This optimum fraction appears to be ~5 vol %. Larger β-phase volume fractions result in wider β regions, so that these regions can deform by dislocation creep. Since Si is much more soluble in the β phase, it does not tend to form atmospheres around dislocations in this phase. Thus, the β-phase regions should have lower resistance to dislocation creep. If they are very thin, plastic constraint from the adjacent α-phase plates retards creep deformation. This description of the effect of β-phase volume fraction, while somewhat speculative, does account for the apparent maximum in creep strength that is observed at small β-phase volume fractions.

The Future of TMP

The foregoing has made it clear that TMP is used routinely in existing commercial practice to enhance the properties of high-strength Al and Ti alloys. So what does the future hold? While we do not claim to be clairvoyant, several logical directions for continued development seem obvious. These include the following:

- Combination of TMP with net shapemaking through more sophisticated die design.
- Use of microprocessors to provide better control of TMP process.
- Inclusion of texture control in commercial practice.
- Combination of powder consolidation with TMP.

The benefits of net shapemaking are based on a reduction in the amount of material which must be machined away from a forging in order to produce a component of proper final dimensions. This is sometimes described as an improved buy: fly ratio. Two benefits accrue from such an improvement: the amount of wasted material is reduced, and machining costs are lowered. In the case of Ti alloys the cost savings can easily reach $45 to $50 per pound of forging weight eliminated. This breaks down into cost reductions of $15 to $20 for the forging and $25 to $30 per pound in machining costs. Thus, the benefits are real and attractive, but such shapemaking cannot be achieved at the expense of introducing nonuniform ε-T-t profiles at various locations in the forging in an uncontrolled manner. Recognizing this, Gegel[131] has made considerable progress in improving die design using the principles of large plastic flow and improved die-lubrication technology. In fact, a major program of study in this area has been in progress for several years. One of the exciting aspects of this activity is that it demonstrates that ε-T-t profiles can be controlled at various locations in a forging so as to obtain, through establishment of controlled microstructural gradients, desired variations in properties from location to location in the forging. Although these techniques are not currently used in commercial practice, the benefits which they appear to offer certainly make them attractive.

One deterrent to implementation of net-shape technology exists — the frequent requirement for 100% sonic inspection of forgings made for critical applications. Such applications require a "sonic envelope" which is free of sharp corners and re-entrant angles that can give spurious reflections. Although there exist computer-based sonic inspection systems which are capable of dealing with this problem, the economics of their implementation is still unsatisfactory. Thus, the sonic-envelope requirement presently places a limit on the extent to which net shapes can be usefully employed.

The use of microprocessors to control and reproduce TMP cycles is another aspect of production technology which has been demonstrated in the laboratory but which has not been implemented in commercial practice. The benefits of this technology appear to be substantial, especially when alloys sensitive to modest variations in the ε-T-t TMP cycle are being processed. The ability to provide improved properties on a reproducible basis is indeed attractive. The advancing technology of robotics would make the commercialization of this capability seem reasonably near at hand. Moreover, this capability complements the shapemaking technology mentioned earlier.

The ability to produce materials with controlled textures is reliant on the availability of knowledge regarding the relations between the ε-T-t profile and the resulting texture. For most alloys this is now fairly well known. Thus, implementation of microprocessor controls for forging appear to make control of texture feasible. However, the variations in local flow patterns which accompany net shapemaking may interfere with the achievement of the desired texture. Here again, an examination of the trade-offs which exist between conflicting goals must be examined on a case-by-case basis.

A strong renewal of interest in high-performance components made by powder metallurgy techniques has led to the promulgation of new alloys and technologies for making and consolidating powder. To this end, the prospect of making alloys by powder techniques to avoid the damaging segregation that can occur in ingots is attractive. In addition, the higher solidification rates obtainable in powders offer the prospect of novel microstructures. However, the current practices of very-high-temperature (on a homologous scale) compaction tend to negate some of these benefits. Thus, concurrent consolidation and TMP at lower temperatures is an attractive proposition. Work is currently being conducted in this area at a number of laboratories. The outcome of this work may be a new technology for combining the benefits of TMP with those of new alloy compositions which can only be made by powder metallurgy techniques. This, indeed, appears interesting, but assessment of its true benefits must await further work.

Summary

In this article we have attempted to discuss the factors which lead to variations in microstructure and texture. We also have summarized the effects of microstructure and texture on properties of Al and Ti alloys. The information presented has made it clear that TMP is an integral part of modern processing technology for high-performance materials. Finally, we have closed by speculating about some fruitful areas for further development of TMP.

Acknowledgments

We wish to acknowledge several of our collaborators for allowing us to cite their work in this paper—specifically, Dr. M. Peters, Dr. G. Terlinde and Dr. J. Allison. One of us (JCW) also gratefully acknowledges partial support by the Office of Naval Research during the preparation of this article.

References

1. G. Gottstein, *Textures of Materials,* edited by G. Gottstein and K. Lücke, Vol 1, Springer-Verlag, Berlin, 1978, p 93.
2. P. Cotterill and P. R. Mould, *Recrystallization and Grain Growth in Metals,* John Wiley & Sons, New York, 1976.

3. P. L. Morris and M. D. Ball, "Recrystallization and Growth of Multiphase and Particle Containing Materials," Risø Conference, 1980, p 97.

4. N. Ryum, *J. Inst. Met.*, Vol 94, 1966, p 191.

5. G. Beghi, M. Grin and G. Piatti, *Mem. Sci. Rev. Met.*, Vol 64, 1967.

6. E. Henry Chia and E. A. Starke, Jr., *Met. Trans.*, Vol 8A, 1977, p 825.

7. C. Zener, private communication to C. S. Smith, *Trans. AIME*, Vol 175, 1949, p 15.

8. E. Nes, Recrystallization and Grain Growth of Multiphase and Particle Containing Materials, Risø. Conference, 1980, p 85.

9. S. S. Rystad and N. Ryum, *Aluminium*, Vol 53, 1977, p 193.

10. E. A. Starke, Jr., T. H. Sanders, Jr., and I. G. Palmer, *J. Met.*, Vol 33, No. 8, 1981, p 24.

11. A. S. Argon, J. Im and R. Safoglu, *Met. Trans.*, Vol 6A, 1975, p 625.

12. R. Sandstrom and J. Lindgren, *Met. Sci. Eng.*, 1980.

13. H. J. McQueen, Concordia University, private communication on unpublished research, 1982.

14. John A. Wert, *J. Met.*, Vol 34, No. 9, 1982, p 35.

15. F. J. Humphreys, *Acta Met.*, Vol 15, 1977, p 1323.

16. John A. Wert, N. E. Paton, C. H. Hamilton and M. W. Mahoney, *Met. Trans.*, Vol 12A, 1981, p 1267.

17. H. Hu, R. S. Cline and S. R. Goodman, *Recrystallization, Grain Growth and Texture*, ASM, Metals Park, OH, 1966, p 295.

18. P. A. Beck and H. Hu, *Recrystallization, Grain Growth and Texture*, ASM, Metals Park, OH, 1966, p 393.

19. I. L. Dillamore and W. T. Roberts, *Met. Rev.*, Vol 10, No. 39, 1965, p 271.

20. W. B. Hutchinson, *Met. Sci.*, Vol 8, 1974, p 185.

21. H. P. Stüwe, *Z. Metallk.*, Vol 52, 1961, p 34.

22. H. Hu, *Textures in Research and Practice*, edited by J. Grewen and G. Wassermann, Springer-Verlag, Berlin, 1969, p 200.

23. I. L. Dillamore and H. Katoh, *Met. Sci.*, Vol. 8, 1974, p 73.

24. C. D. Graham and R. W. Cahan, *Trans. AIME, J. Met.*, Vol 8, 1969, p 517.

25. Victor W. C. Kuo, unpublished research, Fracture and Fatigue Research Laboratory, Georgia Institute of Technology, Atlanta, GA, 1982.

26. D. L. Robinson, unpublished work, Alcoa Laboratories, 1982.

27. N. E. Paton and A. W. Sommer, "Influence of Thermomechanical Processing Treatments on Properties of Aluminum Alloys," Proceedings of the 3rd International Conference on Strength of Metals and Alloys, Cambridge, England, 1973, p 101.

28. I. J. Polmear, *Light Alloys: Metallurgy of the Light Metals*, Edward Arnold, Ltd., London, England, 1981.

29. E. DiRusso, M. Conserva, M. Buratti and F. Gatto, *Mater. Sci. Eng.*, Vol 14, 1974, p 23.

30. J. Waldman, H. Sulinski and H. Markus, *Met. Trans.*, Vol 5, 1974, p 573.

31. B. K. Park and J. E. Vruggink, *Thermomechanical Processing of Aluminum Alloys*, edited by J. G. Morris, TMS-AIME, Warrendale, PA, 1978, p 25.

32. N. E. Paton, C. H. Hamilton, J. Wert and M. Mahoney, *J. Met.*, Vol 34, No. 8, 1982, p 21.

33. B. Fowler, C. P. Cutler and J. W. Edington, *J. Inst. Met.*, Vol 101, 1973, p 332.

34. J. W. Edington, K. N. Melton and C. P. Cutler, *Prog. Mater. Sci.*, Vol 21, No. 3, 1976, Pergamon Press, Oxford, England.

35. Oleg D. Sherby and Jeffrey Wadsworth, "Development and Characterization of Fine-Grained Superplastic Materials," *Deformation, Processing and Structure*, ASM Materials Science Seminar, St. Louis, October, 1982.

36. Roy E. Crooks, Ph.D. Thesis, Georgia Institute of Technology, August, 1982.
37. E. O. Hall, *Proc. Phys. Soc.*, Vol 64, 1951, p 741.
38. N. J. Petch, *J. Iron Steel Inst.*, Vol 197, 1953, p 25.
39. A. W. Thompson, *Met. Trans.*, Vol 8A, 1977, p 833.
40. R. B. Nicholson, *Strengthening Methods in Crystals*, edited by A. Kelly and R. B. Nicholson, Applied Science Publishers, Ltd., London, England, 1971, p 535.
41. R. E. Sanders, Jr., and E. A. Starke, Jr., in *Thermomechanical Processing of Aluminum Alloys*, edited by J. G. Morris, TMS-AIME, Warrendale, PA, 1979, p 50.
42. R. D. Carter, E. W. Lee, E. A. Starke, Jr., and C. J. Beevers, "An Experimental Investigation of the Effects of Microstructure and Environment on Fatigue Crack Closure of 7475" (to be published).
43. J. Wert, in *Strength of Metals and Alloys*, edited by R. C. Gifkins, 6th International Conference on the Strength of Metals and Alloys, Vol 1, Pergamon Press, Sydney, Australia, 1982, p 339.
44. R. Penelle, in *Textures of Materials*, edited by G. Gottstein and K. Lücke, Vol II, Springer-Verlag, Berlin, 1978, p 129.
45. I. G. Palmer, R. E. Lewis and D. D. Crooks, *Aluminum-Lithium Alloys*, edited by T. H. Sanders, Jr., and E. A. Starke, Jr., The Metallurgical Society of AIME, Warrendale, PA, 1981.
46. G. Lütjering, T. Hamajima and A. Gysler, *Fracture 1977*, Vol 2, ICF4 Waterloo, Canada, p 7.
47. T. H. Sanders, Jr. and E. A. Starke, Jr., *Acta Met.*, Vol 30, 1982, p 927.
48. G. Terlinde and G. Lütjering, *Met. Trans.*, Vol 13A, 1982, p 1283.
49. A. Gysler, R. Crooks and E. A. Starke, Jr., *Aluminum-Lithium Alloys*, edited by T. H. Sanders, Jr., and E. A. Starke, Jr., The Metallurgical Society of AIME, Warrendale, PA, 1981, p 263.
50. E. A. Starke, Jr., and F. S. Lin, "The Influence of Grain Structure on the Ductility of the Al-Cu-Li-Mn-Cd Alloy 2020," *Met. Trans. A* (in press).
51. F. S. Lin, S. B. Chakrabortty and E. A. Starke, Jr., *Met. Trans.*, Vol 13A, 1982, p 1401.
52. Fu-Shiong Lin, "The Effect of Grain Structure on the Fracture Behavior and Tensile Properties of an Al-Li-Cu Alloy," *Scripta Met.* (in press).
53. R. E. Sanders, Jr., and E. A. Starke, Jr., *Met. Trans.*, Vol 9A, 1978, p 1087.
54. Edgar A. Starke, Jr., and Gerd Lütjering, *Fatigue and Microstructure*, edited by M. Meshii, ASM, Metals Park, OH, 1979, p 205.
55. Fu-Shiong Lin and E. A. Starke, Jr., *Mater. Sci. Eng.*, Vol 43, 1980, p 65.
56. J. C. Williams and M. J. Blackburn, "A Comparison of Phase Transformations in Three Commercial Titanium Alloys," *ASM Quart. Trans.*, Vol 60, 1967, p 373.
57. J. C. Williams, "Phase Transformations in Titanium Alloys: A Review," in *Titanium Science and Technology*, Vol 3, edited by R. I. Jaffee and H. M. Burte, Plenum Press, 1973, p 1433.
58. J. C. Williams, "Precipitation in Titanium Base Alloys," in *Precipitation Processes in Solids*, edited by K. C. Russell and H. I. Aaronson, AIME, New York, 1976, p 191.
59. J. C. Chesnutt, C. G. Rhodes and J. C. Williams, "The Relationship Between Mechanical Properties, Microstructure and Fracture Topography in $\alpha + \beta$ Titanium Alloys," ASTM STP 600, ASTM, Philadelphia, 1976, p 99.
60. J. C. Williams, "Phase Transformations in Ti Alloys – A Review of Recent Developments," Critical Review, *Proc. 3rd Int. Conf. on Titanium*, Moscow, USSR, edited by J. C. Williams and A. F. Belov, Plenum Press, 1982, p 1477-1498.
61. M. Peters, Ph.D. Thesis, Ruhr-University Bochum, FRG, 1980.

62. M. Peters, G. Luetjering and G. Ziegler, *Z. Metallk.* (in press).
63. J. C. Williams, F. H. Froes, C. F. Yolton and I. M. Bernstein, "The Influence of Thermomechanical Processing on the Microstructure of Metastable β-Ti Alloys," *Proc. 4th Int. Conf. on Strength of Metals and Alloys,* Vol 2, 1976, p 639.
64. F. H. Froes, C. F. Yolton, J. M. Capenos, M. G. H. Wells and J. C. Williams, "The Relationship Between Microstructure and Age Hardening Response in the Metastable Beta Titanium Alloy Ti-11.5Mo-6Zr-4.5Sn (β-III)," *Met. Trans.,* Vol 11A, 1980, p 21-31.
65. J. C. Williams, F. H. Froes and C. F. Yolton, "Some Observations on the Structure of Ti-11.5Mo-6Zr-4.5Sn (Beta III) as Affected by Processing History," *Met. Trans.,* Vol 11A, 1980, p 356-358.
66. T. W. Duerig, Ph.D. Thesis, Carnegie-Mellon University, Pittsburgh, PA, 1980.
67. T. W. Duerig, G. T. Terlinde and J. C. Williams, "Phase Transformations and Tensile Properties of Ti-10V-2Fe-3Al," *Met. Trans.,* Vol 11A, 1980, p 1987.
68. F. R. Larson and A. Zarkades, "Properties of Textured Ti Alloys," MCIC-74-20, June 1974.
69. A. W. Bowen, *Acta Met.,* Vol 23, 1975, p 1401.
70. C. A. Stubbington, *AGARD Conf. Proc.* #185, 1976, p 3-1.
71. P. J. E. Forayth and C. A. Stubbington, *Met. Tech.,* April, 1975, p 158.
72. S. F. Frederick, AFML TR-73-265, 1973.
73. See for example: S. S. Manson, *Thermal Stresses and Low Cycle Fatigue,* McGraw-Hill, New York, 1966, p 132 et seq.
74. S. M. Copley and J. C. Williams, *Alloy Design,* edited by J. Tien and G. Ansell, Academic Press, 1976, p 3.
75. N. E. Paton, G. P. Rauscher and J. C. Williams, *Titanium Science and Technology,* edited by R. I. Jaffee and H. M. Burte, Plenum Press, Vol 2, 1973, p 1049.
76. J. C. Williams and A. W. Thompson, "Strengthening of Metals and Alloys," in *Metallurgical Treatises,* edited by J. K. Tien and J. F. Elliot, TMS-AIME, Warrendale, PA, 1981, p 487.
77. M. J. Blackburn and J. C. Williams, "Strength, Deformation Modes and Fracture in Ti-Al Alloys," *ASM Quart. Trans.,* Vol 62, 1969, p 398.
78. N. E. Paton, J. C. Williams, J. C. Chesnutt and A. W. Thompson, *AGARD Conf. Proc.* #185, 1976, p 4-1.
79. R. B. Sparks and J. R. Long, AFML-TR-73-301, February 1974.
80. J. C. Chesnutt, A. W. Thompson and J. C. Williams, AFML-TR-78-68, 1978.
81. R. R. Boyer and R. Bajoraitis, Boeing Document D6-48016, July 1978.
82. Cf. "Facts About Ti-6Al-4V," RMI Company Publication, Niles, OH, 1967.
83. M. M. Allen, AFML TR-71-78, February 1976.
84. K. Okazaki and H. Conrad, "Deformation Mechanisms," Critical Review, *Proc. 3rd Int. Conf. on Titanium,* Moscow, USSR, edited by J. C. Williams and A. F. Belov, Plenum Press, 1982, p. 429-466.
85. J. C. Williams, unpublished research, 1974.
86. A. W. Sommer and M. Creager, AFML-TR-76-222, January 1977.
87. C. C. Chen and R. R. Boyer, *J. Met.,* Vol 31, 1979, p 33.
88. G. T. Terlinde, T. W. Duerig and J. C. Williams, "The Effect of Heat Treatment on Microstructure and Tensile Properties of Ti-10V-2Fe-3Al," *Titanium '80, Science and Technology,* edited by H. Kumura and O. Izumi, Vol 2, TMS-AIME, 1981, p 1571.
89. F. H. Froes, J. C. Chesnutt, C. G. Rhodes and J. C. Williams, "Relationship of Fracture Toughness and Ductility to Microstructure and Fractographic Features in Advanced Deep Hardenable Titanium Alloys," in *Toughness and Fracture Behavior of Titanium,* ASTM STP 651, July 1978, p 115.

90. J. C. Williams, F. H. Froes, J. C. Chesnutt, C. G. Rhodes and R. G. Berryman, "Development of High Fracture Toughness Titanium Alloys," in *Toughness and Fracture Behavior of Titanium,* ASTM STP 651, July 1978, p 64.

91. R. R. Boyer, *J. Met.,* Vol 32, 1980, p 61.

92. H. Margolin, J. C. Chesnutt, G. Luetjering and J. C. Williams, "Fracture Fatigue and Wear: Critical Review," in *Titanium '80, Science and Technology,* edited by H. Kimura and O. Izumi, Vol 1, TMS-AIME, 1981, p 169.

93. A. W. Bowen, RAE TR 76150, 1976.

94. J. C. Chesnutt and J. C. Williams, unpublished research, 1973.

95. B. Tomkins, *Proc. Storrs Conf. on Corrosion Fatigue,* 1971.

96. M. Peters, A. Gysler and G. Luetjering, in *Titanium '80, Science and Technology,* edited by H. Kimura and O. Izumi, Vol 1, TMS-AIME, 1981, p 1777.

97. J. J. Lucas, in *Titanium Science and Technology,* edited by R. I. Jaffee and H. M. Burte, Plenum Press, 1973, p 2081.

98. P. C. Kelly and J. C. Williams, unpublished research, 1977.

99. C. A. Stubbington and A. W. Bowen, *J. Mater. Sci.,* Vol 9, 1974, p 941.

100. J. C. Williams and G. Luetjering, in *Titanium '80, Science and Technology,* edited by H. Kimura and O. Izumi, Vol 1, TMS-AIME, 1981, p 671.

101. G. R. Leverant, B. S. Langer, A. Yuen and S. W. Hopkins, *Met. Trans.,* Vol 10A, 1979, p 251.

102. R. A. Sprague, D. L. Ruckle and M. P. Smith, in *Titanium Science and Technology,* R. I. Jaffee and H. M. Burte, Plenum Press, 1973, p 2069.

103. C. M. Gilmore and M. A. Imam, Report 3-B, The George Washington University, Washington, DC, August 1976.

104. A. W. Bowen, in *Titanium Science and Technology,* edited by R. I. Jaffee and H. M. Burte, Plenum Press, 1973, p 1271.

105. D. Eylon and C. M. Pierce, *Met. Trans.,* Vol 7A, 1976, p 111.

106. J. C. Williams, N. E. Paton, P. J. Stocker and H. L. Marcus, "Space Shuttle Materials," SAMPE, 1971, p 643.

107. N. E. Paton, J. C. Williams, J. C. Chesnutt and A. W. Thompson, *AGARD Conf. Proc.* #185, 1976, p 4-1.

108. G. R. Yoder, L. A. Cooley and T. W. Crooker, *Met. Trans.,* Vol 8A, 1977, p 1737.

109. G. R. Yoder, L. A. Cooley and T. W. Crooker, *J. Eng. Mater. Tech.,* October 1977, p 313.

110. J. C. Chesnutt, A. W. Thompson and J. C. Williams, "Fatigue Crack Propagation and Fracture of Titanium Alloys," in *Titanium '80, Science and Technology,* edited by H. Kimura and O. Izumi, Vol 3, 1981, p 1875.

111. G. R. Yoder, L. A. Cooley and T. W. Crooker, in *Titanium '80, Science and Technology,* edited by H. Kimura and O. Izumi, Vol 3, TMS-AIME, 1981, p 1865.

112. P. E. Irving and C. J. Beevers, *Mater. Sci. Eng.,* Vol 14, 1974, p 229.

113. See for example: R. O. Ritchie, *Met. Sci.,* August/September 1977, p 368.

114. P. C. Paris and F. Erdogan, *Trans. ASME,* Series D, Vol 85, 1963, p 528.

115. G. T. Gray, J. C. Williams and A. W. Thompson, "Roughness-Induced Crack Closure: An Explanation for Micostructurally Sensitive Fatigue Crack Growth," *Met. Trans.,* 1982, (in press).

116. J. E. Allison, Ph.D. Thesis, Carnegie-Mellon University, Pittsburgh, 1982.

117. J. E. Allison and J. C. Williams, *Proceedings of ICSMA Conference,* Australia, 1982.

118. G. M. Ludtka, Ph.D Thesis, Carnegie-Mellon University, Pittsburgh, 1981.

119. S. B. Chakrabortty and E. A. Starke, Jr., "Fatigue Crack Propagation of Metastable Beta Titanium-Vanadium Alloys," *Met. Trans.,* Vol 10A, 1979, p 1901.

120. G. A. Alers, unpublished research, Rockwell Science Center, 1970.

121. M. J. Blackburn and J. C. Williams, in *Proc. of Conf. on Fundamental Aspects of Stress Corrosion Cracking*, edited by R. W. Staehle et al., NACE, Houston, TX, 1969, p 620.

122. M. J. Blackburn, W. H. Smyrl and J. A. Feeney, in *Stress Corrosion Cracking in High Strength Steels and in Titanium and Aluminum Alloys*, edited by B. F. Brown, Naval Research Laboratory, Washington, DC, 1972, p 246.

123. D. N. Fager and W. F. Spurr, Trans. ASM, *245*, 283 (1968).

124. H. R. Gray, *Corrosion*, Vol 25, 1969, p 337.

125. C. C. Chen, Wyman-Gordon Company Report RD-88-110, October 1977.

126. C. C. Chen, Wyman-Gordon Company Report RD-77-111, October 1977.

127. N. E. Paton, M. W. Mahoney and J. C. Williams, Final Report SC532.8FR, July 1973.

128. M. W. Mahoney and N. E. Paton, Final Report SC5003.7FR, September 1975.

129. M. W. Mahoney, N. E. Paton, W. M. Parris and J. A. Hall, AFML-TR-77-56, May 1977.

130. N. E. Paton, M. W. Mahoney and J. C. Williams, Final Report SC564.7FR, August 1974.

131. H. L. Gegel, private communication, 1982.

Development and Characterization of Fine-Grain Superplastic Materials

Oleg D. Sherby and Jeffrey Wadsworth*

Abstract

Superplasticity was observed 70 years ago, and it remained a scientific curiosity until about 20 years ago. In recent years, however, a number of fine-grain superplastic materials have been developed and have found commercial application. The remarkable formability of superplastic materials is due to their high strain-rate sensitivity. The high strain-rate sensitivity is associated with grain-boundary sliding as the principal mechanism of plastic flow. The microstructural features that are important for developing superplastic structures include: (*a*) size of grains; (*b*) amount, strength, size and distribution of second phase; (*c*) nature, mobility and resistance to tensile separation of grain boundaries; and (*d*) grain shape. Phenomenological equations for fine-structure superplastic flow, incorporating temperature, grain-size and stress dependences, are well developed and are described. Methods for enhancing superplastic flow are discussed, and it is shown that optimization of superplastic properties of fine-grain materials can be achieved by promoting grain-boundary sliding and by inhibiting slip.

I.   Introduction

A principal characteristic of polycrystalline metallic solids is their ability to behave plastically. That is, they can deform permanently, under external forces, with negligible formation of cracks or voids. This capacity for plastic flow is associated with three discrete mechanisms that can occur at the atomic level. These mechanisms are (*a*) slip by dislocation movement, (*b*) sliding of adjacent grains along grain boundaries, and (*c*) directional diffusional flow. Examples of photomicrographs that reflect each of these mechanisms of plastic flow are presented in Fig. 1 (twinning is yet another mechanism of plastic flow but will

*O. D. Sherby is Professor at the Department of Materials Science and Engineering, Stanford University, Stanford, CA 94305, and J. Wadsworth is Research Scientist, Lockheed Palo Alto Research Laboratory, 3251 Hanover St., Palo Alto, CA 94304.

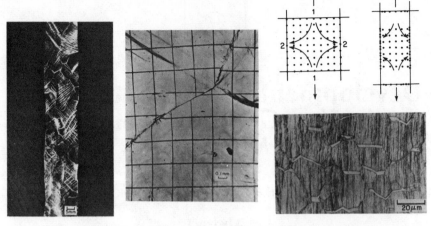

Fig. 1. Photomicrographs showing the three principal mechanisms of plastic flow observed in polycrystalline metallic solids. (Left) Slip bands in coarse-grain pure molybdenum deformed at 1000°C. (Center) Grain-boundary sliding in high-purity polycrystalline aluminum deformed at 200°C.[1] (Right) Zones denuded of precipitates (magnesium hydride) in polycrystalline magnesium after high-temperature tensile deformation;[2] these denuded zones are believed to result as a consequence of plastic deformation by diffusional flow, as illustrated by the schematic diagram above the photomicrograph.

not be covered here). All three mechanisms are generally considered to occur independently of one another. For the case of large plastic strains, these mechanisms are all thermally activated and are controlled by the diffusion of atoms. Studies in the last two decades have led to the development of constitutive equations to describe each of these mechanisms of plastic flow.[3,4]

Thus, the flow stress, σ (stress to cause plastic flow), can be described as a function of strain rate, $\dot{\varepsilon}$, and absolute temperature, T, in the following functional form:

$$\sigma = \left[\dot{\varepsilon} \, \exp \frac{Q}{RT} \right]^{m} E \, f(S) \tag{1}$$

where Q is the activation energy for plastic flow; R is the gas constant; m is the strain-rate-sensitivity exponent; E is the dynamic, unrelaxed, elastic modulus; and f(S) is some function of structure (principally reflecting the influences of grain size, subgrain size and dislocation density). Each mechanism of plastic flow has specific values of m and Q by which the mechanism can be defined uniquely. For example, plastic flow by slip is associated with a low strain-rate sensitivity ($m \simeq 0.1$) and an activation energy which can be related either to lattice diffusion, Q_L, or to dislocation pipe diffusion, Q_P. Plastic deformation by grain-

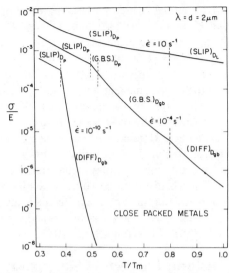

Fig. 2. Modulus-compensated yield strength as a function of homologous temperature at three different strain rates for a fine-grain, close-packed polycrystalline metal (d = grain diameter, λ = subgrain diameter). Breaks in the curves indicate changes in deformation mechanisms.

boundary sliding is characterized by a high value of strain-rate sensitivity (m ≃ 0.5) and an activation energy which is either equal to Q_L or to the activation energy for grain-boundary diffusion, Q_{gb}. Plastic deformation by diffusional flow is characterized by a strain-rate-sensitivity value of unity and an activation energy equal to Q_L or Q_{gb}.

Superplasticity refers to large tensile elongations, typically 500%, that can be achieved in polycrystalline materials under certain conditions of strain rate and temperature. These high elongations are directly related to the high strain-rate-sensitivity exponents exhibited by superplastic materials. The high values of strain-rate sensitivity result in the inhibition of the growth of "necks" during tensile deformation. Typical values of m for superplastic materials are 0.5 or greater, thus placing them in the category where grain-boundary sliding or diffusional flow dominates the deformation process.[*] Constitutive equations can be used to determine the ranges of strain rate and temperature over which superplasticity can be expected. An example illustrating this principle is given in Fig. 2. Here the flow stress (yield strength), normalized by the elastic modulus, is plotted as a function of the homologous temperature, T/T_m, for a typical

---

*There is another type of superplasticity which is based on a slip mechanism and relies on the development of internal stress. This is known as "internal stress" or "environmental" superplasticity and will not be discussed here.

fine-grain close-packed material. Three curves, representing three strain-rate conditions, are shown: $\dot{\varepsilon} = 10$ s^{-1}, 10^{-4} s^{-1} and 10^{-10} s^{-1}. The breaks shown in each curve represent changes in the mechanism of plastic flow. Regions where grain-boundary sliding (m = 0.5) or diffusional flow (m = 1.0) dominate the deformation process are regions where superplasticity can be expected.

This paper deals with the development and characterization of superplastic fine-grain materials. The following two sections (II and III) give a brief historical review of superplasticity followed by some commercial examples of superplastic materials. The next two sections (IV and V) discuss the mechanics and structural requirements for superplasticity. This is followed by a review of phenomenological equations for describing superplastic materials (Sec. VI) and then by a review of the physical mechanisms proposed for superplastic flow (Sec. VII). An important problem in utilizing superplastic materials is the need to maximize the strain rate at which superplastic flow can occur, and methods of achieving this goal are discussed in the last section (Sec. VIII).

II. Historical Aspects of Superplasticity

Pearson[5] dramatically demonstrated in 1934 that unusually large elongations could be achieved in certain fine-structure, two-phase materials. As a result, he is often credited with having first demonstrated superplasticity. The first observation of the phenomenon of superplasticity, however, was much earlier than 1934. In a written response, appended to a paper by Rosenhain and Ewen[6] on the "amorphous cement" theory in 1912, Bengough reported that "a certain special brass which he had examined pulled out to a fine point, just like glass would do, having an enormous elongation." Examination of the original work by Bengough,[7] also published in 1912, reveals that the "special brass" was in fact an α/β brass and exhibited an elongation of nearly 200% at 700°C. As far as the present authors can establish, this work is the first reference to the phenomenon of superplasticity. Other observations regarding viscous-like behavior in fine-grain metals can occasionally be found in the early literature (see, for example, Rosenhain, Haughton and Bingham[8] in 1920, and Jenkins[9] in 1928). In addition, there are early references to extreme plasticity during deformation through a phase change (see, for example, Sauveur[10] in 1924).

It was in the USSR that Bochvar and Sviderskaya[11] first coined the term "sverhplastichnost" (ultrahigh plasticity) in their studies on superplastic Al-Zn alloys in 1945. Although occasional papers appeared on the subject after this date, a significant interest in superplasticity did not develop until after Underwood's[12] review article in 1962 on work in the USSR. Since that time, research on the development of superplastic alloys and their commercial applications has been extensive. A number of monographs, three by Western scientists[13-15] and six by Soviet researchers,[16-21] have been published. In addition, numerous review articles have been written.[12,22-38] In 1982 a conference was held on "Superplastic Forming of Structural Alloys" and the published proceedings[39] (containing 19 invited papers and 58 submitted abstracts) demonstrate the great interest, both from academic as well as commercial viewpoints, in the subject of fine-structure superplasticity.

The increase in commercial interest in superplastic materials reflects the fact that such materials exhibit low resistance to plastic flow (in specific temperature and strain-rate regimes) as well as high plasticity. This combination of properties is ideal because it provides the potential for a complex shape to be formed with a minimum expenditure of energy. In addition, the fine structure is often of benefit to service properties in the finished product. The feasibility of the commercial application of superplasticity has been recently reviewed in Ti-,[40] Ni-,[41] Al-[42,43] and Fe-base[44] alloys.

III.   Commercial Examples of Superplasticity

An example of superplastic forming carried out at Rockwell International using a Ti-6Al-4V alloy is shown in Fig. 3.[45] The component shown is a nacelle center-beam frame. (A number of such parts form a structure in the B-1 aircraft.) In this example, a single superplastic forming and diffusion bonding operation was designed to replace a production route which had involved the forming of eight separate pieces of the same alloy which then had to be joined together with 96 fasteners. Estimated cost savings of 55% and weight savings of 33% were estimated using this fabrication route compared with the conventional production technique.

A similar and equally impressive superplastic forming operation, developed by Pratt and Whitney, Florida,[46] is shown in Fig. 4. In this case, a superplastic, nickel-base superalloy (IN 100), prepared either by conventional or powder metallurgy processing, is formed into the desired shape using a two-step forming operation. In the example of Fig. 4, the billet is made by hot isostatic pressing (HIP'ing) of powders, made fine-structured by liquid atomization. As a first step, the billet is warm pressed in a die to the configuration of a disk (Fig. 4, center). In the second step, this intermediate part is superplastically formed into the complex disk with turbine blades shown at right in Fig. 4. The fine structure of the IN 100 is an advantage in the body of the disk during operation but a disadvantage in the blades, where a high temperature is encountered in service. For this reason, the part of the disk containing the blades is selectively heat treated after forming, in a manner which creates a coarse grain size in the blades. Such a treatment considerably improves the creep-resistance characteristics of the blades but leaves a tough, fatigue-resistant fine-grain material in the body of the disk.

An example of superplastic forming of an ultrahigh-carbon (UHC) steel is shown in Fig. 5. These steels have been examined in detail in recent years for superplastic formability and for room-temperature properties.[47–59] In the present

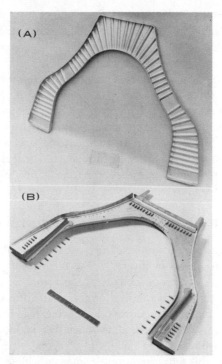

Fig. 3. Nacelle center-beam frame of alloy Ti-6Al-4V, demonstrating the unique forming capabilities of a superplastic material. Eight such frames form the structure separating the engines on the B-1 aircraft. Formerly (B), each frame was made by forming eight separate pieces of the alloy which then had to be joined with 96 fasteners. The superplastic forming route, A, (carried out at 920°C), which also uses diffusion bonding, results in considerable weight and cost savings. *Courtesy of H. Hamilton and E. Weisert, Rockwell International, Thousand Oaks, CA.*

example, a steel casting, containing 1.6 wt % C, was liquid atomized, and the resulting fine-structure powders were warm pressed into a billet at 815°C. The billet was warm forged to a plate shape, and the plate was then superplastically formed to the final shape shown in Fig. 5. In the case of UHC steels, the fine structure developed for superplasticity is also of considerable benefit to ambient-temperature properties. This is demonstrated in Fig. 6, in which a UHC steel, processed to have two different microstructures, is compared both at intermediate temperature and at low temperature. This first microstructure has a ferrite grain size of about 1 μm, contains cementite particles 0.1 to 0.5 μm in diameter and is superplastic at warm temperatures. The other microstructure, which is not superplastic at intermediate temperatures, has a grain size of about 20 μm and contains cementite particles about 2 μm in diameter. The fine-structure steel

Fig. 4. Net shapeforming of an ultrafine-grain-size, nickel-base alloy by superplastic forming in two stages. (Left) Original powder metallurgy IN 100 billet. (Center) Powder metallurgy billet pressed into disk shape. (Right) Disk-shape billet superplastically pressed into disk and turbine blades. *Courtesy of J. Moore and R. Athey, Pratt and Whitney, Florida (AF Contract F33615-72-C-2177).*

Fig. 5. An ultrahigh-carbon (UHC) steel containing 1.6% C was superplastically formed at 815°C to the shape shown above. The processing procedure consisted of warm consolidation (800°C) of liquid-atomized powders and forging the resulting billet into plate shape. The plate shape was then superplastically formed to the final shape shown at top. Bottom photo is a cross section of the part.

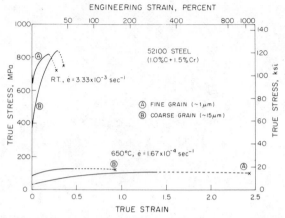

Fig. 6. Comparison of a 52100 bearing steel, made superplastic by a thermomechanical processing route, with an "as-received" 52100 steel at 650°C and at room temperature. The fine structure of the processed steel (ferrite grain size ~1 μm, carbide size ~0.1 to 0.5 μm) leads to considerably better properties than those of the as-received steel (ferrite grain size ~20 μm, carbide size ~2 μm) at both intermediate and ambient temperatures.

exhibits ideal characteristics in that it is weak and superplastic at intermediate temperatures (650°C) and strong and ductile at room temperature (specimens marked "A" in Fig. 6). This is in contrast to the coarse-structure steel which is strong and not very ductile at warm temperature, and relatively weak at room temperature (specimens marked "B" in Fig. 6).

Aluminum alloys have also been commercially utilized for superplastic forming operations in recent years. A wide variety of parts have been manufactured using Al-Cu-Zr alloys,[42,60,61] an Al-Ca-Zn alloy has been developed,[42,62] and high-strength 7000 series aluminum alloys have been made superplastic.[43,63]

In the next section, the phenomenon of superplasticity is described from a mechanics viewpoint.

IV.   Mechanics Description of Superplasticity

In this section, the phenomenon of fine-structure superplasticity is described from the viewpoint of applied mechanics.

Superplastic metallic alloys have one primary property in common. The strength (resistance to plastic flow) of such materials is highly strain-rate sensitive — that is, σ, the flow stress, increases rapidly with $\dot{\varepsilon}$, the strain rate. A measure of this rate sensitivity of strength is given by the relation

$$\sigma = K\dot{\varepsilon}^m \tag{1}$$

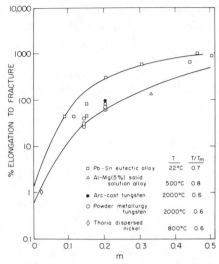

Fig. 7. Metals and alloys show high strain-rate sensitivity at high temperatures. This rate sensitivity is usually measured by the exponent m in the phenomenological equation $\sigma = K\dot{\varepsilon}^m$. The graph above shows that ductility (measured by elongation to fracture) increases as the strain-rate-sensitivity exponent increases. Superplastic metals and alloys generally exhibit strain-rate-sensitivity exponents ranging from 0.4 to 0.6 (m $\simeq$ 0.5). *(Data from Ref 65–69.)*

where K is a material constant and m is the strain-rate-sensitivity exponent.* The value of m determines the rate at which the neck progresses after localized plastic flow starts. It is an experimental fact that as m, the strain-rate-sensitivity exponent, increases, the elongation to failure increases.[64] Data illustrating the relationship between m and elongation are plotted in Fig. 7 for a variety of materials.[65-69] The elongation to failure is found to increase to large values as m approaches a value of about 0.5. When the relation between σ and $\dot{\varepsilon}$ is linear (m = 1), the material is said to behave in a Newtonian-viscous manner: hot glass, tar and well-masticated chewing gum all obey the Newtonian-viscous relationship, and these materials can be classified as ideally superplastic. Most superplastic metals, however, have values of m equal to about 0.5.

Applied mechanics can be used to describe the change in profile of a tensile specimen with strain once a neck forms, provided that the strain-rate-sensitivity

---

*Oftentimes the stress exponent, n, is used to describe the relation between the strain rate (creep rate) and the flow stress. In this case, the expression $\dot{\varepsilon} = K'\sigma^n$, where K' is a material constant, is used. The stress exponent, n, is equal to $1/m$, where m is the strain-rate-sensitivity exponent (Eq 1). When data are plotted as log σ versus log $\dot{\varepsilon}$, the resulting slope is equal to m. When data are plotted as log $\dot{\varepsilon}$ versus log σ, the resulting slope is equal to n. In this paper, both m and n are used.

exponent is known. When m is low, the increase in stress at the neck will lead to a large increase in strain rate in that region. Thus, the neck will grow sharply, leading to sudden failure and a low elongation to fracture. Conversely, when m is large, the strain rate increases slowly due to the increased stress in the neck region. As a result, the neck forms gradually. Various methods[70-75] of applied mechanics (a macroscopic approach) have been applied to predict elongation to fracture for materials that fail by necking. Four of these methods are outlined below:

- Rossard[70] has shown that the strain at the start of necking is given by $\varepsilon_{neck} = N/(1 - 2m)$, where N is the strain-hardening exponent in $\sigma = K''\varepsilon^N$. His theory would predict infinite plasticity at an m value of 0.5 provided that N has a finite positive value.
- Morrison[71] demonstrated that the dimensions of the specimen will influence the total elongation observed and showed that

$$\% \text{ elongation to fracture} = bm^2\left(\frac{d_0}{L_0}\right) \times 100$$

where b is a material constant, and d_0 and L_0 are the initial diameter and initial gage length of the specimen, respectively.

- Avery and Stuart[72] took into account the possible tapered shape of the specimen. Their equation is given as

$$\% \text{ elongation} = \left[\frac{1 - \beta^{1-m}}{1 - \alpha^{1-m}}\right] \times 100$$

where α is the ratio of the minimum to maximum area at the start of the test and β is the same ratio at some arbitrary stage of the test at which the elongation is measured.

- Burke and Nix[73] analyzed the influence of strain-rate sensitivity on neck growth using a finite-element method. Their results were modified by Ashby (quoted in Ref 74) to predict elongation to failure as a function of strain-rate sensitivity, as follows:

$$\% \text{ elongation} = \left[\exp\left(\frac{2m}{1 - m}\right) - 1\right] \times 100.$$

Figure 8 illustrates the predictive ability of three of the four relations described above. As can be seen, the values of elongation to fracture for a number of high-carbon and ultrahigh-carbon steels[47,76] agree well with the predictive relations based on a knowledge of the strain-rate-sensitivity exponent, m.

We thus see that, from an applied mechanics approach, a high strain-rate-sensitivity exponent, m, leads to high ductility. When m = 0.5, the material is generally superplastic (>400% elongation). A high value of m is a necessary condition for superplasticity. It is, however, not a sufficient condition. Material-embrittling characteristics, such as grain-boundary separation, cavitation at inter-

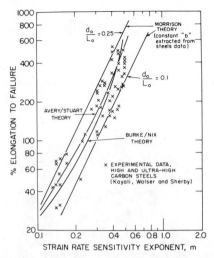

Fig. 8. Predicted elongation to fracture as a function of the strain-rate-sensitivity exponent, m, based on various analytical expressions (using an applied mechanics approach). The curves predicted by the Morrison relation were calculated using b = 100. The elongation and m-value data are for high- and ultrahigh-carbon steels.[47,75]

phase boundaries and other premature failure modes, can lead to early failure even when the material is highly strain-rate sensitive.

Fine-structure superplasticity thus refers to the high elongations obtainable in metallic alloys that have ultrafine grain sizes. Typically, the grain size is less than about ten microns (10 μm). The mode of plastic deformation is believed to be primarily grain-boundary sliding accommodated by slip near the grain boundary. An example of a fine-grain, superplastically deformed ultrahigh-carbon steel[51] is shown in Fig. 9. As can be seen, an elongation of about 1200% was achieved before failure occurred. The transmission electron micrograph in Fig. 9 reveals equiaxed ferrite grains about 2 μm in diameter in the deformed area of the specimen. This micrograph also reveals fine particles of iron carbide (cementite, Fe_3C) distributed along the ferrite grain boundaries. The cementite particles inhibit the growth of the ferrite grains during superplastic deformation.

It is now quite well established that fine-structure superplastic metallic alloys exhibit m values of about 0.5 (n = 2).* Theories have been developed to explain

---

*Oftentimes high elongations are obtained for solid-solution alloys where the grain size is fairly coarse (~100 μm). The elongation observed (~200 to 300%) can be attributed to the presence of a solute-dislocation drag mechanism where the stress exponent is three.[77–80] The respectably high value of the strain-rate-sensitivity exponent (m = 0.33) can explain the high ductilities obtained. Such an explanation has recently been suggested[81] to rationalize the high values of elongation to failure observed in a commercial niobium-base alloy strengthened by solute drag by hafnium additions. These highly ductile solid-solution alloys should not be classified in the same category as fine-structure superplastic alloys where m ≈ 0.5.

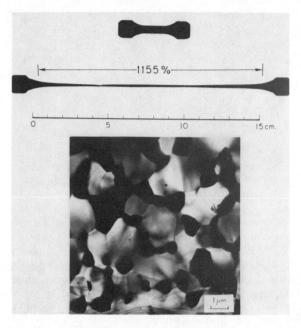

Fig. 9. Specimen of 1.6% C UHC steel deformed in tension at 650°C at an initial strain rate of 1%/min (elongation, 1200%) and micrograph showing that the structure after testing consists of a matrix of fine ferrite grains (~2 μm) and fine cementite particles (~0.1 to 0.5 μm).

this stress dependence of superplastic flow,[82-85] and it is generally accepted that superplastic flow is associated primarily with grain-boundary sliding processes. An example of a curve of log stress versus log strain rate for a nickel-base superplastic alloy is shown over a wide range of strain rate in Fig. 10. This figure illustrates the wide range of strain rate (designated region II) over which m = 0.5 (n = 2) and a second strain-rate range (designated region III) at high strain rates where m decreases to a low value (typically m ≈ 0.15). In this high strain-rate range, plastic flow by diffusion-controlled dislocation creep is known to prevail. In recent years, considerable attention has been directed toward the behavior of superplastic alloys at very low stresses (designated region I). Whereas some researchers observe high values of strain-rate sensitivity (m ≈ 1) in region I,[86,87] others observe low values of strain-rate sensitivity in region I.[88-91] Yet another group[92] believes that if proper control of experimental conditions is observed then there is no region I but an extension of region II. This subject has recently been fiercely debated,[93,94] and it is clear that further work will be required to resolve fully the controversy.

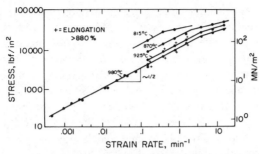

Fig. 10. Flow stress–strain rate relationship (log-log plot) for a two-phase nickel-iron-chromium alloy at elevated temperatures (Ni-39%Cr-10%Fe-1.75Ti-1%Al). The strain-rate-sensitivity exponent is constant and high (m ≃ 0.5) over a wide range of strain rate. Elongations exceeding 880% (extent of machine travel) were commonly observed.[85]

V. Structural Prerequisites for Superplasticity

In this section, the prerequisites that alloys must possess in order to exhibit fine-structure superplasticity are discussed.

1. Fine Grain Size

One of the major requirements for fine-structure superplasticity is that the grain size, d, should be small (although it should be noted that this is not a universally accepted view[76–79]). Typically, values of grain size less than 10 μm are required for superplastic behavior. Almost invariably, it is very difficult to observe superplasticity in single-phase materials because grain growth is too rapid. Despite this observation, most of the models for superplasticity assume, for simplicity, a single-phase structure. As will be seen in the section on the development of phenomenological equations for superplastic flow, the strain rate is very sensitive to grain size. More specifically, the dependences of strain rate on grain size can be grouped into those exhibiting inverse second power ($\dot{\varepsilon} \propto d^{-2}$) and those exhibiting inverse third power ($\dot{\varepsilon} \propto d^{-3}$). At a given stress, the strain rate invariably increases, therefore, with a decrease in grain size. Grain-size refinement thus represents an obvious method of increasing the strain-rate range for superplastic forming of alloys. Quantitative predictions based on this observation will be described in the next section.

2. Presence of Second Phase

In order to maintain a fine grain size at the superplastic forming temperature, and for the times involved during superplastic forming, the presence of a second phase at the matrix grain boundaries is required. For this reason, most, but certainly not all, superplastic materials are based on eutectoid (e.g., Fe-C), eutectic (e.g., Al-Ca) or monotectoid (e.g., Zn-Al) compositions. These materials can be thermomechanically processed to develop fine, two-phase structures. Inhibition of grain growth is usually improved if the quantity of the second phase is increased, provided that the size of the second phase remains fine and its distribution remains uniform.[52] In recent years a number of aluminum alloys have been made superplastic by the addition of small amounts of Zr ($\sim$0.5 wt %). In these cases (Al-6Cu-0.5Zr,[95-98] Al-6Mg-0.4Zr,[99] Al-9Zn-1Mg-0.2Zr[100-102] and, very recently, Al-3Li-0.5Zr and Al-4Cu-3Li-0.5Zr[103]), a very fine distribution of $ZrAl_3$ is formed which prevents grain growth during recrystallization of a heavily deformed matrix.

3. Strength of the Second Phase

There is evidence to suggest that the relative strengths of the matrix and second phase constitute an important parameter in the control of cavitation during super-plastic flow. For example, the deliberate addition of hard particles of Ag_3Sn to the superplastic Pb-Sn alloy (which does not normally cavitate) was found to cause cavitation nucleated exclusively at the matrix/particle interfaces.[104,105]

From these and other observations, it can be deduced that the strength of the second phase should be similar to the strength of the matrix phase. The evidence for cavitation at matrix/hard particle or matrix/second phase interfaces is quite extensive. In addition to the example quoted above, this type of behavior has been observed for Ti(C,N) particles in microduplex stainless steels,[106-109] for inclusions in α/β brasses,[106,110] at deliberate additions of coarse Fe particles to α/β brasses,[111] and at the β-phase/matrix interfaces in Cu-Zn-Ni alloys.[106,110,112] It should be pointed out, however, that when the second-phase particles are fine and similar in strength to the matrix phase, e.g., in ultrahigh-carbon steels,[34,53] cavitation is not observed. The question of cavitation in superplastic materials has been the subject of recent reviews[31,113] and remains an area of superplasticity that requires further study.

4. Size and Distribution of the Second Phase

If the second phase is considerably harder than the matrix phase it should be distributed uniformly and in fine particle form within the matrix. In the form of fine, but hard, particles, cavitation during superplastic flow can in fact be inhibited by various recovery mechanisms occurring in the vicinity of the particle. Chung and Cahoon[114] have shown how hard silicon particles, when in fine particle form, can minimize cavitation during superplastic flow of a fine-grain Al-Si eutectic alloy. Coarse particles, on the other hand, can lead to cavitation. For example, in white cast irons processed by a hot-and-warm rolling procedure, cavitation occurs during superplastic flow at the interfaces between the matrix and large eutectic carbides (>10 μm).[55] These large carbides cannot be refined by further mechanical processing. One way of eliminating this cavitation is to manufacture white cast irons using rapid solidification technology, because fine grains and fine distributions of second-phase particles can be achieved in powders obtained by rapid solidification technology (RST) as a result of the high rate of nucleation of solid phases with little time for growth. In white cast irons produced in this way, virtually no cavitation is observed after superplastic deformation.[115] Such a result confirms the important effects of a fine microstructure on high-temperature ductility and fracture of materials at intermediate temperatures.

5. Nature of the Grain Boundary

The grain boundaries between adjacent matrix grains should be of a high-angle nature (i.e., disordered), because grain-boundary sliding is the predominant mode of deformation during superplastic flow. Low-angle boundaries, such as those obtained by extensive warm working, do not slide under shearing stresses. In some cases — in eutectoid-composition steels, for example — structures containing low-angle grain boundaries are not superplastic but can be made superplastic by converting the low-angle grain boundaries to high-angle ones by use of an appropriate thermal or thermomechanical heat treatment.[34,75] Similar results have been obtained in a tool steel composition.[58]

Large-strain deformation at intermediate temperatures can change low-angle subgrain boundaries into high-angle grain boundaries. It has been shown[116] that the misorientation angle of subgrains in a ferritic stainless steel is a function of the amount of deformation. When the amount of mechanical deformation is very large ($\varepsilon = 15$, for example), the misorientation angle can approach 15°. These large strains were obtained in torsion testing and it was shown that, as a misorientation angle was increased, the material simultaneously became weaker at intermediate temperatures. This is in agreement with the concept that grain-boundary sliding is facilitated by an increase in the misorientation angles between grains. This result suggests a method of preparing a superplastic material, albeit a difficult one, because such large strains ($\varepsilon \simeq 15$) are clearly not readily achieved in most metalworking operations.

6. Mobility of Grain Boundaries

Grain boundaries in superplastic alloys should be mobile. During grain-boundary sliding, stress concentrations develop at triple points and other obstructions along the grain boundaries. The capacity of grain boundaries to be mobile (i.e., grain-boundary migration) permits reduction of these stress concentrations. The fact that the grains of superplastic materials remain essentially equiaxed even after very large deformations is indirect evidence that grain-boundary migration is occurring.

The ability to influence the behavior of metals in regions immediately adjacent to grain boundaries has been utilized in a remarkable investigation by Wyon and his colleagues.[117,118] They made an aluminum particulate composite superplastic near room temperature by diffusion of gallium into the grain-boundary regions. In their work, aluminum containing 10% alumina particles was heavily worked and recrystallized to develop a fine grain size (~ 1 μm). This material exhibited about 8% tensile elongation at 50°C. At this temperature, as expected, the strain-rate-sensitivity exponent is low because slip processes dominate deformation. If the Al-Al$_2$O$_3$ specimen is immersed in liquid gallium at 50°C, the gallium diffuses along the grain boundaries, forming a very thin film at such regions. If the specimen is then soaked for 50 h at 50°C, the gallium diffuses into a narrow region adjoining the grain boundary, creating an Al-Ga solid-solution mantle. The mantle is at a high homologous temperature. The material is now superplastic, exhibiting 300% elongation at 50°C. Its strain-rate sensitivity is high ($m \approx 0.5$), and it is weak. In this structural state, grain-boundary sliding is possible, because the grain boundary is embedded in the low-melting-point Al-Ga solid-solution region. Thus, even though the temperature is only 50°C ($T/T_m \simeq 0.35$ for pure aluminum), the homologous temperature at the grain-boundary region is probably around $0.5T_m$. After superplastic forming, the material can be strengthened by a high-temperature heat treatment which disperses the gallium throughout the aluminum matrix.

7. Grain Shape

The shapes of grains should be equiaxed in order that the grain boundaries can be subjected to shear stresses allowing grain-boundary sliding to occur. Materials with elongated grains, even though of fine grain size in a direction transverse to the applied stress, would not be expected to exhibit very much grain-boundary sliding if tested longitudinally. Testing in a transverse direction, however, could lead to extensive grain-boundary sliding and superplasticity. Rotation of grains also has to take place during superplastic flow, and this is clearly also more difficult with elongated grains than with equiaxed grains.

8. Grain Boundaries and Their Resistance to Tensile Separation

Grain boundaries in the matrix phase should not be prone to ready tensile separation. This factor may be the reason why fine-grain polycrystalline ceramics are not superplastic despite sometimes having strain-rate-sensitivity exponents of nearly unity. The tensile separation at grain boundaries may be a result of the high grain-boundary energy inherent in ceramics.[119] The inability of grain boundaries in polycrystalline ceramics to migrate readily also may contribute to the difficulty of making these materials superplastic (see Sec. 6, above).

VI. Phenomenological Equations for Describing Superplastic Flow in Fine-Grain Materials

The factors that influence superplastic flow have been investigated extensively. Several theories of superplastic flow have been proposed and developed in a quantitative manner.[76,82–85,120] None of these theories, however, appears to have resolved clearly the collective contributions of the three major factors that influence superplasticity, namely: (*a*) the temperature dependence of superplastic flow, (*b*) the effect of grain size on superplastic flow, and (*c*) the stress dependence of superplastic flow. An extensive study[34,35,121] of the phenomenology of superplastic flow has been made on a large number of systems. This study has made it possible to determine the influences of temperature, grain size and stress on superplastic flow, and these factors will be described in the following sections.

1. Temperature Dependence of Superplastic Flow

It is generally believed that the activation energy for superplastic flow, Q_{spf}, of fine-structure materials is equal to that for grain-boundary diffusion. Figure 11(a) illustrates this equality for a number of materials. It can be seen, however, that some fine-structure materials (represented by open symbols in Fig. 11a) yield activation energies that are higher than those for grain-boundary diffusion. These materials, in fact, exhibit an activation energy for superplastic flow which is about equal to that for lattice self-diffusion (Fig. 11b).

Because some superplastic materials exhibit activation energies for plastic flow that are nearly equal to those for lattice diffusion whereas others exhibit activation energies nearly equal to those for grain-boundary diffusion (Fig. 11a and b), it is tempting to suggest that an effective diffusion coefficient should be used to correlate all superplasticity data. Such an effective diffusion coefficient has previously been suggested to describe the temperature dependence of creep rate, where lattice and grain-boundary diffusion are believed to be rate-controlling as in Nabarro-Herring-Coble creep:[84]

$$D_{eff} = D_L f_L + c D_{gb} f_{gb} \qquad (2a)$$

where f_L and f_{gb} are the fractions of atoms associated with lattice and grain-boundary diffusion, respectively, and $c = 1$. Typically, $f_L \approx 1$, and $f_{gb} = \pi w/d$, where w is the grain-boundary width (which often is given as 2b, where b is the Burgers vector) and d is the grain size. Equation 2a can thus be rewritten as

$$D_{eff} = D_L + D_{gb}\left(\frac{\pi w}{d}\right) \qquad (2b)$$

This type of effective diffusion coefficient apparently works well in correlating diffusional creep data (i.e., Nabarro-Herring-Coble creep; see, for example, Ashby and Verrall[84]). The temperature dependence of superplastic materials, however, does not follow the prediction given by Eq 2b. This is shown in Fig. 12, where $D_L f_L$ is plotted as a function of $D_{gb} f_{gb}$ (where $f_{gb} = 2\pi b/d$) for a large number of superplastic metals. The dashed line in this figure is for the case $D_L f_L = D_{gb} f_{gb}$. Equation 2b would predict that all lattice-diffusion-controlled superplastic materials should fall above this line ($D_L f_L > D_{gb} f_{gb}$) and that all grain-boundary-diffusion-controlled superplastic materials should fall below it ($D_{gb} f_{gb} > D_L f_L$). As can be seen, a D_{eff} concept as given by Eq 2b cannot explain

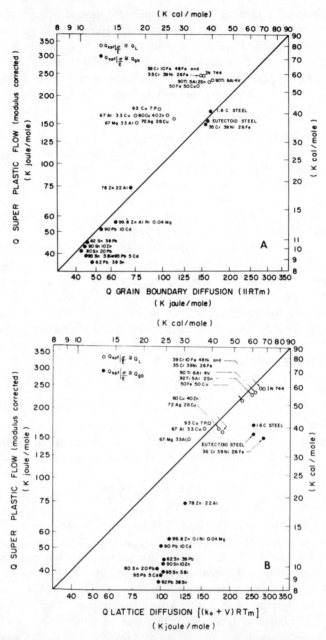

Fig. 11. Activation energy for superplastic flow is equal to that for grain-boundary diffusion (a) or to that for lattice diffusion (b).[121]

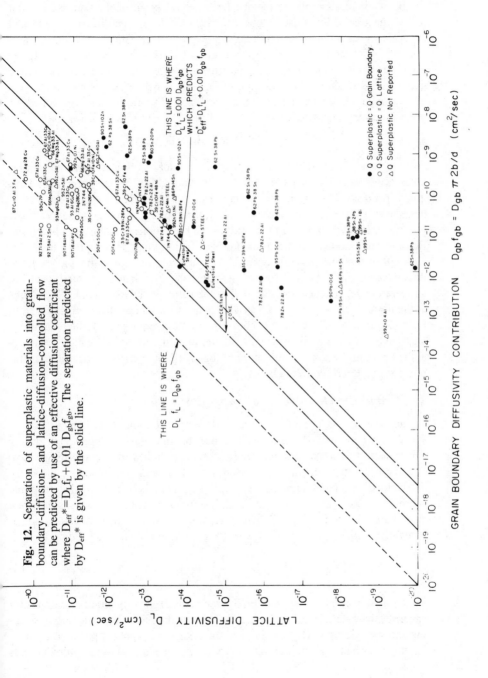

Fig. 12. Separation of superplastic materials into grain-boundary-diffusion- and lattice-diffusion-controlled flow can be predicted by use of an effective diffusion coefficient where $D_{eff}* = D_L f_L + 0.01 D_{gb} f_{gb}$. The separation predicted by $D_{eff}*$ is given by the solid line.

the behavior of superplastic metals, because nearly all of the data plotted fall below the line. Nevertheless, one can see that the lattice-diffusion-controlled superplastic materials (open symbols) are grouped separately from the grain-boundary-diffusion-controlled superplastic materials (solid symbols). In fact, it is possible to give a modified effective diffusion coefficient which will correlate the data. This is given by the solid line in Fig. 12, which separates well the superplastic materials controlled by lattice diffusion (D_L) from those controlled by grain-boundary diffusion (D_{gb}). The modified D_{eff} can thus be given as

$$D_{eff}{}^* = D_L f_L + 0.01 D_{gb} f_{gb} \tag{3}$$

where f_L and f_{gb} are as previously defined. Equation 3 states that the grain-boundary term is 100 times less effective than predicted by the normal D_{eff} term given in Eq 2b. No physical explanation of the phenomenological relation obtained can be offered at the present time. It is tempting, however, to suggest that, because extensive grain-boundary sliding and migration occur during superplastic flow, normal models for grain-boundary diffusion must be inappropriate.

We illustrate the validity of Eq 3 for superplastic flow with the example illustrated in Fig. 13. This figure illustrates the excellent predictability of the temperature dependence of a 35Cr-39Ni-26Fe alloy[122] by means of Eq 3. Whereas the normal D_{eff} (where c in Eq 2a equals unity) would predict a transition from D_{gb} to D_L at near 1600°C (220°C above the melting temperature), the relation for $D_{eff}{}^*$ given in Eq 3 predicts a transition temperature of 900°C, which is exactly at the break in the slope of the curve of log creep rate versus reciprocal temperature for the 35Cr-39Ni-26Fe alloy.

2. Grain-Size Dependence of Superplastic Flow

The grain-size effect on superplastic flow has been reported to be of the form $\dot{\varepsilon} \propto d^{-p}$, where p is the grain-size exponent and d is the grain size during superplastic flow. Experimental and theoretical studies of superplasticity have indicated that p is in the range from 1 to 3. Phenomenologically, p can be determined for a large number of superplastic materials by plotting the creep rate (normalized as $\dot{\varepsilon}/D_{eff}{}^*$ at a given modulus-compensated stress) versus the grain size. This is shown in Fig. 14. Although some scatter in the data can be noted, a negative slope of 2 represents well the correlation obtained, yielding

$$\dot{\varepsilon} \propto \frac{D_{eff}{}^*}{d^2} f\left(\frac{\sigma}{E}\right) \tag{4}$$

Equation 4, in combination with Eq 3, predicts the following grain-size trends for superplastic materials. When lattice diffusion is the rate-controlling process during superplastic flow, $D_{eff}{}^* = D_L$, and the grain-size exponent p should equal 2. On the other hand, when grain-boundary diffusion is the rate-controlling process

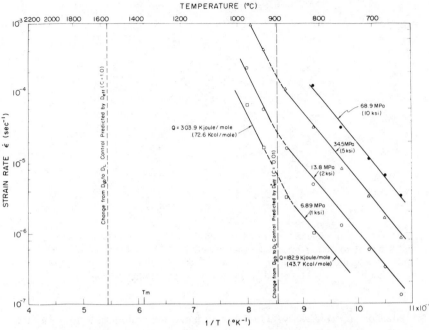

Fig. 13. Arrhenius plot of data from Goodell, Floreen and Hayden[122] for a 35Cr-39Ni-26Fe alloy, showing the good agreement between the temperature where the change from grain-boundary-diffusion-controlled to lattice-diffusion-controlled superplastic flow occurs and that predicted by Eq 2a with c = 0.01. The predicted temperature for this change, as given by D_{eff} with c = 1, is also shown.

during superplastic flow, $D_{eff}^* \propto D_{gb}d^{-1}$, the grain-size exponent p should equal 3. These predictions are consistent with the grain-size exponents reported in the literature for p = 2[122–126] and p = 3.[122,127–129]

3. Stress Dependence of Superplastic Flow

The stress dependence of fine-structure superplastic flow is well established, with $\dot{\varepsilon} \propto \sigma^n$ where n = 2 (i.e., m = 0.5). An example of this stress dependence was shown earlier in Fig. 11. It is now possible to write a quantitative expression for the superplastic flow rate as a function of the modulus-compensated flow stress (σ/E) utilizing Eq 3 and 4 and the data of Fig. 14.

The quantitative relations to be presented have to be modified in one small way. All quantitative analyses of creep data for superplastic materials (Fig. 12 and 14) were based on grain-size measurements carried out by the mean-linear-intercept ($\bar{L}$) method. (This is the most common method of reporting grain sizes

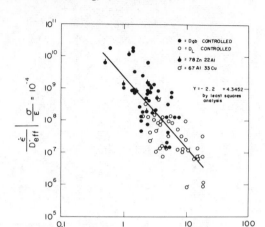

FINAL GRAIN SIZE d (microns)

Fig. 14. Dependence of grain size on the diffusion-compensated creep rate, at constant σ/E, for all superplastic materials studied (56 different investigations). Data from R. White.[121] The resulting correlation reveals that $\varepsilon \propto d^{-2}$. (The term d is actually the linear intercept grain size.)

in studies on superplasticity.) On the other hand, the grain-size term, d, as used in Eq 2b, usually refers to the true spatial grain size. The terms are related by the expression $d = 1.776\overline{L}$.[82] In the following, quantitative relations for super-plastic flow rate are given in terms of $\overline{L}$. Thus,

$$\dot{\varepsilon} = 2 \times 10^9 \frac{D_{eff}{}^*}{\overline{L}^2} \left(\frac{\sigma}{E}\right)^2 \tag{5}$$

This general equation can be written for the case where fine-structure superplastic flow is controlled by grain-boundary diffusion and where fine-structure super-plastic flow is controlled by lattice diffusion. Thus, when grain-boundary diffusion controls the superplastic flow process, $2 \times 10^{-2} D_{gb}\pi b/\overline{L} \gg D_L$ and therefore $D_{eff}{}^* = 2 \times 10^{-2} D_{gb}\pi b/\overline{L}$. Substituting this expression for $D_{eff}{}^*$ in Eq 4, we obtain:

$$\dot{\varepsilon} \simeq 10^8 \left(\frac{\sigma}{E}\right)^2 \left(\frac{bD_{gb}}{\overline{L}^3}\right) \text{ (grain-boundary-diffusion controlled)} \tag{5a}$$

In the case when lattice diffusion controls the superplastic flow process, $D_L \gg 2 \times 10^{-2}\pi D_{gb}\pi b/\overline{L}$ and therefore $D_{eff}{}^* = D_L$. Substituting this expression for $D_{eff}{}^*$ in Eq 4, we obtain:

$$\dot{\varepsilon}_{spf} = 2 \times 10^9 \left(\frac{\sigma}{E}\right)^2 \left(\frac{D_L}{\overline{L}^2}\right) \text{ (lattice-diffusion controlled)} \tag{5b}$$

Figures 15 and 16 summarize the ability of Eq 5 to predict the creep behavior of superplastic materials. Figure 15 is a plot of creep data for superplastic materials that are controlled by grain-boundary diffusion. Figure 16 gives data for superplastic materials that are controlled by lattice diffusion. The predicted curves from Eq 5a and 5b are shown by dashed lines. The correlations are quite good and predict the creep behavior of all superplastic materials within an order of magnitude of creep rate.

VII.   Physical Mechanisms for Superplastic Flow

Three principal modes of deformation have been considered in explaining the creep behavior of fine-structure superplastic materials. These are: (a) diffusional flow (Nabarro-Herring-Coble), (b) grain-boundary sliding accommodated by diffusional flow (Ashby-Verrall) and (c) grain-boundary sliding accommodated by slip. We consider the last, i.e., grain-boundary sliding (g.b.s.) accommodated by slip, to be the most likely mechanism and believe that such a mechanism is described by the phenomenological equations 5a and 5b. A comparison of the relations between strain rate and stress predicted by the three different modes of deformation mentioned above, with experimental data, is presented in Fig. 17. The data in this figure are for a Zn-Al monotectoid-composition alloy and are from eleven separate investigations (the references are cited in the figure). Creep testing was done at intermediate temperatures and at fine grain sizes ($\overline{L}$ = 0.5 to 3.5 μm). Under such conditions the superplastic creep rate is controlled by grain-boundary diffusion. As can be seen, the shapes of the curves represented by diffusional creep (Coble creep)[156] and by g.b.s. accommodated by diffusional flow (Ashby-Verrall)[84] do not fit the experimental data. On the other hand, g.b.s. accommodated by slip, as represented by Eq. 5a, does fit the data rather well. These correlations lead us to favor the concept of g.b.s. occurring in a mantlelike region within and adjacent to the grain boundaries, similar to the "mantle and core" theory proposed by Gifkins.[76] Thus, one can consider plastic flow as arising from two independent processes in fine-grain superplastic materials. One is due to g.b.s. accommodated by slip occurring in the mantle region, and the other is due to slip processes occurring within the core of each grain. When the former process dominates deformation, superplasticity can occur; and when the latter dominates deformation, normal ductility is expected.

Other information exists to suggest that g.b.s. accommodated by slip is the dominant mode of deformation during superplastic flow. For example, a Zn-Al eutectic[157] processed to a fine grain size with a strong crystallographic hcp texture exhibited the following behavior. During superplastic deformation in the range where m = 0.5, the originally round specimen developed an elliptical shape. This directionality in plastic flow indicates that slip along specific crys-

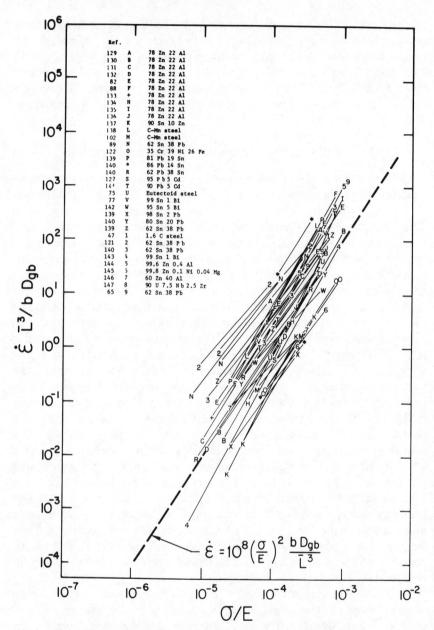

Fig. 15. Diffusion- and grain-size-compensated strain rate as a function of modulus-compensated stress for polycrystalline materials where superplastic flow is controlled by grain-boundary diffusion.[121]

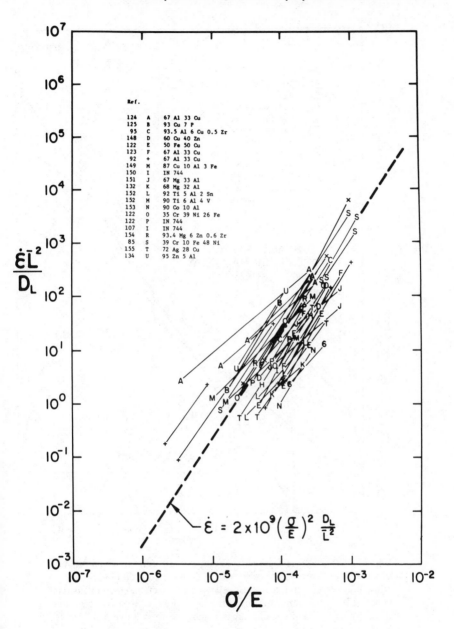

Fig. 16. Diffusion- and grain-size-compensated strain rate as a function of modulus-compensated stress for polycrystalline materials where superplastic flow is controlled by lattice diffusion.[121]

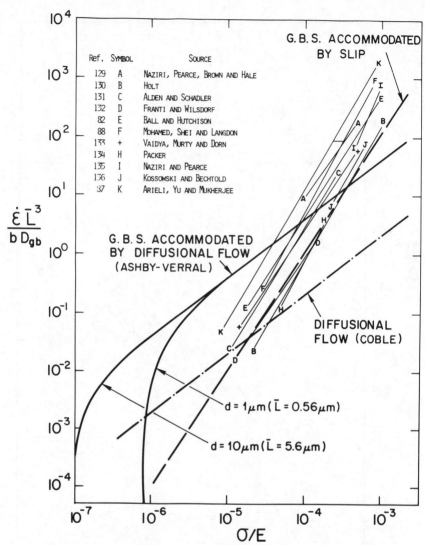

Fig. 17. Diffusion- and grain-size-compensated strain rate as a function of modulus-compensated stress for a Zn-Al monotectoid-composition alloy. Predicted creep rate – stress relations are given based on diffusional creep, g.b.s. accommodated by diffusional flow, and g.b.s. accommodated by slip. The threshold stress values are from Table 5 in Ref 14.

tallographic planes occurs in the vicinity of the grain-boundary region contributing to directionality in grain-boundary sliding. A detailed study of an Al-9Zn-1Mg alloy by Matsuki, Morita, Yamada and Murakami[102] also led to the conclusion that slip processes occur during grain-boundary sliding. Another example is a fine-grain-size cadmium specimen[158] deformed at a temperature (250°C) and within a strain-rate range (10^{-3} to 10^{-4} s^{-1}) where m = 0.5. In this case it was shown that the strength of the fine-grain cadmium polycrystal was dictated by the crystallographic texture existing in the specimen. Specimens with grains oriented for easy basal slip required lower stresses for g.b.s. than did specimens in which the grains were oriented so that basal slip was difficult.

An understanding of the mode of deformation during superplastic flow can be an important guide in developing new methods of enhancing the superplastic properties of fine-grain materials. It is our conclusion that grain-boundary sliding and migration, accommodated by slip processes at regions adjoining the grain boundaries, constitute the most likely mechanism of deformation during super-plastic flow of fine-grain-size materials. This concept has support from a number of theories, besides the "mantle and core" theory of Gifkins, which invoke g.b.s. accommodated by slip.[82,83,120,138,159-161]

VIII. Optimizing the Rate of Superplastic Flow in Fine-Structure Materials

For a given superplastic material, and at a given temperature, there is a range of strain rate where superplastic flow by grain-boundary sliding can be expected. Above a certain specific strain rate, grain-boundary sliding is no longer the dominant process and another mode of deformation becomes important, namely, diffusion-controlled dislocation creep (slip creep). The maximum strain rate at which grain-boundary sliding remains rate-controlling is typically on the order of 10^{-4} to 10^{-3} s^{-1}, a rate considerably lower than those used in commercial forming operations (e.g., 10^{-1} to 1 s^{-1}). From a technological viewpoint it would be highly desirable to increase the maximum strain rate for superplastic flow. Because the factors that influence grain-boundary sliding differ from those that influence slip creep, the approach to attainment of this goal is straightforward. One must select variables that will enhance grain-boundary sliding but make slip creep more difficult. We illustrate this schematically in Fig. 18. In this graph, the logarithm of the strain rate is plotted against the logarithm of the flow stress. The two separate processes contributing to grain-boundary sliding and to slip creep are represented as straight lines, and the point of intersection (marked 1) represents the maximum strain rate for superplastic flow for a given microstructural condition. The dashed lines on the figure indicate how a new set of microstructural conditions can increase the maximum strain rate for superplastic flow by making grain-boundary sliding more facile and by inhibiting the slip-creep rate. Such a change then leads to an increase in the maximum rate for superplastic flow, as indicated by symbol 2 in Fig. 18.

The major microstructural feature influencing superplastic flow is the grain size. When grain-boundary diffusion is rate-controlling the superplastic flow rate is inversely proportional to the cube of the grain size (Eq 5a). An example of the

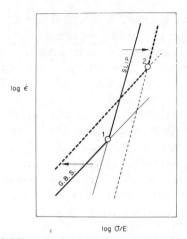

Fig. 18. The solid lines depict the creep rate of a fine-structure super-plastic material as a function of stress at a given structural state. Point 1 on the curve represents the maximum strain rate where superplastic behavior can be expected. If the structural state is changed such that grain-boundary sliding is enhanced and slip creep is made more difficult, as shown by the dashed lines, then the maximum strain rate for superplastic flow is increased (point 2).

predicted enhancement of superplastic flow by control of grain size is shown for an ultrahigh-carbon steel[53] in Fig. 19. If the grain size is decreased from 2 to 0.4 μm, it is predicted that the maximum strain rate for superplastic flow is enhanced by a factor of about a thousand. Maintaining a grain size of less than one micron during superplastic flow is not a trivial problem and represents an important area of research for the development of commercially useful super-plastic alloys.

Diffusion-controlled dislocation creep (slip creep) is influenced by a number of microstructural variables.[162-164] The slip-creep rate can be influenced by changing the crystal structure[162] (bcc, hcp, fcc, diamond cubic, ordered structures, etc.), by adjusting texture,[165] by substructure strengthening,[166] or by decreasing the stacking-fault energy.[167]

We illustrate in Fig. 20 the possible enhancement of superplastic behavior in nickel as influenced by stacking-fault energy. The creep rate of nickel in the slip-creep region is decreased by making solid-solution alloying additions of tungsten[168] so as to lower the stacking-fault energy. It is predicted that the maximum superplastic strain rate of a 2-μm-grain-size nickel alloy will be enhanced by a factor of 20 by the addition of 6 at.% tungsten to the nickel. A possible complication arising from such a prediction is that the rate of grain-boundary sliding may also be inhibited by a decrease in stacking-fault energy. This is because grain-boundary-sliding mechanisms appear to be linked to slip

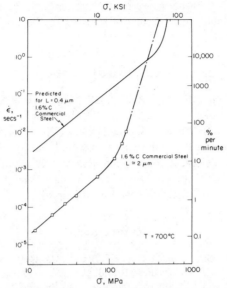

Fig. 19. Graph illustrating the strain rate – flow stress relationship for a 1.6% C steel at two grain sizes. Superplastic flow can be expected to occur at strain rates as high as 5000%/min at L = 0.4 μm.[53]

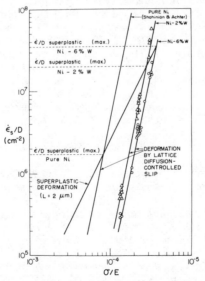

Fig. 20. Tungsten enhances the creep resistance of nickel in the slip-creep region by decreasing the stacking-fault energy. Thus, the extent of superplastic flow may be broadened by solute additions.[168]

mechanisms in the grain-boundary mantle region, as discussed earlier. Studies to assess such variables as the effect of stacking-fault energy on superplasticity are needed.

Acknowledgments

This work was supported by DARPA, ARO and ONR at Stanford University. The authors would like to thank Drs. L. Jacobson and S. Fishman (DARPA), Dr. G. Mayer (ARO) and Dr. B. MacDonald (ONR) for their guidance and encouragement. One of the authors (J. W.) would like to acknowledge support from the Lockheed Palo Alto Independent Research Program. The contributions from Dr. Richard White are gratefully acknowledged, and discussions with him on various aspects of this paper proved most helpful.

References

1. B. Fazan, O. D. Sherby and J. E. Dorn, *Trans. TMS-AIME*, Vol 200, 1954, p 919.
2. R. L. Squires, R. T. Weiner and M. Phillips, *J. Nucl. Mater.*, Vol 8, 1963, p 77.
3. B. Walser and O. D. Sherby, *Scripta Met.*, Vol 16, 1982, p 213.
4. M. F. Ashby and H. J. Frost, in *Constitutive Relations in Plasticity*, edited by A. Argon, M.I.T. Press, 1975.
5. C. E. Pearson, *J. Inst. Metals*, Vol 54, 1934, p 111.
6. W. Rosenhain and W. Ewen, *J. Inst. Metals*, Vol 8, 1912, p 149.
7. G. D. Bengough, *J. Inst. Metals*, Vol 7, 1912, p 123.
8. W. Rosenhain, J. L. Haughton and K. E. Bingham, *J. Inst. Metals*, Vol 23, 1920, p 261.
9. C. H. M. Jenkins, *J. Inst. Metals*, Vol 40, 1928, p 21.
10. A. Sauveur, *Iron Age*, Vol 113, 1924, p 581.
11. A. A. Bochvar and Z. A. Sviderskaya, *Izvest. Akad. Nauk SSSR, Ordel Tekhn. Nauk*, Vol 9, 1945, p 821.
12. E. E. Underwood, *J. Metals*, Vol 14, 1962, p 914.
13. S. Tang, *Mechanics of Superplasticity*, Kriegar Publishing Co., New York, 1979.
14. J. W. Edington, K. N. Melton and C. P. Cutler, *Prog. Mater. Sci.*, Vol 21, 1976, p 61.
15. K. A. Padmanabhan and G. T. Davies, *Superplasticity*, Springer-Verlag, Berlin, Germany, 1980.
16. M. H. Shorshorrov, A. S. Tichonov, C. I. Bulat, K. P. Gurov, N. I. Nadirashvili and V. Z. Antipov, *Superplasticity in Metallic Materials*, Izdatelsvo, "Nauk", Moscow, 1973.
17. A. S. Tichonov, *Superplastic Effect in Metals and Alloys* (questions on theory and on application), Izdatelsvo, "Nauk", Moscow, 1978.
18. O. M. Smirnov, *Working of Metals Under Pressure in the Superplastic State*, Machinostroyenia, Moscow, 1979.
19. I. I. Novikov and V. K. Portnoy, *Superplasticity in Alloys with Ultrafine Grains*, Metallurgiya, Moscow, USSR, 1981.
20. A. A. Presnyakov, *Sverkhplastichrost' Metallov i Splavov*, publ. "Nauka", Alma Ata, USSR, 1969.
21. O. A. Kaibyshev, *Plastichnost' i Sverkhplastichrost Metallov*, publ. Metallurgiya, Moscow, USSR, 1975.

22. M. W. Grabski, Nadplastyczność Strukturalna Metali, publ. "Slgsk", Katowice, Poland, 1973.
23. O. D. Sherby, *Sci. J.,* Vol 5, 1969, p 75.
24. J. J. Burke and V. Weiss, *Ultrafine Grain Metals,* Syracuse University Press, Syracuse, New York, 1970.
25. R. H. Johnson, *Met. Rev.,* Vol 15, 1970, p 115.
26. G. J. Davies, J. W. Edington, C. P. Cutler and K. A. Padmanabhan, *J. Mater. Sci.,* Vol 5, 1970, p 1092.
27. R. B. Nicholson, in *Electron Microscopy and Structures of Materials,* edited by G. Thomas, R. M. Fulrath and R. M. Fisher, University of California Press, Berkeley, CA, 1972, p 689.
28. A. K. Mukherjee, in *Treatise on Materials Science and Technology,* edited by R. J. Arsenault, Academic Press, New York, 1975, Vol 6, p 163.
29. T. H. Alden, in *Fundamental Aspects of Structural Alloy Design,* edited by R. T. Jaffee and B. A. Wilcox, Plenum Press, 1977, p 411.
30. J. Gittus, *Creep Viscoelasticity and Creep Fracture in Solids,* Applied Science Publishers, Ltd., London, 1975, p 509.
31. D. M. R. Taplin, G. L. Dunlop and T. G. Langdon, *Ann. Rev. Mater. Sci.,* Vol 9, 1979, p 151.
32. A. K. Mukherjee, *Ann. Rev. Mater. Sci.,* Vol 9, 1979, p 191.
33. J. W. Edington, *Metals Technol.,* Vol 3, 1976, p 138.
34. J. Wadsworth, T. Oyama and O. D. Sherby, *6th Inter-American Conference on Materials Technology,* San Francisco, August 1980, Vol II, edited by I. LeMay, ASME, New York, 1980, p 29.
35. O. D. Sherby, R. D. Caligiuri, E. S. Kayali and R. A. White, in *Advances in Metal Processing,* edited by J. J. Burke, R. Mehrabian and V. Weiss, Plenum Press, 1981, p 133.
36. O. Izumi, *J. Japan Soc. Tech. Plasticity,* Vol 16, 1975, p 1015.
37. J. P. Poirier, *Plasticité à Haute Température des Solides Cristallins*, Eysolles, Paris, France, 1976, p 191.
38. M. Suéry and B. Baudelet, *Rev. Phys. Appl.,* Vol 13, 1978, p 53.
39. *Superplastic Forming of Structural Alloys,* edited by N. E. Paton and C. H. Hamilton, AIME, Warrendale, PA, 1982.
40. C. H. Hammond, ibid., p 131.
41. H. F. Merrick, ibid., p 209.
42. D. J. Lloyd and D. M. Moore, ibid., p 147.
43. C. H. Hamilton, C. C. Bampton and N. E. Paton, ibid., p 173.
44. N. Ridley, ibid., p 191.
45. C. H. Hamilton and G. W. Stacher, *Metal Progr.,* Vol 109, 1976, p 34.
46. J. B. Moore, J. Tequesta and R. L. Athey, U.S. Patent No. 3,519,503, 1970.
47. O. D. Sherby, B. Walser, C. M. Young and E. M. Cady, *Scripta Met.,* Vol 9, 1975, p 569.
48. O. D. Sherby, C. M. Young, B. Walser and E. M. Cady, U.S. Patent No. 3,951,967, 1976.
49. B. Walser, S. Kayali and O. D. Sherby, *Proceedings of the 4th International Conference on the Strength of Metals and Alloys,* Laboratoire de Physique du Solide, E.N.S.M.I.M., Nancy, France, 1976, Vol 1, p 413.
50. J. Wadsworth and O. D. Sherby, *J. Mech. Work Tech.,* Vol 2, 1978, p 53.
51. J. Wadsworth and O. D. Sherby, *J. Mater. Sci.,* Vol 13, 1978, p 2645.
52. T. Oyama, J. Wadsworth, M. Korchynsky and O. D. Sherby, *Proceedings of the 5th International Conference on the Strength of Metals and Alloys,* Aachen, W. Germany, edited by P. Hassen, V. Gerold and G. Kostovz, Vol 1, 1979, p 351.

53. B. Walser and O. D. Sherby, *Met. Trans. A,* Vol 10A, 1979, p 1461.
54. J. Wadsworth and O. D. Sherby, *Foundry Manag. Technol.,* Vol 106, 1978, p 59.
55. J. Wadsworth, L. E. Eiselstein and O. D. Sherby, *Mater. Eng. Appl.,* Vol 1, 1979, p 143.
56. O. D. Sherby, J. Wadsworth, R. D. Caligiuri, L. E. Eiselstein, B. C. Snyder and R. T. Whalen, *Scripta Met.,* Vol 13, 1979, p 941.
57. E. S. Kayali, H. Sunada, T. Oyama, J. Wadsworth and O. D. Sherby, *J. Mater. Sci.,* Vol 14, 1979, p 2688.
58. J. Wadsworth, J. H. Lin and O. D. Sherby, *Metals Technol.,* Vol 8, 1981, p 190.
59. O. A. Ruano, J. Wadsworth and O. D. Sherby, *Met. Trans. A,* Vol 13A, 1982, p 355.
60. R. Grimes, C. Baker, M. J. Stowell and B. M. Watts, *Aluminium,* Vol 51, 1975, p 720.
61. R. Grimes, M. J. Stowell and B. M. Watts, *Metals Technol.,* Vol 3, 1976, p 154.
62. D. M. Moore and L. R. Morris, *Mater. Sci. Eng.,* Vol 43, 1980, p 85.
63. J. A. Wert, N. E. Paton, C. H. Hamilton and M. W. Mahoney, *Met. Trans. A,* Vol 12A, 1981, p 1265.
64. D. A. Woodford, *Trans. ASM,* Vol 62, 1969, p 291.
65. D. H. Avery and W. A. Backofen, *Trans. ASM,* Vol 58, 1965, p 551.
66. A. W. Mullendore and N. J. Grant, *Trans. TMS-AIME,* Vol 200, 1954, p 973.
67. J. B. Conway and P. N. Flagella, "Creep-Rupture Data for the Refractory Metals at High Temperatures," Nuclear Systems Programs Technical Report GEMP-685 (R-69-NSP-9), General Electric Co., 1969.
68. W. Green, *Trans. TMS-AIME,* Vol 215, 1959, p 1057.
69. B. A. Wilcox and A. H. Clauer, *Trans. TMS-AIME,* Vol 236, 1966, p 570.
70. C. Rossard, *Revue de Metallurgie,* Vol 63, 1966, p 225.
71. W. B. Morrison, *Trans. ASM,* Vol 61, 1968, p 423.
72. D. H. Avery and J. M. Stuart, in *Surfaces and Interfaces II, Physical and Mechanical Properties,* edited by J. J. Burke, N. L. Reed, and V. Weiss, Syracuse University Press, Syracuse, New York, 1968, p 371.
73. M. A. Burke and W. D. Nix, *Acta Met.,* Vol 33, 1975, p 793.
74. R. W. Lund and W. D. Nix, *Acta Met.,* Vol 24, 1976, p 469.
75. W. A. Backofen, I. R. Turner and D. H. Avery, *Trans. ASM,* Vol 57, 1964, p 980.
76. E. S. Kayali, Ph.D. Thesis, Stanford University, Stanford, CA, 1976.
77. R. C. Gifkins, *Met. Trans. A,* Vol 7A, 1976, p 1225.
78. T. H. Alden, *Trans. TMS-AIME,* Vol 236, 1966, p 1633.
79. P. Griffith and C. Hammond, *Acta Met.,* Vol 20, 1972, p 935.
80. J. C. Wei and W. D. Nix, *Scripta Met.,* Vol 13, 1979, p 1017.
81. J. Wadsworth, C. A. Roberts, and E. H. Rennhack, *J. Mater. Sci.,* Vol 17, 1982, p 2539.
82. A. Ball and M. M. Hutchinson, *Metal. Sci. J.,* Vol 3, 1969, p 1.
83. T. G. Langdon, *Phil. Mag.,* Vol 22, 1970, p 689.
84. M. F. Ashby and R. A. Verrall, *Acta Met.,* Vol 21, 1973, p 149.
85. G. W. Hayden, R. C. Gibson, H. P. Merrick and J. H. Brophy, *Trans. ASM,* Vol 60, 1967, p 3.
86. S. C. Misro and A. K. Mukherjee, *Rate Processes in Plastic Deformation of Materials,* edited by J. C. M. Li and A. K. Mukherjee, American Society for Metals, Metals Park, OH, 1975, p 434.
87. A. Arieli, A. K. S. Yu and A. K. Mukherjee, *Met. Trans. A,* Vol 11A, 1980, p 181.
88. F. A. Mohamed, S. Shei and T. G. Langdon, *Acta Met.,* Vol 23, 1975, p 1443.
89. F. A. Mohamed and T. G. Langdon, *Phil. Mag.,* Vol 32, 1975, p 697.

90. D. Grivas, Report No. LBL-7375, Lawrence Berkeley Laboratory, University of California, Berkeley, 1978.
91. S. H. Vale, D. J. Eastgate and P. M. Hazzledine, *Scripta Met.*, Vol 113, 1979, p 1157.
92. G. Rai and N. J. Grant, *Met. Trans. A*, Vol 6A, 1975, p 385.
93. D. Grivas, J. W. Morris, Jr., and T. G. Langdon, *Scripta Met.*, Vol 15, 1981, p 229.
94. A. Arieli and A. K. Mukherjee, *Scripta Met.*, Vol 15, 1981, p 237.
95. B. M. Watts, M. J. Stowell, B. L. Baikie and D. G. E. Owen, *Metal Sci.*, Vol 10, 1976, p 189.
96. B. M. Watts, M. J. Stowell, B. L. Baikie and D. G. E. Owen, *Metal Sci.*, Vol 10, 1976, p 198.
97. R. H. Bricknell and J. W. Edington, *Acta Met.*, Vol 27, 1979, p 1303.
98. R. H. Bricknell and J. W. Edington, *Acta Met.*, Vol 27, 1979, p 1313.
99. K. Matsuki, Y. Yetani, M. Yamada and Y. Murakami, *Metal Sci.*, Vol 10, 1976, p 235.
100. K. Matsuki and M. Yamada, *J. Japan Inst. Metals*, Vol 37, 1973, p 448.
101. K. Matsuki, Y. Ueno and M. Yamada, *J. Japan Inst. Metals*, Vol 38, 1974, p 219.
102. K. Matsuki, H. Morita, M. Yamada and Y. Murakami, *Metal Sci.*, Vol 11, 1977, p 156.
103. J. Wadsworth, I. G. Palmer and D. D. Crooks, *Scripta Met.*, Vol 17, 1983, p 347.
104. C. W. Humphries and N. Ridley, *J. Mater. Sci.*, Vol 12, 1977, p 851.
105. D. W. Livesey and N. Ridley, *J. Mater. Sci.*, Vol 13, 1978, p 825.
106. N. Ridley, C. W. Humphries and D. W. Livesey, *Proceedings of 4th International Conference on the Strength of Metals and Alloys*, Laboratoire de Physique du Solide, E.N.S.M.I.M., Nancy, France, 1976, Vol 1, p 433.
107. C. I. Smith, B. Norgate and N. Ridley, *Metal Sci.*, Vol 10, 1976, p 182.
108. C. W. Humphries and N. Ridley, *J. Mater. Sci.*, Vol 9, 1974, p 1429.
109. C. W. Humphries and N. Ridley, *J. Mater. Sci.*, Vol 13, 1978, p 2477.
110. N. Ridley and D. W. Livesey, *Proceedings of 4th International Conference on Fracture*, 1977, Vol 2, p 533.
111. S. Sagat and D. M. R. Taplin, *Acta Met.*, Vol 24, 1976, p 307.
112. D. W. Livesey and N. Ridley, *Met. Trans. A.*, Vol 9A, 1978, p 519.
113. M. J. Stowell, *Superplastic Forming of Structural Alloys*, edited by N. E. Paton and C. H. Hamilton, TMS-AIME, Warrendale, PA, 1982, p 321.
114. S. W. Chung and J. R. Cahoon, *Metal Sci.*, Vol 13, 1979, p 635.
115. O. A. Ruano, L. E. Eiselstein and O. D. Sherby, *Met. Trans. A*, Vol 13A, 1982, p 1785.
116. C. G. Schmidt, C. M. Young, B. Walser, R. H. Klundt and O. D. Sherby, *Met. Trans. A*, Vol 13A, 1982, p 447.
117. S. K. Marya and G. Wyon, *Proceedings of the 4th International Conference on the Strength of Metals and Alloys*, Laboratoire de Physique du Solide, E.N.S.M.I.M., Nancy, France, 1976, Vol 1, p 438.
118. F. Weill and G. Wyon, *Proceedings of the 5th International Conference on the Strength of Metals and Alloys*, Aachen, West Germany, 1979, Vol 1, p 387.
119. J. J. Gilman, "Mechanical Behavior of Crystalline Solids," National Bureau of Standards Monograph #59, 1963, p 79.
120. A. K. Mukherjee, *Mater. Sci. Eng.*, Vol 8, 1971, p 83.
121. R. A. White, Ph.D. Thesis, Stanford University, Stanford, CA, 1978.
122. H. W. Hayden, S. Floreen, and P. D. Goodell, *Met. Trans.*, Vol 3, 1972, p 833.
123. P. A. Holt and W. A. Backofen, *Trans. ASM*, Vol 59, 1966, p 755.
124. M. J. Stowell, J. L. Robertson and B. M. Watts, *Metal. Sci.*, Vol 3, 1969, p 41.

125. G. Herriot, M. Suery and B. Baudelet, *Scripta Met.*, Vol 6, 1972, p 657.
126. H. W. Hayden and J. H. Brophy, *Trans. ASM*, Vol 61, 1968, p 542.
127. T. H. Alden, *Trans. ASM*, Vol 61, 1968, p 559.
128. C. M. Packer and O. D. Sherby, *Trans. ASM*, Vol 60, 1967, p 21.
129. H. Naziri, R. Pearce, M. Brown and K. F. Hale, *Acta Met.*, Vol 23, 1975, p 489.
130. L. Holt, *Trans. TMS-AIME*, Vol 242, 1968, p 25.
131. T. H. Alden and H. W. Schadler, *Trans. TMS-AIME*, Vol 242, 1968, p 825.
132. G. W. Franti and H. G. F. Wilsdorf, Report MS-3557-102-75, University of Virginia, Feb 1975.
133. M. L. Vaidya, K. L. Murty, and J. E. Dorn, *Acta Met.*, Vol 21, 1973, p 1615.
134. C. M. Packer, Ph.D. Thesis, Stanford University, Stanford, CA, 1967.
135. H. Naziri and R. Pearce, *J. Inst. Metals*, Vol 101, 1973, p 197.
136. R. Kossowsky and J. H. Bechtold, *Trans. TMS-AIME*, Vol 242, 1968, p 716.
137. R. J. Prematta, P. S. Venkatesan and A. Pense, *Met. Trans. A.*, Vol 7A, 1976, p 1235.
138. H. W. Schadler, *Trans. TMS-AIME*, Vol 242, 1968, p 1281.
139. H. E. Cline and T. H. Alden, *Trans. TMS-AIME*, Vol 239, 1967, p 710.
140. S. W. Zehr and W. A. Backofen, *Trans. ASM*, Vol 61, 1968, p 300.
141. P. Chaudhari and S. Mader, *Applied Polymer Symposium*, Vol 1, 1969, p 1.
142. T. H. Alden, *Acta Met.*, Vol 15, 1967, p 469.
143. M. A. Clark and T. H. Alden, *Acta Met.*, Vol 21, 1973, p 1195.
144. H. Naziri and R. Pearce, *J. Inst. Met.*, Vol 98, 1970, p 71.
145. J. D. Lee and P. Niessen, *Met. Trans.*, Vol 4, 1973, p 949.
146. K. N. Melton and J. W. Edington, *Scripta Met.*, Vol 9, 1975, p 559.
147. D. L. Bly, Ph.D. Thesis, Stanford University, Stanford, CA, 1973.
148. B. Sagat, P. Blenkinsop and D. M. R. Taplin, *J. Inst. Metals*, Vol 100, 1972, p 268.
149. G. L. Dunlop, E. Sapiro, D. M. R. Taplin and P. Crane, *Met. Trans.*, Vol 4, 1973, p 2039.
150. D. Weinstein, *Trans. TMS-AIME*, Vol 245, 1969, p 2041.
151. D. Lee, *Acta Met.*, Vol 17, 1969, p 1057.
152. D. Lee and W. A. Backofen, *Trans. TMS-AIME*, Vol 239, 1967, p 1034.
153. H. E. Cline, *Trans. TMS-AIME*, Vol 239, 1967, p 1906.
154. A. Karim and W. A. Backofen, *Met. Trans.*, Vol 3, 1972, p 709.
155. H. E. Cline and D. Lee, *Acta Met.*, Vol 18, 1970, p 315.
156. R. L. Coble, *J. Appl. Phys.*, Vol 34, 1963, p 1969.
157. R. H. Johnson, C. M. Packer, L. J. Anderson and O. D. Sherby, *Phil. Mag.*, Vol 18, 1968, p 1309.
158. Shu-en Hsu, G. R. Edwards and O. D. Sherby, *Acta Met.*, Vol 31, 1983, p 763.
159. J. H. Gittus, *J. Eng. Mater. Tech.*, Vol 99, 1977, p 244.
160. J. H. Gittus, *Phil. Trans. Roy. Soc.*, Vol A288, 1978, p 121.
161. R. C. Gifkins, *Scripta Met.*, Vol 7, 1973, p 27.
162. O. D. Sherby and P. M. Burke, *Prog. Mater. Sci.*, Vol 13, 1967, p 325.
163. A. K. Mukherjee, J. E. Bird and J. E. Dorn, *Trans. ASM*, Vol 62, 1969, p 155.
164. J. Weertman, *Trans. ASM*, Vol 61, 1968, p 681.
165. G. R. Edwards, T. R. McNelley and O. D. Sherby, *Scripta Met.*, Vol 8, 1974, p 475.
166. O. D. Sherby, R. H. Klundt and A. K. Miller, *Met. Trans. A*, Vol 8A, 1977, p 843.
167. C. R. Barrett and O. D. Sherby, *Trans. TMS-AIME*, Vol 223, 1965, p 1116.
168. W. R. Johnson, C. R. Barrett and W. D. Nix, *Met. Trans.*, Vol 3, 1972, p 963.

Question from Dr. U. F. Kocks,
Argonne National Laboratory, Argonne, IL

Your contention that grain-boundary sliding is an independent process has led you to conclude that making slip harder should extend the regime of super-plasticity. Do you know of any experimental evidence to support this surprising prediction?

Authors' Reply

We believe that methods of inhibiting slip deformation are a powerful means of enhancing superplastic behavior of fine-grain materials. This can be understood from our model, which is similar to the "core and mantle" concept of Gifkins. In this model, two independent modes of deformation can occur in polycrystalline material. One mode is diffusion-controlled dislocation creep and occurs in the "core" of each grain. The other mode is grain-boundary sliding accommodated by slip and occurs in the "mantle" regions. An example was shown in Fig. 19 of how the region of superplasticity can be extended by inhibiting deformation of the core through a decrease in grain size. Grain refinement (which decreases the core size) is known to inhibit slip. This inhibition of slip causes the yield strength to be increased, following the well-established Hall-Petch relation. On the other hand, grain refinement increases the contribution of grain-boundary sliding to plastic flow, because the amount of "mantle" material is increased. Thus, the region of high strain-rate sensitivity is extended over a wider range of strain rate by a decrease in grain size for two reasons: because of "core" strengthening and because of an increase in the number of "mantles." An example of possible enhancement in the strain-rate range of superplasticity by "core strengthening" alone was presented by us in Fig. 20. In this figure, we illustrate the inhibition of slip in nickel by the addition of tungsten, a solid-solution-strengthening element. If grain-boundary sliding is essentially unaffected by the presence of tungsten in the "mantle" region, then one can expect the range of superplastic flow to be extended to higher-than-normal strain rates. This prediction requires experimental verification, because the contribution of tungsten to slip accommodation in the "mantle" region is unknown.

Transformation Plasticity and the Stability of Plastic Flow

G. B. OLSON

Massachusetts Institute of Technology
Cambridge, MA

Abstract

Deformation stimulates martensitic transformations via *stress-assisted* nucleation on the same sites responsible for the transformation on cooling, and *strain-induced* nucleation on new sites produced by plastic strain. Quantitative models for the kinetics of each of these transformation modes allow prediction of constitutive flow relations for the transformation plasticity that accompanies transformation under stress. The flow stress is influenced by both the *static hardening* contribution of the transformation product, and the *dynamic softening* effect of the transformation as a deformation mechanism. These combined effects can distort the true stress-strain curve to an upward-curving shape approximating an ideal *exponential hardening* behavior which imparts maximum stability of the macroscopic flow behavior. Application of transformation plasticity has achieved enhanced ductility and toughness and other unusual mechanical properties in a variety of materials including metals, ceramics, and biological materials.

Introduction

Transformation plasticity, the macroscopic plastic strain associated with phase transformations under stress, has allowed the achievement of unusual mechanical properties in a wide range of materials. The phenomenon has been observed not only in the well-known case of martensitic transformations, but also in diffusional allotropic transformations and precipitation reactions, including carbide precipitation during the tempering of steels.[1-6] In the case of diffusional reactions, transformation plasticity appears to be adequately explained in terms of stress biasing of microscopic plastic accommodation around transforming particles, triggered by the transformation volume change.[3] For such cases, repeated cycling through the transformation under stress can produce large uniform deformations similar to superplasticity. The maximum strain obtained in one cycle, however, is generally on the order of 1%. In contrast, strains an order of magnitude higher can be obtained during martensitic transformations under stress. Transformation plasticity arises in this case not only from biasing of accommodation slip, triggered by both the volume change and the transformation shear,[5] but from the martensitic transformation shape strain itself as a result of stress biasing of the martensitic orientation variants.[4] The relative contributions of these two factors vary with the plastic-flow properties of the parent and martensitic phases in a given material. The large transformation plasticity accompanying martensitic transformation allows substantial alteration of mechanical behavior under monotonic as well as cyclic loading conditions. A dramatic illustration of the potential of this phenomenon is given by the experimental high-strength TRIP (TRansformation-Induced-Plasticity) steels,[7] which exhibit an extraordinary combination of strength, ductility, and toughness.

Transformation plasticity provides a significant departure from the classical view of structure/property relationships in materials. The traditional role of transformation kinetics is in the control of materials processing to achieve desired properties associated with the structure of the transformation product. Via transformation plasticity, one can design metastable materials to exploit the properties of the transformation itself, i.e., properties of structural *change*. Under these circumstances, knowledge of transformation kinetics can be directly applied to the prediction of the constitutive flow relations underlying observed macroscopic mechanical behavior of materials during service as well as during processing. These principles are illustrated here for the case of martensitic transformation plasticity, with special attention to the behavior of steels.

Kinetics of Deformation-Induced Transformation

Deformation can stimulate the kinetics of solid-state phase transformations through both the thermodynamic effect of applied stress and the production of new catalyzing defects by plastic strain. For martensitic transformations, heterogeneous nucleation on the same sites responsible for transformation on cooling but assisted by stress is termed *stress-assisted*, whereas nucleation on new sites produced by plastic strain is termed *strain-induced*.[8] Although many details are still under investigation, the essential mechanism of heterogeneous martensitic nucleation is now well understood[9,10] and is believed to be the same for both stress-assisted and strain-induced transformations. Observed nucleation sites, both those initially present and those produced by plastic strain, generally consist of dislocationlike linear defects with long-range stress fields (e.g., extrinsic interfacial defects and dislocation pile-ups). The energetics of classical heterogeneous martensitic nucleation at such defects has been equivalently treated using both dislocation-dissociation descriptions[11,12] and more traditional continuum descriptions[13-15] of the nucleus/defect elastic interaction. Under the thermodynamic driving forces at which these transformations occur, this interaction can lead to "barrierless" nucleation in which the kinetics are controlled by the propagation of the nucleus interface. As the net driving force acting on the interface increases with particle size, the interface accelerates to high velocities during growth. The rate-controlling initial growth start-up taking a nucleus to the size where rapid growth ensues has been termed "operational nucleation".[16]

Direct measurement of the intrinsic mobility of martensitic interfaces in single-interface transformation experiments[17] reveals that interfacial motion at low velocities is thermally activated in the same manner as for the motion of slip dislocations. For operational nucleation in steels at low temperatures, this accounts for an experimentally observed linear dependence of the nucleation activation energy Q on the transformation molar free-energy change (driving force) ΔG, expressed as:

$$Q = A + B\Delta G \tag{1}$$

where A and B are constants.* The constant B corresponds to the "activation volume" for thermally activated motion of the nucleus interface, and shows good agreement with available data for the kinetics of slip in similar alloys.[18,19] The intrinsic thermally activated (isothermal) character of initial martensitic nucleation events can sometimes be obscured by solute-drag interaction with mobile interstitials,[20] termed "dynamic stabilization",[18] or by the influence of autocatalysis[21] (stimulation of nucleation by growth events), which may in some cases lead to "overdriven" nucleation.

For transformation under conditions of imposed macroscopic deformation, the temperature regimes in which transformation initiates predominantly from either stress-assisted or strain-induced nucleation are depicted in the schematic stress-temperature diagram of Fig. 1. Spontaneous transformation triggered by existing nucleation sites occurs on cooling to the M_s temperature. Stress-assisted nucleation on the same sites will occur at the stress denoted by the solid line indicated. At a temperature designated M_s^{σ}, this stress reaches the yield stress σ_y for slip in the parent phase. Above M_s^{σ} new potent nucleation sites introduced by plastic strain trigger strain-induced nucleation at the stress level depicted by the indicated solid curve. The temperature M_s^{σ} thus defines an approximate boundary between the temperature regimes where the two modes of nucleation dominate; near M_s^{σ} both modes will operate. Due to transformation plasticity, the observed yield stress follows the stress for stress-assisted transformation below M_s^{σ}. A reversal of the temperature dependence of the yield stress thus provides a convenient determination of the M_s^{σ} temperature. Above the temperature M_d no trans-

---

*Equation 1 approximates the activation-energy behavior observed over the range of experimentally measurable activation energies. When Q is expressed as energy per mole of events, and ΔG as the (volume) free-energy change per mole of atoms, B is a dimensionless quantity representing the number of atoms in the activation volume controlling thermally activated nucleus interfacial motion. With ΔG expressed per unit transformed volume, B has dimensions of volume; the term $-\Delta G$ then represents a force per unit area acting on a planar interface, and can thus be regarded as a "thermodynamic driving force".[17]

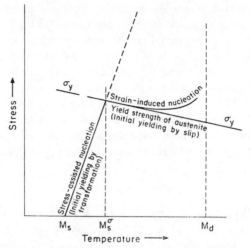

Fig. 1. Schematic representation of interrelationships between stress-assisted and strain-induced martensitic transformation.[8]

formation occurs on deformation to fracture. As will be discussed later, the diagram of Fig. 1 represents the behavior for a particular stress state.

Although the origin of the nucleation site need not influence the morphology of the resulting transformation product, the condition of general plastic flow prevailing in the strain-induced transformation regime apparently inhibits martensitic growth with a resulting change in product morphology. Martensite morphologies observed in a TRIP steel in the annealed condition are shown in Fig. 2.[22] Plate martensite formed during isothermal holding at 78 K without applied stress is shown in Fig. 2(a). A similar plate morphology formed by stress-assisted transformation at the same temperature is depicted in Fig. 2(b). Strain-induced transformation at 245 and 298 K (Fig. 2e and f) produces small martensite laths at shear-band intersections. Such intersections have been identified as the major nucleation sites for strain-induced transformation. Growth beyond the intersection regions is generally very restricted. At intermediate temperatures of 145 and 198 K, near M_s^σ, mixtures of the two morphologies are observed, as shown in Fig. 2(c) and (d).

Typical results of simultaneous measurements of stress-strain (σ-ε) and volume fraction martensite versus strain (f-ε) curves are shown in Fig. 3 for an annealed TRIP steel similar to that represented in Fig. 2.[22] The strong influence of transformation on strain-hardening behavior is evident from the qualitative similarity between the shapes of the two sets of curves. Details of these interrelationships will be discussed later. At higher temperatures where strain-induced transformation is dominant, the f-ε curves have a characteristic sigmoidal shape. At

Fig. 2. Morphology of martensites formed in 0.27C TRIP steel solution treated at 1200°C. (a) Martensite formed at −196°C without applied stress. (b-f) Martensite formed by deformation at various temperatures: (b) −196°C; (c) −128°C; (d) −75°C; (e) −28°C; (f) 25°C.[22]

low temperatures, below M_s^σ ($\sim$ −33°C), the f-ε curves associated with stress-assisted transformation show an initially linear behavior:

$$f = k\varepsilon \tag{2}$$

For sufficiently low temperatures where transformation occurs at stresses well below that for general slip in the parent phase, the constant k reaches a fixed value characteristic of "pure" transformation plasticity. Such linear behavior is expected when deformation is controlled entirely by the transformation.

The detailed kinetics of stress-assisted transformation can be predicted by simply adapting existing theory of the spontaneous (stress-free) transformation kinetics to take into account the additional thermodynamic contribution of applied stress. The latter has been estimated by calculating the interaction work of the resolved stresses acting through the shape strain of the most favorably oriented martensitic variant. For an invariant-plane shape strain with shear and normal components γ_0 and ε_0, the work is then[23]

$$W = \tau\gamma_0 + \sigma_n\varepsilon_0 \tag{3}$$

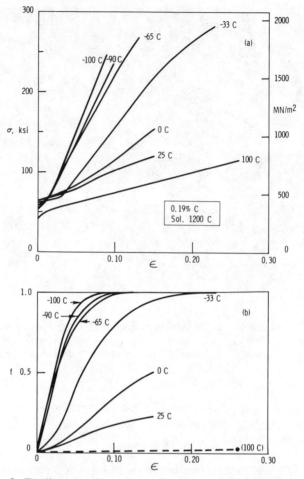

Fig. 3. Tensile properties and transformation curves for 0.19C TRIP steel solution treated at 1200°C. (a) σ-ε curves. (b) f-ε curves.[22]

where τ and σ_n are the shear and normal stresses resolved in the planes and directions of γ_0 and ε_0. Relevant Mohr's circle diagrams for calculation of W for the most favorably oriented variant in a random polycrystal are shown in Fig. 4 for stress states associated with uniaxial compression, uniaxial tension, and an elastic crack tip. Defining θ as the angle between the martensitic habit plane normal and the axis of the maximum magnitude principal stress σ_1, and expressing the resolved stresses in terms of the maximum and minimum principal stresses, Eq 3 becomes:

$$W = \tfrac{1}{2}\gamma_0 |\sigma_1 - \sigma_3| \sin 2\theta + \tfrac{1}{2}\varepsilon_0 [\sigma_1 - \sigma_3 + |\sigma_1 - \sigma_3| \cos 2\theta] \qquad (4)$$

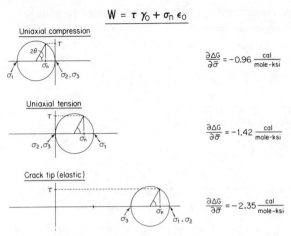

Fig. 4. Mohr's circle diagrams and thermodynamic assist of applied stress for stress states corresponding to uniaxial compression, uniaxial tension and elastic crack tip.[24]

Setting $dW/d\theta = 0$, the habit orientation for maximum W is given by $\tan 2\theta = \gamma_0/\varepsilon_0$. For the shape strain of ferrous plate martensites, $\gamma_0 = 0.20$, $\varepsilon_0 = 0.04$, and 2θ is $79°$. Expressing stress in terms of von Mises equivalent tensile stress $\overline{\sigma}$, the driving-force contributions of stress $\partial\Delta G/\partial\overline{\sigma}$ for the three cases in Fig. 4 are then -0.58 J/mole-MPa (-0.96 cal/mole-ksi), -0.86 J/mole-MPa (-1.42 cal/mole-ksi) and -1.42 J/mole-MPa (-2.35 cal/mole-ksi), respectively.[23,24]* These estimates are found to be in reasonable agreement with the stress-state dependence of the conditions for stress-assisted transformation (i.e., the stress-assisted transformation line in Fig. 1). As the $\overline{\sigma}$ for slip in the parent phase is relatively insensitive to stress state, the M_s^σ temperature (and M_d as well) will vary with stress state. A precise treatment of the kinetics of stress-assisted transformation will be developed during discussion of constitutive flow relations.

A quantitative model for the sigmoidal f-ε curves observed during *strain-induced* transformation in steels has been developed based on the assumption that shear-band intersections constitute the primary strain-induced nucleation site.[25] Such nucleation at intersections has been observed for shear bands consisting of hcp ε-martensite, mechanical twins, dense stacking-fault bundles, and slip bands.[26-29] The volume fraction of shear bands f^{sb} is assumed to be related to plastic strain by an expression of the form:

$$f^{sb} = 1 - \exp(-\alpha\varepsilon) \qquad (5)$$

---

*The unit J/mole-MPa is equivalent to $10^{-6} m^3$/mole.

where the rate (with respect to strain) of shear-band formation is determined by the single dimensionless parameter α. For an average shear-band volume $\bar{v}^{sb}$, the number of shear bands per unit volume N_v^{sb} is $f^{sb}/\bar{v}^{sb}$. The number of intersections N_v^I is assumed to be related to N_v^{sb} by a simple power law:

$$N_v^I = K(N_v^{sb})^n \tag{6}$$

where K is a geometric constant and $n \geq 2$. The number of martensitic units $N_v^{\alpha'}$ of average volume $\bar{v}^{\alpha'}$ is related to N_v^I by the probability p that an intersection will act as a nucleation site:

$$dN_v^{\alpha'} = p\,dN_v^I \tag{7}$$

These assumptions lead to an expression for the volume fraction of strain-induced martensite as a function of plastic strain of the form:

$$f^{\alpha'} = 1 - \exp\{-\beta[1 - \exp(-\alpha\varepsilon)]^n\} \tag{8}$$

where $\beta = p[\bar{v}^{\alpha'}K/(\bar{v}^{sb})^n]$. This relation gives sigmoidal transformation curves which saturate below $f^{\alpha'} = 1.0$. The saturation level is determined by the parameter β, while the rate of approach to saturation is controlled by both α and β.

Temperature dependence of the strain-induced transformation kinetics arises from both the α and β parameters. Relative to homogeneous slip, shear-band formation is favored by low intrinsic stacking-fault energy and high strain rate. The α parameter will be temperature dependent through the temperature dependence of stacking-fault energy. The β parameter, proportional to p, will be temperature dependent through the dependence of p on thermodynamic driving force, ΔG. Assuming a Gaussian distribution of the potency of intersections as nucleation sites, p will have the form of a Gaussian cumulative function with respect to ΔG or temperature T (when ΔG and T are linearly related). Figure 5 shows a fit of Eq 8 to available data[30] for strain-induced transformation kinetics during tensile deformation of a type 304 stainless steel, using $n = 4.5$. The temperature dependence of the corresponding α and β parameters is shown in Fig. 6 along with the associated nucleation-site potency distribution with respect to temperature, $n_v(T)$. The model accounts quite well for the shape of individual curves as well as the form of their temperature dependence. The strong temperature dependence of β near room temperature can pose a problem for the practical application of strain-induced transformation, as will be discussed later. In addition to the behavior in uniaxial tensile deformation depicted in Fig. 5 and 6, recent work[31] has demonstrated that the model applies equally well for other stress states provided that plastic strain is expressed as equivalent strain, $\bar{\varepsilon}$. The β parameter is stress-state dependent through the stress-state dependence of ΔG as developed in Fig. 4.

The temperature sensitivity of transformation kinetics just mentioned is of concern for both stress-assisted and strain-induced modes of transformation, and arises from both thermodynamic and kinetics factors which are subject to some

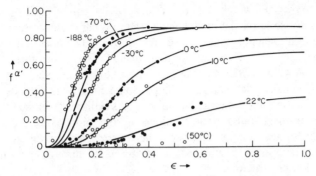

Fig. 5. Comparison of calculated strain-induced transformation curves with available data[30] for a 304 stainless steel. Experimental data are indicated by points. The solid curves represent the best fit of Eq 8 for each temperature, consistent with the assumed temperature dependence of the α and β parameters.[25]

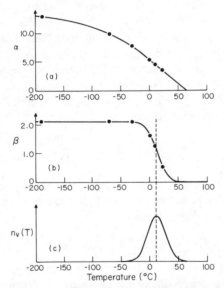

Fig. 6. Temperature dependence of the kinetic parameters derived from the data of Fig. 5. (a) α parameter, representing the rate of shear-band formation. (b) β parameter, representing the probability of an intersection forming a martensitic embryo. (c) $n_v(T)$, the Gaussian distribution of intersection potencies with respect to temperature giving the best fit for the derived dependence of β in (b).[25]

degree of control. Both modes of transformation are strongly influenced by the chemical contribution to the thermodynamic driving force, ΔG_{ch}. The temperature dependence of ΔG_{ch} is controlled by the transformation entropy change

$\Delta S_{ch} = -\partial \Delta G_{ch}/\partial T$, which can be influenced by composition modification based on alloy solution thermodynamics. The primary temperature-dependent kinetic factor influencing strain-induced transformation kinetics is the α parameter through its dependence on stacking-fault energy. The temperature dependence of intrinsic stacking-fault energy has been shown to be controlled by the fcc-hcp entropy difference, which is also subject to some control via composition modification guided by alloy thermodynamics.[11] As will be developed in the next section, the primary kinetic factor influencing the temperature dependence of stress-assisted transformation kinetics is the B parameter (activation volume) of Eq 1, which may also be subject to some influence through control of composition and microstructure.

Constitutive Flow Relations

Detailed comparison of σ-ε and f-ε curves such as those depicted in Fig. 3 reveals two major factors controlling the plastic-flow behavior during deformation-induced martensitic transformation.[22] In addition to the well-known *static hardening* contribution of the transformation product, a *dynamic softening* contribution arises from the operation of the transformation as a deformation mechanism. The latter effect is most amenable to quantitative treatment in the case of stress-assisted transformation at sufficiently low temperatures (stresses) where plastic flow is entirely controlled by the transformation kinetics.

When the nucleation activation energy is of the form of Eq 1, the initial rate of transformation $\dot{f}$ can be expressed as[32]

$$\dot{f} = n_s V \nu \, \exp - \frac{A + B\Delta G}{RT} \qquad (9)$$

where n_s is the density of nucleation sites, V is the instantaneous mean martensitic plate volume, and ν is the nucleation-attempt frequency. The critical ΔG required to achieve a given $\dot{f}$ is then given by

$$\Delta G_{crit}(\dot{f}) = -\frac{1}{B}\left(A + RT \ln \frac{\dot{f}}{n_s V \nu} \right) \qquad (10)$$

which shows a linear dependence on temperature, T. When plastic deformation arises from stress-assisted transformation, yielding in a constant strain-rate test will occur when the rate of transformation plasticity matches the imposed strain rate. From Eq 2, this corresponds to an imposed transformation rate of $\dot{f} = k\dot{\varepsilon}$. Calculation of the total thermodynamic driving force (chemical plus stress contributions) at the yield stress observed[22] at temperatures below M_s^σ for a high-strength TRIP steel verifies the linear $\Delta G_{crit}(T)$ relation of Eq 10.[24] This linear relation then predicts the critical stress for stress-assisted transformation σ_t depicted by the solid curve in Fig. 7 and compared against measured stresses for

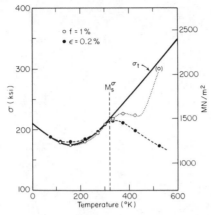

Fig. 7. Observed temperature dependence of 0.2% yield stress and stress at which 1% martensite is detected in high-strength TRIP steel.[22] Solid curve (σ_t) represents theoretical (tensile) stress for stress-assisted transformation.[24]

0.2% plastic strain and 1% transformation. In contrast to the linear behavior represented in Fig. 1, $\sigma_t(T)$ is in this case a curve which passes through a minimum at 158 K. The curvature arises from nonlinearity of $\Delta G_{ch}(T)$ at low temperatures, and the minimum is consistent with the behavior of relatively stable alloys which exhibit suppressible isothermal ("C-curve") transformation kinetics. Agreement between the yield stress and transformation stress at temperatures below M_s^σ in Fig. 7 verifies that yielding is controlled by stress-assisted transformation in this regime. Observation of transformation at stresses below σ_t at temperatures above M_s^σ is consistent with the contribution of strain-induced transformation. In the stress-assisted regime, Eq 2 and 10 also predict a strain-rate sensitivity of flow stress of the form:[24]

$$\frac{\partial \sigma}{\partial \ln \dot{\varepsilon}} = -\frac{RT}{B \dfrac{\partial \Delta G}{\partial \sigma}} \tag{11}$$

In line with the close relation between B and the activation volume for slip, the rate sensitivity of transformation plasticity is very similar to that for conventional plastic deformation.

Using the $\partial \Delta G / \partial \sigma$ values determined for the three stress states depicted in Fig. 4, the corresponding σ_t curves calculated for the alloy of Fig. 7 are compared in Fig. 8. Also shown are 0.2% yield stress measurements for uniaxial compression. An unusually large strength-differential (S-D) effect, comparing flow stress in compression and tension, is expected in the stress-assisted regime.

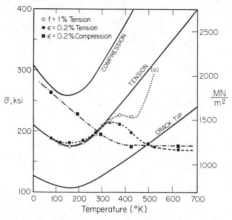

Fig. 8. Calculated transformation stresses and measured yield stresses for the TRIP steel of Fig. 7 under different stress states.[24]

Although an anomalous reverse S-D effect is observed near 300 K, the expected effect is observed at low temperatures, suggesting that flow is controlled by transformation in tension and slip in compression. The prediction for the crack-tip case suggests that transformation will control flow over a wider temperature range and that the effective flow stress near 300 K is significantly reduced relative to uniaxial tension. As will be discussed further, this stress-state sensitivity may be important to the achievement of improved strength/toughness combinations, allowing a high-tensile-strength material to benefit from the crack-tip plasticity of a softer material.

A knowledge of the kinetics of isothermal martensitic transformations can also be used to predict the shape of the σ-ε curve during stress-assisted transformation by taking into account the variation in the number of nucleation sites (n_s in Eq 10) with progress of the transformation.[24] A model which satisfactorily accounts for the course of stress-free isothermal martensitic transformation describes the nucleation-site density by an expression of the form:[32,33]

$$n_s = (n_i + pf - N_v)(1 - f) \tag{12}$$

Here n_i is the initial-site density, p is an "autocatalytic factor" accounting for new sites produced during transformation, and N_v is the number of martensitic plates per unit volume (accounting for sites which have already operated); the $(1 - f)$ factor takes into account potential sites which have been "swept up" by transformation. While the p factor can cause n_s to increase in the initial stages of transformation, a decreasing average plate volume $\overline{V}$ due to "partitioning" can cause the $-N_v$ term to reduce n_s at later stages. A saturation level of transformation is reached when $n_s \rightarrow 0$. Substituting Eq 12 into Eq 10, the flow stress during stress-assisted transformation can be expressed as:[24]

$$\sigma_t(f, \dot{f}, T) = -\left(B\frac{\partial \Delta G}{\partial \sigma}\right)^{-1}\left[A + B\Delta G_{ch} + RT \ln \frac{\dot{f}}{(n_i + pf - N_v)(1 - f)V\nu}\right]$$

(13)

Substitution of Eq 2 then provides a complete constitutive relation for $\sigma_t(\epsilon, \dot{\epsilon}, T)$. The shape of the σ-ϵ curve is determined by the behavior of the denominator in the last term of Eq 13. The yield stress is controlled by n_i, a stress drop is produced by the pf autocatalytic term, and approach to saturation causes rapid hardening as site depletion brings the denominator toward zero. Using the A and B parameters defined by the temperature dependence of σ_t in Fig. 7, and typical experimental values of the other kinetic parameters, a calculated σ-ϵ curve is compared with that measured at 158 K for the same TRIP steel in Fig. 9. The yield drop and flat portions of the curves are in good agreement, but the calculated final hardening (saturation) stage is too abrupt. This arises from the approximation of a singly activated process. The distribution of nucleation-site potencies is known to give a distribution of nucleation activation energies,[21] and this would account for a more gradual approach to saturation. It is important to note that the flow stresses depicted in Fig. 9 are all well below the stress for general yielding by slip. Relative to the "normal" slip-controlled flow behavior of this material, the phenomena thus far considered are entirely "softening" contributions. The high strain hardening associated with the saturation stage is best understood as the cessation of a softening phenomenon rather than the onset of a hardening mechanism. Once the transformation plasticity phenomenon (dynamic softening) associated with stress-assisted transformation is essentially complete, the higher strength of the transformation product makes its "static hardening" contribution to the flow stress.

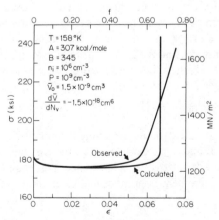

Fig. 9. Comparison of calculated and observed[22] true σ-ϵ curves for the TRIP steel of Fig. 7.[24]

While control of plasticity by transformation kinetics allows rather precise prediction of flow behavior in the stress-assisted transformation regime, the simultaneous operation of general slip and transformation in the strain-induced regime makes the quantitative treatment of flow behavior more difficult. Although much attention has been given to the static-hardening aspects of the two-phase mixture formed during strain-induced transformation, careful comparison of σ-ε curves has demonstrated that dynamic softening is also an important contribution. A recent study of the transformation and flow behavior of a metastable austenitic stainless steel, summarized in Fig. 10, has allowed a quantitative separation of the static hardening and dynamic softening contributions during strain-induced transformation.[34] The measured σ-ε curve of the metastable steel is labeled σ_{exp}. The dashed $\sigma_{\alpha'}$ and σ_{γ} curves represent the corresponding σ-ε curves of the martensite and stable austenite measured on specially designed similar steel compositions. One estimate of the static hardening behavior of the two-phase mixture is obtained from a simple "rule of mixtures" using the $\sigma_{\gamma}(\varepsilon)$, $\sigma_{\alpha'}(\varepsilon)$ data and the measured $f(\varepsilon)$ curve shown at the bottom of the figure; this estimate is depicted by the dashed curve labeled RM. Because the transformation plasticity contribution arising from biasing of the transformation shape strain does not contribute to the strain in either phase, the RM curve represents an upper limit to the static-hardening behavior. A strain-corrected rule of mixtures is obtained using σ_{γ} and $\sigma_{\alpha'}$ values at a strain of ε-αf, where αf corrects for the shape-strain contribution to the measured ε. An upper-limit estimate of the α coefficient is $\alpha = k^{-1}$, taking k (Eq 2) from the measured stress-assisted transformation behavior at low temperatures. This gives the solid curve depicted just below the dashed RM curve. The open points represent static flow-stress estimates determined by prestrain experiments. After the transforming material had been deformed to a particular strain, the static flow stress of the resulting two-phase mixture (in the absence of dynamic transformation softening) was determined from its flow stress subsequently measured at a higher temperature where the austenite is stable, correcting for its temperature dependence. These estimates and the RM estimate bracket the strain-corrected rule of mixtures estimate which is therefore taken as a reasonable approximation of the flow stress arising from static hardening, σ_s. This is expressed as:

$$\sigma_s = [1 - f] \cdot \sigma_{\gamma}(\varepsilon - \alpha f) + f \cdot \sigma_{\alpha'}(\varepsilon - \alpha f) \tag{14}$$

As indicated in Fig. 10, the dynamic softening increment $\Delta\sigma_d$ is then taken as the difference between σ_s and σ_{exp}. Correlation of the $\Delta\sigma_d$ increment thus obtained with f-ε behavior observed over a wide temperature range indicates that the fractional softening increment is proportional to $df/d\varepsilon$:

$$\frac{\Delta\sigma_d}{\sigma_s} = \beta \cdot \frac{df}{d\varepsilon} \tag{15}$$

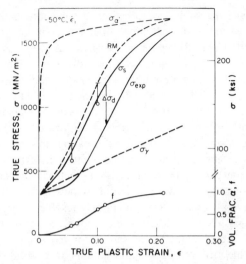

Fig. 10. Experimental flow stress, σ_{exp}, and volume fraction martensite, f, vs plastic strain, ε, for metastable austenitic steel at $-50°C$, $\dot{\varepsilon}_1 = 2.2 \times 10^{-4}$ s^{-1}. Dashed curves represent the stable austenite flow stress, σ_γ, the martensite flow stress, $\sigma_{\alpha'}$, and the prediction of the rule of mixtures for two-phase hardening, RM. Solid curve, σ_s, is prediction of strain-corrected rule-of-mixtures model.[34]

where $\beta = 5.3 \times 10^{-2}$. This then gives a constitutive flow relation for strain-induced transformation of the form:*

$$\sigma = \{[1 - f] \cdot \sigma_\gamma(\varepsilon - \alpha f) + f \cdot \sigma_{\alpha'}(\varepsilon - \alpha f)\} \cdot \left[1 - \beta \cdot \frac{df}{d\varepsilon}\right] \qquad (16)$$

Through Eq 16, $\sigma(\varepsilon, \dot{\varepsilon}, T)$ can be predicted from a knowledge of f, σ_γ and $\sigma_{\alpha'}$ as functions of ε, $\dot{\varepsilon}$ and T.

As in the case of stress-assisted transformation, dynamic softening is the dominant factor at low strains, causing the flow stress σ_{exp} of the transforming material in Fig. 10 to fall below that of the stable austenite, σ_γ. The static hardening becomes dominant at high strains. The combined effect of these two factors delays the maximum hardening rate $d\sigma/d\varepsilon$ to a higher strain than that where $df/d\varepsilon$ is maximum. The maximum $d\sigma/d\varepsilon$ arises from both the static hardening (proportional to $df/d\varepsilon$) and the diminution of dynamic softening (proportional to $-d^2f/d\varepsilon^2$) as $df/d\varepsilon$ decreases. Again, the cessation of a softening phenomenon provides a major contribution to the net rate of hardening.

---

*The α and β parameters in Eq 14 to 16 are distinct from those employed in Eq 5 to 8.

Before we leave the subject of constitutive behavior, some rather esoteric phenomena related to transformation/slip interactions in the vicinity of the M_s^σ temperature should be briefly mentioned. As evinced by the very limited growth of martensitic units in the strain-induced transformation regime, concurrent (or prior) slip apparently exerts a retarding influence on the motion of martensitic interfaces. The additional free-energy dissipation associated with interfacial motion during concurrent slip has been invoked to account for an observed negative rate-sensitivity of the flow stress at temperatures just below M_s^σ where flow is controlled primarily by stress-assisted transformation but where attendant accommodation slip is extensive;[35] the anomalous rate sensitivity has been demonstrated to be separate from interstitial-diffusion-controlled dynamic strain aging. At temperatures just above M_s^σ, a "pretransformation strengthening" effect is sometimes observed. Comparison of the critical resolved shear stress for slip in austenitic Fe-Ni-Cr alloys of differing thermodynamic stabilities indicates that just above its M_s^σ temperature, a metastable alloy has a higher yield stress than a stable alloy at the same temperature, despite its lower solid-solution strengthening indicated by higher temperature properties.[36] The anomalous reverse strength-differential effect near 300 K in Fig. 8 mentioned earlier may represent the same phenomenon, because comparison of behavior in two stress states is thermodynamically equivalent to comparison of two compositions of differing stability. The effect occurs in a temperature region where detectable transformation does not accompany yielding, and it has been proposed to arise from reduced slip-dislocation mobility due to possible core rearrangements resembling a coherent bcc structure as the thermodynamic conditions are approached for spontaneous martensitic nucleation by defect dissociation.[24] Such a strengthening mechanism would constitute an interaction of slip with metastability, rather than with transformation, *per se*.

Stability of Plastic Flow

Having developed constitutive flow relations for transformation plasticity based on the transformation kinetics, we may now consider the associated macroscopic flow behavior as predicted by continuum plasticity theory. The ability of transformation plasticity to dramatically alter the shape of a σ-ε curve naturally raises the question of the optimum or "ideal" curve shape for stability of plastic flow. The minimum strain-hardening rate required to maintain stable flow during tensile deformation is expressed by[37]

$$\frac{d\sigma}{d\varepsilon} = \sigma \tag{17}$$

A consequence of strain hardening is that the hardening rate necessary to maintain stability is increased. Excessive hardening therefore makes it more difficult to maintain stability at higher strains. It follows that the most efficient use of a hardening increment $\Delta\sigma$ is to distribute it with respect to strain in such a way that $d\sigma/d\varepsilon$ increases with ε. The ideal hardening curve, which provides the minimum hardening rate necessary to maintain stability at all strains, is given by the solution of the differential equation represented by Eq 17:

$$\sigma = \sigma_0 \exp \varepsilon \tag{18}$$

where σ_0 corresponds to the yield stress. In contrast to the usual downward curvature associated with the structural hardening and recovery processes controlling deformation by slip, the ideal flow relation of Eq 18 possesses upward curvature.[38] The "delayed hardening" behavior associated with the interaction of dynamic softening and static hardening phenomena in transformation plasticity provides an effective means of increasing $d\sigma/d\varepsilon$ with ε to achieve this curvature.

The ideal hardening behavior of Eq 18 is contrasted with the conventional downward-curving behavior, approximated by a power law $\sigma = K\varepsilon^n$, in Fig. 11. The open circle indicates the point where the power-law curve reaches the condition for flow localization by tensile necking, i.e., where $d\sigma/d\varepsilon$ falls below σ (Eq 17). From a local viewpoint which considers the conditions at the necking strain, one would attribute the necking to insufficient hardening, and would propose to increase uniform ductility by increasing the hardening exponent, n. From a perspective which considers the σ-ε curve as a whole, the necking can be equally well ascribed to excessive hardening at low strains. In the region where the $K\varepsilon^n$ curve lies above the $\sigma_0 \exp \varepsilon$ curve, the superposition of a *softening* mechanism which lowers the flow stress to the $\sigma_0 \exp \varepsilon$ curve would double the uniform ductility for the case depicted in Fig. 11. This illustrates the potential of the dynamic softening phenomenon in transformation plasticity for

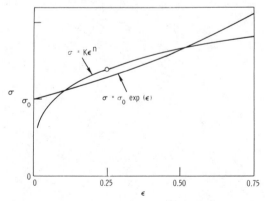

Fig. 11. Comparison of ideal exponential hardening behavior with power-law hardening. Open point indicates necking strain.

the enhancement of flow stability. At higher strains, where the $K\varepsilon^n$ curve falls below the $\sigma_0 \exp \varepsilon$ curve, superposition of a hardening mechanism, such as the high-strain static hardening contribution in transformation plasticity, is required to further maintain stable flow. If the ideal $\sigma_0 \exp \varepsilon$ curve is followed, necking will never occur, and uniform ductility will be controlled by fracture rather than plastic instability. Figure 11 illustrates that the dynamic softening contribution to transformation plasticity at lower strains can be equally important as the high-strain static-hardening contribution in promoting stability of plastic flow.

The enhancement of flow stability in metastable austenitic steels via transformation plasticity has been extensively investigated,[7,22,30,31,39-44] and analyzed in terms of Eq 18.[45,46] A continuous local strain measurement technique has allowed determination of true stress–true strain behavior of high-strength TRIP steels during both uniform and localized plastic flow under a variety of conditions.[22,45,46] Figure 12 summarizes such curves obtained for uniaxial tensile deformation of an austenitic TRIP steel of nominal composition Fe-9Cr-8Ni-4Mo-2Si-2Mn-0.3C strengthened by warm working to a 60% reduction at 450°C.[46] Arrows indicate the necking strain or uniform strain, ε_u. The temperature (stability) dependence of the strength and ductility properties is summarized in Fig. 13. The true σ-ε curves in Fig. 12 have a characteristic sigmoidal shape similar to those of the lower-strength annealed TRIP steel of Fig. 3. The temperature dependence of the 0.2% yield strength σ_y in Fig. 13 is similar to that of Fig. 7, indicating an M_s^σ temperature near 280 K. The ultimate tensile strength (UTS) reaches a maximum in the stress-assisted transformation regime, limited

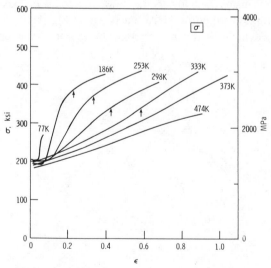

Fig. 12. True stress-strain curves of high-strength TRIP steel. Arrows indicate necking strain.[46]

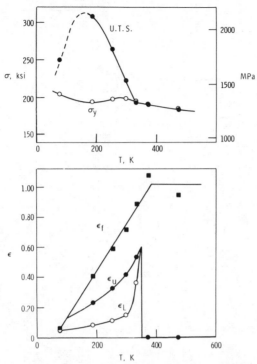

Fig. 13. Tensile properties of TRIP steel of Fig. 12, including Lüders strain ε_L, uniform (necking) strain ε_u, and fracture strain ε_f.

at lower temperatures by fracture. The latter effect is indicated by the intersection of ε_u and the fracture strain ε_f at the lowest temperature. The uniform ductility ε_u reaches a maximum above M_s^σ at a temperature near 350 K, reflecting an optimum over-all rate of deformation-induced transformation. As indicated by the shape of the σ-ε curves, early completion of transformation at low temperatures leads to early necking in the downward-curving portion of the σ-ε behavior that follows. Above 350 K the strain hardening associated with a reduced degree of transformation is insufficient to maintain stable flow in this high-strength material, and the uniform ductility is negligible. The existence of such an optimum stability for maximum uniform ductility in metastable austenitic steels in general is well established.[42-44]

The abrupt drop in ε_u beyond the peak in Fig. 13 is peculiar to the high-strength TRIP steels, arising from the complication of an initial transient flow instability in the form of a Lüders band.[7,45] The origin of the TRIP steel Lüders band phenomenon in the sigmoidal shape of the true σ-ε curves has been analyzed in terms of Eq 17 and 18,[45] and can be understood via the constructions of

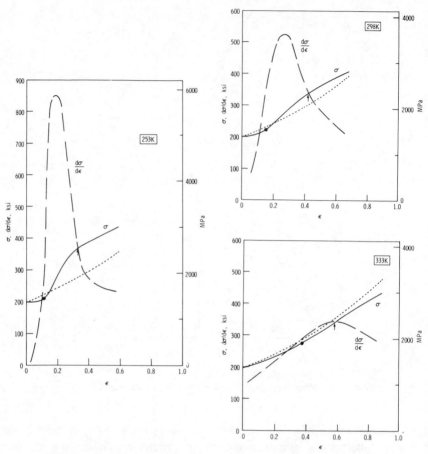

Fig. 14. Flow stress and strain-hardening rate of TRIP steel of Fig. 12 at temperatures of 253, 298 and 333 K. Dotted curves represent ideal exponential hardening behavior. Arrows indicate necking strains; solid points indicate Lüders strains.

Fig. 14, representing three of the σ-ε curves in Fig. 12. Solid curves depict the flow stress σ, dashed curves denote the strain-hardening rate $d\sigma/d\varepsilon$, and the dotted curves represent the ideal exponential hardening curves (Eq 18) consistent with the observed yield stress. Comparison of the σ and $d\sigma/d\varepsilon$ curves shows that the necking condition $d\sigma/d\varepsilon \leq \sigma$ is met not only beyond the necking strain ε_u indicated by the arrows, but also at yielding. Initial flow must therefore be localized in a manner essentially equivalent to final necking. As the strain-hardening rate increases with strain, however, flow can eventually stabilize at a Lüders strain ε_L; once this strain is reached in the initial "neck," flow will "delocalize" by the propagation of a Lüders front sequentially deforming material

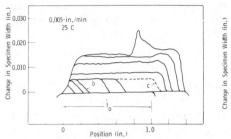 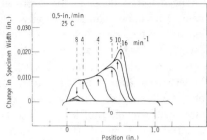

Fig. 15. Specimen width-change profiles during tensile deformation of TRIP steel. (Left) Isothermal deformation. (Right) Nonisothermal deformation. Numbers indicate local strain rate in min^{-1}.[45]

to ε_L. This is illustrated by the sequence of recorded tensile-specimen width-change profiles in Fig. 15 (left). Once propagation has brought the entire specimen gage section to the strain ε_L, simultaneous homogeneous straining of the gage section proceeds until final necking occurs at ε_u. Although deformation *during* Lüders-band formation and propagation is heterogeneous, the final result is a uniform strain ε_u. The Lüders strain can be predicted approximately from the shape of the true σ-ε curve by the constructions of Fig. 14. A lower limit to ε_L is given by the intersection of the dashed $d\sigma/d\varepsilon$ curve and the solid σ curve, since $d\sigma/d\varepsilon \geq \sigma$ is a necessary condition for stable flow. However, because the Lüders band has a reduced cross section relative to the adjacent material, additional hardening is required to compensate for this geometric difference in propagating the plastic flow. The adjacent material is loaded to the yield stress σ_0 when the σ curve crosses the dotted exponential hardening curve.[45] This intersection defines an upper limit to ε_L, as autocatalytic effects and a possible stress concentration at the Lüders front can allow plastic flow at nominal stresses below σ_0. The observed Lüders strains indicated by the closed points in Fig. 14 are seen to be consistently bracketed by these two intersections. As temperature and thermodynamic stability are increased, not only does the necking strain ε_u increase but so does the strain ε_L required for recovery of the initial instability. Eventually ε_u and ε_L become identical, as indicated in Fig. 13; ε_u drops abruptly to a negligible level as initial and final instabilities merge. The dashed $d\sigma/d\varepsilon$ curve for 333 K in Fig. 14 shows that the initial strain-hardening rate under the conditions for optimum stability with respect to uniform ductility is approximately 1000 MPa (150 ksi). Consistent with this hardening rate, the complication of the Lüders phenomenon is not significant for TRIP steels with yield strengths below this level. The curves for 333 K in Fig. 14 further indicate that for the conditions promoting optimum uniform ductility, the combined transformation plasticity effects of dynamic softening and static hardening bring the shape of the σ-ε curve almost exactly to that of the ideal exponential hardening curve.

The deformation behavior discussed thus far reflects isothermal test conditions. The temperature sensitivity alluded to in discussion of transformation kinetics further complicates flow stability under the nonisothermal conditions encountered at higher strain rates. Inhibition of transformation by heating can not only reduce uniform ductility, but the increased local strain rate accompanying Lüders-band formation (e.g., under conditions of fixed specimen boundary displacement rate) can further augment the problem.* Figure 15 (right) shows specimen width-change profiles recorded in a nonisothermal tensile test of the same material as in Fig. 15 (left).[45] A reduced transformation rate accompanying heating during deformation causes the Lüders band to propagate at increasing strains, merging continuously with final necking, and providing essentially no uniform ductility. This interaction of initial localization and nonisothermal transformation kinetics poses a serious problem to the practical application of transformation plasticity in the higher-strength TRIP steels.

The smooth stress-strain curves of Fig. 3, 9, 12 and 14 and the "diffuse" localization events (both Lüders bands and final necks) depicted in the width-change profiles are both macroscopic representations of the plastic flow behavior. On closer inspection, the actual σ-ε curves exhibit fine serrations, and the Lüders bands are actually composed of fine "shear bands" arising from local transient shear instabilities. Localization in a shear mode can arise from the condition:

$$\frac{d\sigma}{d\varepsilon} \leq 0 \qquad (19)$$

Due to the heterogeneous nature of martensitic transformation, particularly during autocatalytic events, the condition of Eq 19 can be locally met in transforming regions leading to clusters of martensitic units arranged in the form of bands, as shown in Fig. 16, which are found to be oriented at an angle of 48° to the tensile axis.[46] Whereas bands of pure shear deformation would be located at 45°, the higher angle is consistent with the additional contribution of the transformation volume change to the transformation plasticity. More macroscopic shear localization is evident when the over-all σ-ε curve shows strain softening as in the extreme case of the stress-assisted transformation behavior at yielding, as depicted in Fig. 6.

In addition to such transient localization effects, shear localization can play an important role in final instabilities, including the mechanism of ductile fracture in some high-strength steels.[47-49] Although the high-strain strain-softening phenomena which bring these materials to the shear-instability condition (Eq 19) are

---

*As discussed in Ref 45, Lüders-band formation in TRIP steels occurs under conditions where both strain hardening and strain-rate hardening are small but positive. This is in contrast to the Lüders bands associated with dynamic strain aging in mild steels, where both strain hardening and strain-rate hardening can be negative. The apparent rate sensitivity of flow-instability behavior in TRIP steels arises predominantly from the temperature sensitivity of the strain hardening associated with transformation plasticity.

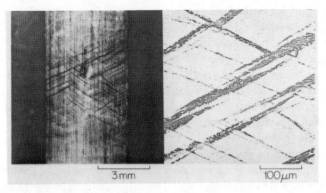

Fig. 16. Localized transformation bands within macroscopic Lüders band formed on yielding of high-strength TRIP steel of Fig. 7. (Left) Surface steps delineating local bands within Lüders band (oblique illumination). (Right) Microstructure of local bands.[46]

not yet precisely known, one would expect that the delayed hardening behavior associated with transformation plasticity could make an important contribution in delaying such final shear instabilities by preferentially increasing $d\sigma/d\varepsilon$ at the high strains where it is most needed. The possible relevance of this to enhancement of fracture toughness will be discussed in the next section.

Applications of Transformation Plasticity

Uniform Ductility

As discussed in the previous section, the flow-stabilizing influence of the σ-ε curve – shaping effects associated with transformation plasticity allows dramatic enhancement of uniform ductility under conditions of optimum thermodynamic stability, and this phenomenon is the most well-established transformation plasticity application. Although temperature sensitivity and Lüders-band behavior limit the useful ductility of high-strength TRIP steels, which depend almost entirely on their transformation behavior for stability of flow, the flow-stabilizing contribution of a moderate rate of transformation in lower-strength metastable austenitic steels has been practically applied for many years. Control of transformation plasticity has led to the development of numerous commercial high-formability stainless steels.

A recent study of the strain-hardening behavior of Hadfield's manganese steel[50] has shown that mechanical twinning in this alloy can be modeled in much the same manner as strain-induced martensitic transformation. The associated effects of dynamic softening and static hardening account for the exceptional stability of plastic flow in this material as well. The mechanical twinning phenomenon may be more amenable to application in higher-strength steels, owing to lower intrinsic temperature sensitivity compared to γ-α' martensitic transformation.

Several investigations have suggested that small amounts of metastable retained austenite in "dual-phase" ferrite/martensite steels may contribute significantly to uniform ductility via transformation plasticity.[51-53] Others have suggested that this contribution is negligible.[54,55] Of particular note are recent studies on the effect of tempering,[56] which demonstrate a significant decrease in uniform ductility after the second stage of tempering where retained austenite is eliminated.

The contribution of retained austenite to the strain-hardening behavior and uniform ductility of dual-phase steels has now been determined quantitatively by a series of controlled experiments using high-magnetic-field cryogenic treatments on an intercritically annealed 0.11C-0.5Si-1.4Mn-0.2Mo-0.1Cr steel.[57] A heat treatment consisting of austenitizing for ½ h at 1000°C and water quenching, followed by intercritical annealing for 7 min at 810°C and oil quenching (cooling rate, 30°C/s), was found to give the best combination of content and stability of retained austenite in this steel. The austenite content was substantially converted to martensite in identically heat treated control specimens by application of a sustained magnetic field of 22 tesla at 77 K for 1 h. The tensile-flow properties of the two materials were then measured at a strain rate of 0.02 min^{-1} and compared. Typical results are plotted in Fig. 17. Curve A represents the true stress-strain curve of the as-heat-treated material, while curve B depicts that of

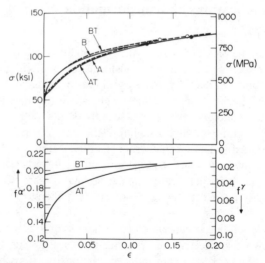

Fig. 17. True stress-strain curves and corresponding martensite contents during tensile deformation of dual-phase steel. Curve A, as heat treated; B, magnetic-field treated; AT and BT, tempered 1/2 h at 180°C. Solid and open points indicate necking strains.[57]

the magnetic-field-treated control material. The solid points denote the necking strain. To eliminate the possible complication of additional mobile dislocations introduced by the magnetically induced martensitic transformation in B, which might be different from those of the as-oil-quenched material A, specimens identically treated as A and B were tempered for 1 h at 180°C to fully "age" the dislocation substructures. Their true stress-strain curves are represented by the dashed curves AT and BT, with the open points denoting the necking strains. Retained austenite content as a function of strain was determined by x-ray diffraction measurements after interrupted straining of identically treated specimens; ferrite content was determined by quantitative metallography. These measurements provided the martensite and austenite volume fraction estimates depicted in Fig. 17 for the AT and BT materials. Very similar transformation behavior was observed for the untempered A and B materials.

The data presented in Fig. 17 allow comparison of the flow behavior of materials with the same UTS and identical martensite contents at UTS, but with a difference in initial austenite content of $\Delta f^\gamma = 0.05$. The transformation plasticity associated with transformation of this amount of austenite provides an additional increment of uniform ductility of 4 to 5% strain. A detailed analysis of the difference between the stress-strain curves has been performed according to the model of Eq 14 to 16, using $\alpha = 0.1$[34] and taking $\partial\sigma_s/\partial f^{\alpha'} = 954.3$ MPa (138.5 ksi) based on a recent determination[58] of the martensite-content dependence of flow stress in dual-phase steels at constant martensite carbon content. The difference between the curves is consistent with a dynamic softening β parameter equal to or slightly less than that determined for fully austenitic stainless steels.[34] The curve-shaping influence of relatively modest austenite contents is clearly a potent mechanism for enhanced control of the stability of plastic flow. Compositional modifications such as manganese increases to provide higher austenite contents will likely lead to a new class of low-alloy "triple-phase" steels with exceptional combinations of strength and formability.

The continuous yielding behavior of dual-phase steels is also of importance to formability. Circumvention of the discontinuous yielding normally encountered in microalloyed steels is commonly attributed to either the mobile accommodation dislocations or residual internal stresses induced by the martensitic transformation during dual-phase processing. Comparison of the AT and BT σ-ε curves in Fig. 17 reveals that discontinuous yielding observed in the BT material is eliminated by the presence of austenite in the AT material. Stress-assisted transformation at low strains, indicated by the shape of the AT $f^{\alpha'}$-ε curve in Fig. 17, is undoubtedly an important mechanism for controlling continuous yielding behavior.

Fracture Toughness

Enhancement of fracture toughness via transformation plasticity in high-strength TRIP steels is well established.[59-61] Quantitative determination of the temperature (thermodynamic stability) dependence of plane-strain toughness in a high-carbon TRIP steel revealed a transformation-toughening increment of $\Delta K_{IC} = 50$ MPa√m.[61] This increment was rationalized in terms of the additional dissipative work of the applied stress acting through the transformation shape strain (Eq 3) in the crack-tip plastic zone.[61] The underlying assumption that the transformation shape strain contributes an additional strain superimposed on the fracture process requires closer scrutiny.

Possible fundamental mechanisms of transformation toughening in metastable austenitic steels can be grouped into two classes of phenomena: (a) "flow delocalization" and (b) "dilatant plasticity." The flow-delocalizing effect of transformation plasticity can operate on two microstructural levels. On the size scale of a relatively coarse martensite plate produced by stress-assisted transformation, the transformation shape strain is a homogeneous deformation which may be superimposed on whatever inhomogeneous flow processes (e.g., void coalescense) control fracture; this allows the "additive strain" assumption of the toughening model just mentioned.[61] On a "continuum" level, the flow stabilizing influence of the σ-ε curve – shaping effect of fine strain-induced martensite might delay fracture-controlling inhomogeneous (localized) flow processes, particularly in the case of shear-instability-controlled ductile fracture. The "dilatant plasticity" aspects of transformation toughening could involve three phenomena: (a) direct $P\Delta V$ energy dissipation from the transformation volume change; (b) modification of the crack-tip stress state and effective stress intensity by the volume change; and, less directly, (c) the phenomenon of the stress-state sensitivity of transformation plasticity, which arises from the volume change (as represented in Fig. 8) allowing the triggering of plasticity (including shear strain) at a lower effective stress, enlarging the plastic zone and lowering crack-tip stresses.

Preliminary results of experiments directed at the mechanism of transformation toughening in austenitic steels are summarized in Fig. 18.[62] Two precipitation-hardening TRIP steels of nominal composition Fe-30Ni-3Ti-1.3Mo-0.2Al-0.3V-0.01B with and without 5% Cr were strengthened by warm rolling to a 40% reduction at 450°C and aging at 720°C to a yield strength of 1240 MPa (180 ksi). Figure 18 compares the temperature dependence of the J-integral toughness measured by multiple-specimen tests on the relatively stable 5% Cr steel, alloy A, and the less-stable Cr-free steel, alloy B. Both steels show a toughness increase in the temperature range where γ-α' transformation at the crack tip is observed (T ≤ 300 K for alloy A, T ≤ 400 K for alloy B) as indicated by the arrows in the figure. Extrapolation of the slope of the high-temperature behavior denoting toughness in the absence of transformation suggests a transformation-toughening increment of $\Delta J_{IC} = 120$ KJ/m² (600 in.–lb/in.²). This corresponds to a significantly higher toughness enhancement compared with the previous measurements on the high-carbon TRIP steel,[61] and may be due to differences in fracture mechanism. Whereas the latter steel failed by cleavage, fracture in the present

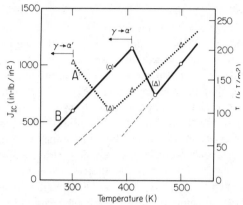

Fig. 18. Temperature dependence of J_{IC} fracture toughness of two precipitation-hardened TRIP steels of differing thermodynamic stability. Arrows indicate region of γ-α' transformation during testing. Points in parentheses estimated from single-specimen tests.[62]

case is predominantly ductile. Rather than by simple void coalescence, however, the ductile failure occurs by shear localization, giving a "zig-zag" mode of tensile fracture. The peak toughness of the transforming steels in Fig. 18 is associated with the formation of fine strain-induced martensite in the shear bands, suggestive of the proposed delay of plastic instability via transformation hardening. The "tearing modulus" defined by $(E/\sigma_0^2)(dJ/d\Delta a)$, corresponding to the slope of the J-integral/crack-advance curve normalized to the yield stress and elastic modulus, is found to double from 10 to 20 when transformation occurs. Further experiments are planned to isolate the toughening role of the transformation volume change by comparing the behavior of similar steel compositions designed to vary the volume change via modification of the austenite magnetic properties.

The possible toughening role of small amounts of retained austenite in martensitic steels remains a controversial area where controlled experiments are needed. Some recent studies suggest toughness improvements in tool steels[63] and carburizing steels[64] associated with retained austenite. The stress-oriented transformation product of interlath retained austenite films may inhibit low-temperature cleavage-crack propagation across the (normally low-angle) lath boundaries in packet martensites.[65] In ceramics, the deformation-induced transformation of dispersed metastable zirconia particles has produced dramatic improvements in fracture toughness.[66] Models proposed for the toughness improvement[67] have been essentially equivalent to the additive dissipation approach previously applied to steels.[61] A more recent calculation has demonstrated that the volume change of transformed particles left in the wake of a moving crack tip can reduce the effective local stress intensity, accounting for toughness enhancement.[68]

Fatigue Properties

Studies on fatigue-crack propagation (FCP) conducted under controlled stress-intensity amplitude (ΔK) conditions indicate that deformation-induced transformation retards crack growth in lower-strength metastable austenites, particularly at low ΔK,[69] and also exerts a beneficial influence in high-strength TRIP steels, although to a much lesser extent.[70] This growth retardation may be due to crack-closure effects arising from the transformation volume change, which may be particularly effective in the fatigue-threshold regime. Smooth-bar fatigue properties appear to be dominated by transformation hardening, which is desirable under stress-control conditions (reducing strain amplitude) but generally undesirable under strain-control conditions (increasing stress amplitude). In lower-strength austenites, transformation reduces fatigue life under conditions of controlled *plastic* strain amplitude;[71] under controlled *total* strain amplitude, transformation is detrimental to low-cycle fatigue life, but a small amount of transformation may be beneficial at high cycles.[72] Similarly, the low-cycle fatigue properties of high-strength TRIP steels are found to be degraded by transformation under controlled total strain amplitude.[73] Under *stress* control, the fatigue life of lower-strength austenites is greatly enhanced by transformation; for a stress ratio ($R = \sigma_{min}/\sigma_{max}$) of 0, fatigue limits in excess of the yield strength are observed.[74] Investigation of the smooth-bar fatigue properties of high-strength TRIP steels at $R = 0.1$, in which thermodynamic stability was varied by heat treatment, also revealed transformation enhancement of fatigue life.[75] Such enhancement allows the achievement of exceptional fatigue strength at high ultimate strength levels, as illustrated by comparison with other high-strength steels in Fig. 19.

Reversible Plasticity

As mentioned earlier, the relative contribution to transformation plasticity from biasing of accommodation slip versus biasing of the transformation shape strain varies among different materials. "Thermoelastic" alloys are distinguished by a virtual absence of accommodation slip,[76] and transformation plasticity in this case arises entirely from biasing of the shape-strain variants produced by stress-assisted transformation. Owing to the high coherency and intrinsic mobility of martensitic interfaces in the absence of extrinsic accommodation slip, reversion of the transformation by reverse interface motion leads to reversal of the associated transformation plasticity. Under conditions where the reversion occurs on unloading, the reversible plasticity* is termed "pseudoelastic" or "superelastic;" when reversion requires heating, the term "shape-memory effect" is applied.

While the monotonic σ-ε behavior of thermoelastic alloys shows the essential features of stress-assisted transformation plasticity as described by Eq 10 to 13,

---

*These deformations are "plastic" in the sense that they arise from intrinsically metastable rather than unstable (elastic) microscopic states.

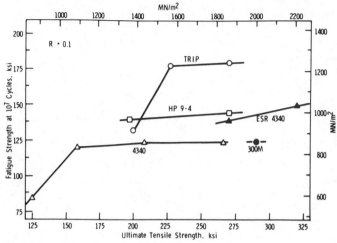

Fig. 19. Fatigue strength at 10^7 cycles (R = 0.1) vs ultimate tensile strength for TRIP steels compared with other high-strength steels.[75]

an additional contribution to transformation (and strain) arises from continued growth and reorientation of variants allowed by the mobile interfaces, and the kinetics cannot be modeled as being entirely nucleation-controlled. The σ-ε behavior associated with the formation of a single martensite plate during pseudoelastic deformation has been modeled[77] based on an analysis originally applied to the thermoelastic transformation on cooling.[76,78] The predicted behavior is summarized in Fig. 20, where the first diagram depicts an ideal case for which interfaces propagate without dissipation. The stress at which the parent and martensitic phases have equal free energies is τ_0. The stress τ_s is required to nucleate a martensite plate. Once nucleated, the plate grows rapidly. After radial growth is stopped by an obstacle (e.g., grain boundary), the plate thickens until a thermoelastic balance is achieved between the forces associated with elastic strain energy and the thermodynamic driving force.[76] This rapid event is depicted by the horizontal line at τ_s in Fig. 20(a), producing a shear strain γ proportional to the plate volume. On continued loading and subsequent unloading, the plate volume and associated strain increase and decrease along the straight line which extrapolates back to τ_0, as the plate thickens and thins (at fixed radius) according to the thermoelastic force balance. Upon unloading to τ_r, a radial force imbalance causes the plate to vanish and the associated strain is fully reversed. Hysteresis in this case arises entirely from the difference between the stresses τ_s and τ_r required for nucleation and reversion. If a finite friction stress τ_f is required to drive interfacial motion, additional hysteresis arises, as depicted in Fig. 20(b). Finally, the macroscopic τ-γ curve arising from progressive multiple nucleation and growth events can have the form of Fig. 20(c). It is evident

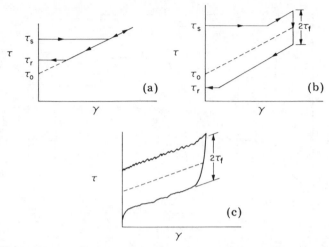

Fig. 20. Stress-strain hysteresis loops for pseudoelastic transformation plasticity in thermoelastic alloys. (a) Ideal behavior associated with a single martensite plate without interfacial friction ($\tau_f = 0$). (b) Behavior of single plate with interfacial friction ($\tau_f > 0$). (c) Macroscopic behavior of multiple plates.[77]

from the behavior depicted in Fig. 20 that thermodynamic conditions which lower τ_0 can cause strain reversal to be incomplete on unloading to $\tau = 0$, providing shape-memory behavior on reversal by heating instead of pseudoelastic behavior. Similar reversible plasticity phenomena can also be obtained by reorientation of variants and internal twin substructures by deformation after stress-free transformation on cooling. Numerous applications of reversible transformation plasticity are being developed, based primarily on Ti-Ni and various Cu-base thermoelastic alloys.[79]

Transformation Plasticity in Biology

A surprising number of biological structures involve periodic or crystalline arrays of protein molecules, some of which perform life functions by undergoing displacive phase transformations.[80] Figure 21 illustrates the example of tail-sheath contraction in the T4 bacteriophage, a virus which infects *E. Coli* bacteria. This virus consists of a DNA-filled icosahedral head or capsid to which is attached a tail assembly consisting of a rigid tail-core, a cylindrical tail-sheath, and a baseplate assembly with six long and short tail fibers. The tail-sheath is a metastable two-dimensional protein crystal closed to form a cylinder. When the tail fibers attach to the appropriate bacterial membrane, they distort the baseplate, which in turn triggers a strain-induced martensitic transformation in the crystalline tail-sheath. The transformation shape strain involves a substantial contraction which drives the rigid tail-core through the bacterial membrane, injecting

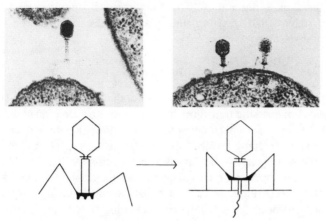

Fig. 21. Bacterial virus tail-sheath contraction.[80]

Fig. 22. Cylindrical crystal structure of virus tail-sheath during contraction by martensitic transformation. Interface is described by partial dislocations which spiral up helical close-packed crystal rows.[80]

the virus DNA into the bacterium. Figure 22 depicts a dislocation model of the martensitic interface transforming the crystal between the structures corresponding to the extended and contracted states. The shape strain involves a shear greater that 100%, which not only shortens the sheath but also causes a large twist about the cylinder axis, helping the tail-core to "bore" through the bacterial cell wall.

Although the example in Fig. 21 represents an irreversible martensitic transformation, reversible stress-assisted transformations producing shape-memory effects in similar cylindrical protein crystals are found to control motion in bacteria.[80] Transformation plasticity associated with martensitic transformations in protein crystals may represent a common mechanism of motion and shape control in living things. The "applications" in viruses and bacteria cited here suggest that transformation plasticity has actually been in use for 3.5 billion years.

Conclusions

Quantitative models have been developed for the kinetics of stress-assisted and strain-induced martensitic transformations. Constitutive flow relations for transformation plasticity have been derived based on these kinetic models, incorporating the static hardening effect of the transformation product and the dynamic softening effect of the transformation as a deformation mechanism. The shape of the predicted true stress-strain curves determines the stability of the macroscopic flow behavior, accounting for the Lüders-band formation and final necking behavior. The transformation softening and hardening effects can distort the stress-strain curve from the usual downward-curving shape to an upward-curving shape which approximates an ideal exponential hardening behavior promoting maximum uniform ductility. Transformation plasticity has been applied to the enhancement of ductility and toughness, and the achievement of other unusual mechanical behavior, in a wide variety of materials including metals, ceramics, and biological materials.

Acknowledgments

The author is grateful to Professors M. Cohen, I. W. Chen and A. S. Argon of MIT and Dr. U. F. Kocks of Argonne National Laboratory for helpful discussions, and to Dr. M. Azrin of AMMRC, Watertown, MA, for permission to use unpublished results obtained at AMMRC. Research at MIT on martensitic transformations and structure/property relationships is sponsored by the National Science Foundation under Grant #DMR-7915196, and by the Office of Naval Research under Contract #N00014-81-K-0013.

References

1. L. F. Porter and P. G. Rosenthal, *Acta Met.*, Vol 7, 1959, p 504.
2. F. W. Clinard and O. D. Sherby, *Acta Met.*, Vol 12, 1964, p 911.
3. G. W. Greenwood and R. H. Johnson, *Proc. Roy. Soc. (London)*, Vol A283, 1965, p 403.
4. C. L. Magee and H. W. Paxton, *Trans. AIME*, Vol 242, 1968, p 1741.
5. R. A. Kot and V. Weiss, *Met. Trans.*, Vol 1, 1970, p 2685.
6. R. L. Brown, H. J. Rack, and M. Cohen, *Mat. Sci. Eng.*, Vol 21, 1975, p 25.
7. V. F. Zackay, E. R. Parker, D. Fahr, and R. Busch, *Trans. ASM*, Vol 60, 1967, p 252.
8. G. B. Olson and M. Cohen, *J. Less-Common Metals*, Vol 28, 1972, p 107.
9. G. B. Olson and M. Cohen, *Ann. Rev. Mater. Sci.*, Vol 11, 1981, p 1.
10. G. B. Olson and M. Cohen, *Proc. Intl. Conf. Solid-Solid Phase Transf.* (Carnegie-Mellon Univ.), AIME, 1982, p 1145.

11. G. B. Olson and M. Cohen, *Met. Trans.*, Vol 7A, 1976, p 1897.
12. G. B. Olson and M. Cohen, *Met. Trans.*, Vol 7A, 1976, p 1905.
13. M. Suezawa and H. E. Cook, *Acta Met.*, Vol 28, 1980, p 423.
14. H. C. Ling and G. B. Olson, *Proc. Intl. Conf. Solid-Solid Phase Transf.* (Carnegie-Mellon Univ.), AIME, 1982, p 1171.
15. A. L. Roitburd, *Dok. Akad. Nauk SSSR*, Vol 256, 1981, p 80.
16. M. Cohen, *Met. Trans.*, Vol 3, 1972, p 1095.
17. M. Grujicic, G. B. Olson, and W. S. Owen, *Proc. Intl. Conf. Mart. Transf. ICOMAT-82*, Leuven, Belgium, *J. de Physique*, Vol 43, 1982, p C4-173.
18. C. L. Magee, in *Phase Transformations*, ASM, 1970, p 115.
19. G. B. Olson and M. Cohen, *Met. Trans.*, Vol 7A, 1976, p 1915.
20. W. S. Owen, F. J. Schoen, and G. R. Srinivason, in *Phase Transformations*, ASM, 1970, p 157.
21. C. L. Magee, *Met. Trans.*, Vol 2, 1971, p 2419.
22. G. B. Olson and M. Azrin, *Met. Trans.*, Vol 9A, 1978, p 713.
23. J. R. Patel and M. Cohen, *Acta Met.*, Vol 1, 1953, p 531.
24. G. B. Olson and M. Cohen, *Met. Trans. A*, Vol 13A, 1982, p 1907.
25. G. B. Olson and M. Cohen, *Met. Trans. A*, Vol 6A, 1975, p 791.
26. J. A. Venables, *Phil. Mag.*, Vol 7, 1964, p 35.
27. R. Lagneborg, *Acta Met.*, Vol 12, 1964, p 823.
28. P. L. Mangonon and G. Thomas, *Met. Trans.*, Vol 1, 1970, p 1577.
29. F. Lecroisey and A. Pineau, *Met. Trans.*, Vol 3, 1972, p 387.
30. T. Angel, *J. Iron Steel Inst.*, Vol 177, 1954, p 165.
31. S. S. Hecker, M. G. Stout, K. P. Staudhammer and J. L. Smith, *Met. Trans. A*, Vol 13A, 1982, p 619.
32. S. R. Pati and M. Cohen, *Acta Met.*, Vol 17, 1969, p 189.
33. V. Raghavan and M. Cohen, *Met. Trans.*, Vol 2, 1971, p 2409.
34. T. Narutani, G. B. Olson and M. Cohen, *Proc. Intl. Conf. Mart. Transf. ICOMAT-82*, Leuven, Belgium, *J. de Physique*, Vol 43, 1982, p C4-429.
35. R. H. Richman and G. F. Bolling, *Met. Trans.*, Vol 2, 1971, p 2451.
36. J. F. Breedis and W. D. Robertson, *Acta Met.*, Vol 11, 1963, p 547.
37. W. A. Backofen, *Deformation Processing*, Addison-Wesley, Reading, MA, 1972, Chap. 10.
38. U. F. Kochs, J. J. Jonas and H. Mecking, *Acta Met.*, Vol 27, 1979, p 419.
39. V. N. Krivobok and A. M. Talbot, *Proc. ASTM*, Vol 50, 1950, p 895.
40. H. C. Fiedler, B. L. Averbach and M. Cohen, *Trans. ASM*, Vol 47, 1955, p 267.
41. G. W. Powell, E. R. Marshal and W. A. Backofen, *Trans. ASM*, Vol 50, 1958, p 478.
42. J. P. Bressanelli and A. Moskowitz, *Trans. ASM*, Vol 59, 1966, p 223.
43. I. Tamura, T. Maki, H. Hato, Y. Tomota and M. Okada, *Proc. 2nd. Intl. Conf. Strength of Metals and Alloys*, Asilomar, CA, 1970, Vol 3, p 894.
44. V. F. Zackay, M. D. Bhandarkar and E. R. Parker, in *Advances in Deformation Processing*, edited by J. J. Burke and V. Weiss, Plenum Press, New York, 1978, p 351.
45. M. Azrin, G. B. Olson and R. A. Gagne, *Mater. Sci. Eng.*, Vol 23, 1976, p 33.
46. G. B. Olson and M. Azrin, unpublished research, AMMRC, Watertown, MA.
47. T. B. Cox and J. R. Low, *Met. Trans.*, Vol 5, 1974, p 1457.
48. J. A. Van den Avyle, Ph.D. Thesis, MIT, Cambridge, MA, 1975.
49. G. B. Olson, M. Azrin and N. Tsangarakis, in *Material Behavior Under High Stress and Ultrahigh Loading Rates*, Proc. 29th Sagamore Army Materials Research Conference, 1982 (in press).
50. P. H. Adler, S. M. Thesis, MIT, Cambridge, MA, Sept 1981.

51. T. Furukawa, H. Morikawa, H. Takecki and K. Koyama, in *Structure and Properties of Dual-Phase Steels*, AIME, New York, 1979, p 281.
52. J. M. Rigsbee and P. J. Vander Arend, in *Formable HSLA and Dual-Phase Steels*, edited by A. T. Davenport, TMS-AIME Conf. Proc., 1977, p 56.
53. A. R. Marder, *ibid.*, p 87.
54. G. T. Eldis, in *Structure and Properties of Dual-Phase Steels*, AIME, New York, 1979, p 202.
55. G. R. Speich and R. L. Miller, *ibid.*, p 145.
56. M. S. Rashid and B. V. N. Rao, *Met. Trans. A*, Vol 13A, 1982, p 1679.
57. T. Narutani, G. B. Olson and M. Cohen, unpublished research, MIT, Cambridge, MA, 1982.
58. A. M. Sarosiek Wojewodski, Sc.D. Thesis, MIT, Cambridge, MA, June 1982.
59. W. W. Gerberich, P. L. Hemmings, M. D. Merz and V. F. Zackay, *ASM Trans. Quart.*, Vol 61, 1968, p 843.
60. W. W. Gerberich, P. L. Hemmings and V. F. Zackay, *Met. Trans.*, Vol 2, 1971, p 2243.
61. S. D. Antalovich and B. Singh, *Met. Trans.*, Vol 2, 1971, p 2135.
62. R. Léal, MIT doctoral research in progress.
63. B. Lou and B. L. Averbach, *Met. Trans. A*, Vol 14A, 1983, p 1899.
64. M. M. Shea, SAE Technical Paper 780772, 1978.
65. J. I. Kim, C. K. Syn and J. W. Morris, Jr., *Met. Trans. A*, Vol 14A, 1983, p 93.
66. R. C. Garvie, R. H. Hannick and R. T. Pascoe, *Nature*, Vol 253, 1975, p 703.
67. A. G. Evans and A. H. Heuer, *J. Amer. Cer. Soc.*, Vol 63, 1980, p 241.
68. B. Budiansky, J. W. Hutchinson and J. C. Lambropoulos, *Intl. J. Solids and Structures*, Vol 19, 1983, p 337.
69. A. G. Pineau and R. M. N. Pelloux, *Met. Trans.*, Vol 5, 1974, p 1103.
70. G. R. Chanani, S. D. Antalovich and W. W. Gerberich *Met. Trans.*, Vol 3, 1972, p 2661.
71. G. Baudry and A. G. Pineau, *Mater. Sci. Eng.*, Vol 28, 1977, p 229.
72. D. Hennessey, G. Steckel and C. Altstetter, *Met. Trans. A*, Vol 7A, 1976, p 415.
73. G. R. Chanani and S. D. Antalovich, *Met. Trans.*, Vol 5, 1974, p 217.
74. R. G. Luther and T. R. G. Williams, *Met. Sci.*, Vol 11, 1977, p 219.
75. G. B. Olson, R. Chait, M. Azrin and R. A. Gagne, *Met. Trans. A.*, Vol 11A, 1980, p 1069.
76. G. B. Olson and M. Cohen, *Scripta Met.*, Vol 9, 1975, p 1247.
77. M. L. Green, M. Cohen and G. B. Olson, *Mater. Sci. Eng.*, Vol 50, 1981, p 109.
78. G. B. Olson and M. Cohen, *Scripta Met.*, Vol 11, 1977, p 345.
79. C. M. Wayman, *J. Metals*, Vol 32, 1980, p 129.
80. G. B. Olson and H. Hartman, *Proc. Intl. Conf. Mart. Transf.* ICOMAT-82, Leuven, Belgium, *J. de Physique*, Vol 43, 1982, p C4-855.

Adiabatic Strain Localization During Dynamic Deformation

Dr. H. C. Rogers

Drexel University
Philadelphia, PA 19104

Abstract

Deformation-induced thermal effects play a major role in strain localization during high-strain-rate deformation of metals. This strain localization generally takes the form of bands of intense local shear. In carbon steels strain localization is frequently further complicated by allotropic transformations occurring on heating. Under some conditions the band that forms is merely a region of very high local deformation; under others it appears to have transformed to an extremely hard, brittle structure. Band generation and band properties are discussed in terms of the deformation conditions as well as microstructure. Material variables that play a significant role in the susceptibility to adiabatic shearing are shown to be rate of strain hardening, thermal softening rate, and over-all strength level. Strain-rate sensitivity and microstructural stability are also discussed. The evidence supporting a critical strain and strain-rate requirement for adiabatic shear-band formation is reviewed and evaluated. Dynamic fracture is discussed in relation to the involvement of adiabatic shear banding. Some thoughts on current problem areas are put forth — in particular, the question of whether or not all the apparent adiabatic shear bands observed are actually so in fact.

Introduction

The term "adiabatic shearing" is applied to localized macroscopic shearing deformation that sometimes results when materials, particularly metals, are deformed at high rates. Rarely if ever can the deformation be considered truly adiabatic; however, the term is loosely applied whenever the localization is strongly thermally assisted. Perhaps the name used by Recht[1] — "catastrophic thermoplastic shear" — would be more appropriate, although the localization is not always catastrophic. The discussion throughout this paper will concentrate on the behavior of metals although some aspects of adiabatic shearing behavior are also applicable to polymers and other nonmetals as well. In metals, plastic

425

deformation takes place by the motion and interaction of dislocations. A small portion of the work of plastic deformation is stored as elastic strain energy. The remaining 90 to 95% is converted into heat. Zener ahd Hollomon[2] recognized in 1944 that localized temperature increases and resulting strain concentrations play a major part in high-speed deformation of metals. The degree of heat retention, the amount and distribution of the associated temperature rise, and the effect of elevated temperatures on the mechanical properties of the material undergoing high-strain-rate deformation comprise some of the major aspects of the adiabatic shearing phenomenology. During dynamic deformation, the amount of heat generated that is lost to the surrounding metal depends both on the strain rate and on the thermal properties of the metal.

No attempt will be made to be comprehensive in the coverage of all aspects of adiabatic shearing, but sufficient background and clarification will be given to put in context the discussion of the influence of material variables on this phenomenon. It is principally this aspect that will be addressed in this paper. Several extensive reviews of this phenomenon[3–5] are available if more information is desired. Short reviews of the pertinent facts of adiabatic deformation also exist in background sections of several recent reports and published articles. Furthermore, the material aspects specifically are discussed in some detail in another recent paper.[6]

A recognition of the magnitudes of the local strains and strain rates that arise because of adiabatic strain localization during dynamic* deformation is necessary to appreciate the difference between this regime and the more familiar quasistatic* deformation. Strains as large as 100 can be developed in microseconds, producing strain rates of 10^7 to 10^8 per second. Moreover, hypervelocity impact is not required to produce such strains and strain rates; they can result from extreme strain localization during what may be relatively low displacement rates, as in shearing of metal plate.

Adiabatic shearing is involved in a wide variety of processes or situations where dynamic deformation occurs. Ordnance problems, in particular, are concerned with adiabatic shearing phenomenology; these include armor penetration, penetrator performance, and explosive fragmentation. Machining, even at conventional speeds, produces high deformation rates in the workpiece ahead of the tool. Impact erosion of metals has now been convincingly shown to involve adiabatic shearing in material-removed mechanisms.[7,8] Other areas of commercial significance in which adiabatic shearing plays a significant role include ore crushing, impact tooling failure, and metal shaping and forming processes.

---

*The terms "static," "quasistatic," "dynamic" and "high" are relative terms the usage of which depends on the concerns of the investigator and the phenomena involved. In this paper, the term "dynamic deformation" will refer to those deformation rates in the general range that will cause adiabatic shear to appear. "Quasistatic" is applied to strain rates that are not static and yet do not cause significant local heating.

Again, these will not be discussed here in detail; if more information is desired, the review articles previously mentioned should be consulted.

One type of adiabatic strain localization is exemplified by the deformed shear band in the aluminum alloy 2014-T6 shown in Fig. 1. Precipitate particles and grain boundaries provide the markers for delineation of strain, which is obviously very high in an extremely thin zone of deformation. There is no evidence from the microstructure that this was an adiabatic shear band; only the fact that the deformation that resulted in the shear-band formation was caused by projectile impact would necessitate consideration of deformation heating as a contributing factor in strain localization. On the other hand, in the adiabatic shearing of carbon steels, a "signature" often described as a "white band" is frequently left (Fig. 2). The "white" appearance of such a band is caused by its failure to provide significant structural detail on etching with nital or other conventional steel etches. In addition to their peculiar etching characteristics, these white shear bands are

Fig. 1. Deformed shear band produced below a flat-nose projectile in aluminum alloy 2014-T6, showing the high degree of shear in the band.[4]

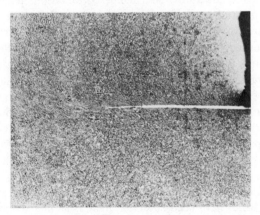

Fig. 2. A sectional plate clearly showing a shear band "propagating" ahead of a projectile in an impacted plate of AISI 1040 steel quenched and tempered at 400°C. Magnification, 100×.

also extremely hard. It was this hardness that led Zener and Hollomon[2] to the conclusion that they are bands of martensite resulting from the deformation heating of the steel. The local temperature rises above that required for austenitization, followed by a rapid quench by the surrounding metal on cessation of deformation. They estimated that the shear strain in the bands observed in mild steel might be as high as 100, but that a shear strain of only 5 would have been sufficient to heat the band to 1000°C if the deformation was adiabatic. That austenitization has occurred has not in fact been unequivocally established; nevertheless, the evidence is sufficiently good and the character of these bands sufficiently different from the normal "deformed" bands that they are referred to as "transformed" bands.

Deformed bands are likely to form in iron, in low-carbon steels and in higher-carbon steels when the structure is coarse. Also, as the rate of deformation decreases, there is a greater tendency for the shear bands to be of the deformed type. Backman and Finnegan[9] studied the tendencies of several different metallurgical structures in AISI 4130 steel to form different types of bands during ballistic impact. When the steel was heat treated to produce either pearlite or Widmanstätten ferrite, only deformed bands were observed. When the steel was heat treated to produce a tempered martensitic or lower bainitic structure, white-etching transformed bands were formed. It should be emphasized, however, that not all pearlitic structures are resistant to transformed band formation. If the structure is fine, such as in patented high-carbon steel wire, transformed bands are frequently observed under adiabatic shearing conditions.

White-etching bands are found in a wide variety of ferrous alloys and under a variety of conditions, as discussed above. Bands appear to form relatively easily in martensitic structures, both tempered to various degrees and untempered. Bainites and fine pearlites are also susceptible. To categorize compositions or microstructures of steels into two types, those that form transformed bands and those that form deformed bands, however, is an oversimplification. The type of band that forms is highly dependent on the specific deformation conditions. Ballistic penetration studies[10] show that within a particular steel an adiabatic shear band may be a transformed band along part of its length and a deformed band along the remainder. A transformed portion extends from the region of most intense shear to a point where the local temperature has not exceeded the transformation temperature. At greater distances the band is deformed only; the hardness varies along the length of the deformed band as the local temperature and maximum strain vary.

Not only during ballistic impact but in almost all other situations where adiabatic shear bands have been observed, there is an interaction between a tool or hard body and a somewhat softer body or workpiece. This generally leads to deformation in the workpiece that is not homogeneous at the outset and becomes increasingly inhomogeneous as the dynamic deformation continues.

There are also other reasons for initially inhomogeneous deformation. For example, Semiatin et al.[11,12] have shown that when titanium alloys are hot upset using cold tooling, thermal gradients are set up in the hot metal as a result of tool-workpiece contact. This in turn produces regions of higher and lower flow stress. This material inhomogeneity, combined with frictional effects, leads to localization of the deformation in narrow zones bordering the cooler regions. Only in the case of the fragmentation of explosively expanded smooth-bore steel cylinders is there no identifiable reason for the deformation to be initially inhomogeneous, although well-defined transformed bands are observed at later stages in the deformation process. Until further information is gained about either the explosive process itself or the identifiable structural inhomogeneities in the metal, band formation in this instance can be treated as a homogeneous nucleation-and-growth process. All other cases must be considered as instances of initially inhomogeneous deformation localizing catastrophically under the imposed dynamic displacements because of the thermal and mechanical properties of the deforming material.

Whether or not adiabatic shear bands are observed during any dynamic deformation process depends strongly on the state of stress that exists during deformation. They are not observed under tensile loading, because, as deformation starts to localize, fracture ensues rapidly by void growth and rupture. This prevents sufficient accumulation of local deformation to raise the temperature significantly. As the hydrostatic component of the stress operative during shearing in any dynamic loading operation becomes less tensile and more compressive, fracture will be increasingly suppressed. The greater local strains required for significant heating can then accumulate. Thus, adiabatic shear bands have been observed in torsional loading but are most commonly seen during dynamic deformation that involves a highly compressive stress state.

Since the conclusion of Zener and Hollomon nearly forty years ago that adiabatic shear bands of the transformed type resulted from deformation-induced heating and subsequent quenching, there have been many investigations into their actual structure, with conflicting results. There is no doubt that their hardness exceeds that obtainable by conventional quenching techniques in the same steel. Rogers and Shastry[10] have shown that the hardness of the transformed bands in a 1040 steel undergoing projectile impact is independent of the impact velocity within the range studied and is also independent of the hardness level of the steel target. The hardness of the bands was observed to increase linearly with the carbon level of the target steel. The substructure in these bands is generally reported to consist of fine, equiaxed grains with diameters in the range of tenths of micrometers. Rogers and Shastry[10] have also shown that the extraordinary hardness of these transformed bands can be accounted for by considering the combined hardening effect of ultrafine grain size and that resulting from the carbon supersaturation in the transformation product (i.e., martensite). They have also shown[13] by use of stain etching techniques that, at least in hypo-eutectoid steels having pearlite-ferrite microstructures, the structure of the white, apparently transformed band is not homogeneous but is a mixture of transformed structure and retained ferrite (Fig. 3). The scale is too fine to check the individual hardness of these two different structures. The fact that the transformed band structure is not always homogeneous may have been the cause of the different reported crystal structures attributed to such bands.

Analyses

There have been many analyses of the adiabatic shearing phenomenon in the past and they continue to appear, usually associated with experimental investigations involving dynamic deformation of one sort or another. These analyses have been commented on previously.[3,4] A more comprehensive discussion presented recently concentrates on the significance of the material variables that influence the tendency to localized adiabatic plastic deformation.[6] In the majority of the instances where adiabatic shearing is observed, the deformation is inhomogeneous from the outset. Most of the analyses, however, consider the initiation of plastic instability in a homogeneous strain field. Stress is considered to be a function of strain, strain rate and temperature. Clifton[14] uses perturbation theory to analyze adiabatic strain localization with results qualitatively similar to those of the instability analyses. Most of the models are variations of the instability analysis, which may be accompanied by a thermal model appropriate to the particular dynamic problem under investigation.

Fig. 3. "Transformed" shear band in 1040 steel with ferrite-pearlite microstructure etched in acidified sodium metabisulfite. Dark parts of band are actually transformed; white parts are ferrite contiguous with the matrix ferrite. Magnification, 500×.

In Recht's model of adiabatic shearing associated with machining, the comparative susceptibilities of two materials to adiabatic strain localization are evaluated in terms of the critical strain rate at which localization occurs in each:

$$\frac{\dot{\varepsilon}_{c_1}}{\dot{\varepsilon}_{c_2}} = \frac{(k\rho C)_1}{(k\rho C)_2} \left[\frac{\left(\frac{\partial \tau}{\partial \varepsilon}\right)_1 \left(\frac{\partial \tau}{\partial T}\right)_2}{\left(\frac{\partial \tau}{\partial T}\right)_1 \left(\frac{\partial \tau}{\partial \varepsilon}\right)_2} \right]^2 \left(\frac{\tau_{y_2}}{\tau_{y_1}}\right)^2 \tag{1}$$

where ρ is mass, C is heat capacity, k is thermal conductivity and τ_y is yield stress.

Using available data, Recht predicts that titanium and its alloys are approximately 1400 times as susceptible as mild steel. From a further analysis of this data it can be shown that, of the critical strain rate ratio of 1400 to 1, a factor of approximately 6 arises because of differences in the thermophysical properties and a factor of 4 is due to differences in the yield strengths of the two materials. The remaining and by far the largest factor of 58 derives from the ratio of two material characteristics—the rate of strain hardening, $(\partial \tau / \partial \varepsilon)$, to the rate of thermal softening, $(\partial \tau / \partial T)$.

Most models consider the effect of the strain-rate sensitivity of materials to be negligible compared with the effects of strain hardening and thermal softening. For example, Staker[15] and Olson et al.,[16] considering the controlled explosive expansion of thick-wall steel cylinders, considered the time to be so short during explosive loading that the process could be considered to be completely adiabatic with no effect on the thermophysical properties. Semiatin et al.,[11,12] however, consider strain-rate sensitivity to be a major factor in determining strain localization during the hot upsetting of titanium alloys either conventionally with cold tools or isothermally. In both instances there is a tendency toward localized flow because of die-workpiece interface friction that leads to restricted flow zones beneath the tools. For nonisothermal forging, chill zones set up by contact with the cold tools enhance the tendency toward localization of the deformation. A parameter, α, that measures a material's tendency toward marked flow localization is developed. This parameter is of the form

$$\alpha = \frac{\eta' - 1}{m}$$

where η' contains both the strain hardening and thermal softening terms while m is the strain-rate sensitivity $[m = (\partial \ln \sigma)/(\partial \ln \dot{\varepsilon})_{\varepsilon,T}]$. The -1 arises from the geometrical hardening that arises in axisymmetric compression. In plane-strain upsetting, shearing is favored and α becomes simply η'/m. There are two obvious differences between the temperature/strain rate regime that is being analyzed by Semiatin et al. and those being analyzed in the majority of the other studies of dynamic deformation: the temperature is high and the strain rate is relatively low. Also in nonisothermal forging, temperature gradients are set up immediately in the workpiece on contact with the tooling, giving an added time or rate dependence.

It is difficult to assess the various analyses and models utilized in terms of their general applicability. Although it appears relatively easy to find some fault with them on mathematical or other grounds, each nevertheless appears to be satisfactory if the application of a particular model is restricted to the range of processing and material conditions for which it was generated. It may simply be asking too much to seek a generalized model that accounts for all the observations of this complex phenomenon over a wide variety of conditions.

Material Properties

Analytical models, supported in general by experimental observation, agree that the material property most needed to stabilize against adiabatic strain localization is a high strain-hardening capacity, generally characterized by the strain-hardening exponent, n. The factor considered to be most conducive to strain localization is a strong tendency toward thermal softening. The role of strain-rate sensitivity is unclear and may depend on the particular process variables, including strain, strain rate, temperature, and geometry of loading.

In addition to these two material variables, another material characteristic that can have a significant effect on the generation of adiabatic shear bands is the basic strength level of the deforming material. Because the heat generated locally comprises the major part of the local specific work done by the deformation, it is obvious that for a given strain increment a higher flow stress causes a greater amount of heat generation and a greater temperature rise. The maximum temperature attained can be calculated from the relation

$$T_{max} = \frac{F}{C} \int_0^{\gamma_f} \tau d\gamma + T_0 \qquad (2)$$

where T_0 is the initial temperature, γ_f is the final shear strain obtained, and the factor F takes into account both the fraction of work converted to heat and the fraction of the heat generated that is lost to the surroundings through thermal conduction, etc. For any set of alloys of a particular base element, both the heat capacity, C, and thermophysical properties are usually similar, so that for equivalent deformation states the greatest amount of heat will be generated in the strongest alloy. This is at least one of the major reasons for the enhanced susceptibility to adiabatic shear-band formation generally exhibited by high-strength quenched-and-tempered steels in comparison with their lower-strength counterparts.

The concept of a critical strain required for the formation of transformed or strongly localized deformed bands such as those that develop in titanium alloys has been put forth recently by a number of authors.[7,17,18] Using totally different techniques and geometries, Culver,[17] Wulf[18] and Timothy and Hutchings[7] all show that a critical strain for shear-band development in Ti-6Al-4V exists and has a value in the range 0.1 to 0.2. The fact that the critical strains determined were roughly comparable despite significant differences in deformation conditions supports the argument that a critical strain required for the appearance of shear bands is a characteristic property of a material. In the Timothy and Hutchings investigation, moreover, several other possible alternative criteria for the appearance of shear bands were ruled out. In this study the microstructural damage associated with the ballistic impact of spherical projectiles on titanium alloy targets was examined. The density, size, and impact velocity of the projectiles were varied. The results show that velocity, kinetic energy and strain rate were unsatisfactory as alternative criteria for shear-band generation. Ongoing studies, at Drexel University, of ballistic impact on steels using stepped, flat-nose projectiles show that a critical depth of penetration of the target is required for the development of transformed adiabatic shear bands. The results to date on 1040 steel quenched and tempered at two different temperatures are shown in Fig. 4 and 5. With both tempers a comparable critical penetration depth is found. A further aspect of these studies is a determination of the rate at which the transformed bands "propagate" into the target ahead of the tip of the projectile. As shown, the length of the band ahead of the projectile appears to vary linearly

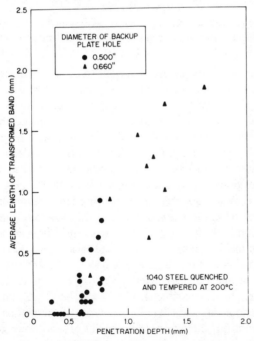

Fig. 4. Critical penetration depth and rate of growth of transformed bands with depth of penetration in steel hardened and tempered at 200°C.

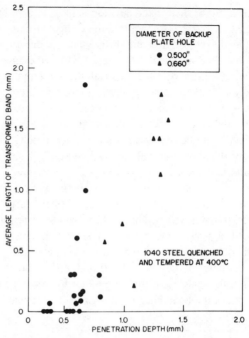

Fig. 5. Critical penetration depth and rate of growth of transformed bands with depth of penetration in steel hardened and tempered at 400°C.

with depth of penetration. By the use of various step heights, the maximum depth of penetration can be controlled while the impact velocity is varied over a considerable range. Obviously, at low velocities the minimum depth of penetration will be variable, depending on velocity, until the full depth of the step is attained. The results in Fig. 4 and 5 were obtained with a range of impact velocities that varied by a factor of two. Experiments are under way to study the behavior of this steel in other heat treated conditions as well as the behavior of other steel compositions under similar ballistic conditions.

Staker[15] has examined the effect of tempering temperature on the susceptibility of 4340 steel to the formation of transformed shear bands in the controlled explosive expansion of thick-wall cylinders. His results indicate that there exists a critical strain for the formation of transformed shear bands that is dependent on the heat treated condition of the steel. Figure 6 shows that the critical shear strain is proportional to a material parameter, $-Cn/(\partial\tau/\partial T)$, derived from a simple instability analysis. In this parameter, C is the volume specific heat, n is the strain-hardening exponent, and $(\partial\tau/\partial T)$ is the rate of thermal softening.

It is obvious from Eq. 2 that a certain amount of work must be done *locally* for the temperature in the deforming band to reach the transformation tempera-

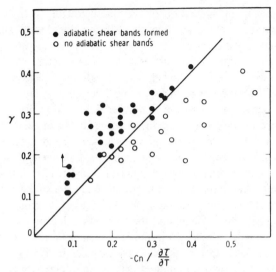

Fig. 6. The relation between measured true shear strain and a material parameter, $Cn(\partial\tau/\partial T)$, defining shear instability.[15]

ture. For a given material this would imply a critical *local* strain but not necessarily a critical macroscopic strain, which might well vary with strain rate and other conditions of dynamic loading. That there also appears to be a macroscopic critical strain in the case of titanium implies a relatively constant relationship between the local and macroscopic strains. When the work at Drexel is complete, the 4340 ballistic results can be compared with those of Staker for explosive loading. Because of the large differences in the strain rates employed in these two studies, the effect of this variable on the critical strain can be compared directly.

The fact that the transformed shear bands formed in steels are clearly identifiable makes measurement of their length and number unambiguous. However, as pointed out earlier, deformed bands can also result from thermally assisted strain localization both in ferrous and nonferrous materials. Because of their relatively diffuse nature, with transverse strain gradients in addition to gradients along their length, bands of this type are difficult to handle quantitatively and are usually not counted in studies of shear-band formation. They are adiabatic shear bands, nevertheless, and must be dealt with when our understanding of the phenomenon improves.

The effects of microstructure and changes in microstructure with strain rate and temperature on the susceptibility to adiabatic shearing are among the least understood aspects of the influence of material characteristics on the adiabatic shearing phenomenon. The term "microstructure" means here the collection and distribution of precipitates or second phases that influence the strength of metals in some degree. This means an interaction of the microstructure with the dislocation substructure to alter its character in terms of dislocation density and/or arrangement. The complexity that arises is that changes in the process variables — strain, strain rate and temperature — can alter both the microstructure and the dislocation substructure as well as modify the behavior of individual dislocations or groups of dislocations. Thus there is a "history effect" during any stage of the adiabatic deformation process, including the time after deformation has ceased. This is clearly illustrated by the results from the ongoing studies of adiabatic shearing at Drexel University shown in Fig. 7 and 8, discussed in more detail previously.[10] Room-temperature microhardness measurements transverse to a transformed band in an impacted 1040 steel target, quenched and tempered at 400°C, show the typical extreme hardness of such a band. On either side of the band are regions of matrix material that have suffered various degrees of deformation strengthening and thermal softening. Generally, the hardness values in the untransformed material decrease with distance from the band center until the matrix hardness is reached. This is also seen for the deformed bands observed in 1018 steel and 304 stainless steel. Thermal softening effects are most clearly identifiable, however, in the zones adjacent to the transformed bands in similarly impacted but more lightly tempered 1040 steel (Fig. 8). The deformation-induced heating, which was high enough in the center of the shear band to cause transformation, was also great enough in the adjacent heat-affected zones to cause a hardness reduction. The fact that such a reduction is observed only in the more lightly tempered steel is considered to be a reflection of its more thermally unstable microstructure. The major point to be made from these results is that the material

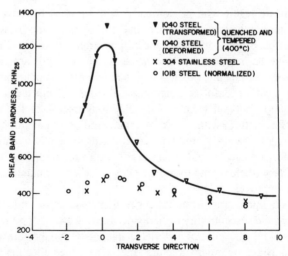

Fig. 7. Profile of the microhardness across adiabatic shear bands in several steels.[10]

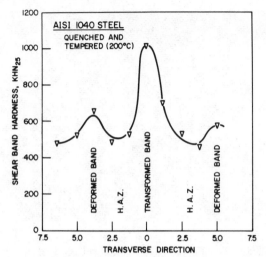

Fig. 8. Profile of the microhardness across a transformed shear band in AISI 1040 quenched and tempered at 200°C.[10]

that is hardest and strongest with respect to post deformation room-temperature mechanical behavior — the material at the center of the band — is precisely the material that was weakest and had the greatest amount of deformation and highest rate of deformation during dynamic loading. The thermal effects during deformation are thus overwhelming the effects of strain and strain rate on dislocation behavior while at the same time producing a dislocation substructure that is more resistant to dislocation motion at subsequent ambient temperatures and quasistatic rates of deformation. Kunze et al. also observed hardness increases of about 50% in deformed adiabatic shear bands in several nonferrous metals.[19]

Quenched-and-tempered steels tend to be particularly susceptible to adiabatic shearing, especially at the higher strength levels that require low tempering temperatures. It is not clear whether this increased susceptibility arises primarily from a lower thermal stability of the microstructures produced or from a lower rate of strain hardening and greater rate of heat generation caused by higher flow stresses. Recent studies have shed some light on this aspect of the problem.[6] A 1018 steel and 70-30 brass were compared in the annealed and 67% cold rolled condition with respect to response to ballistic impact from blunt-nose projectiles. The targets in all cases were 6.35-mm-thick plates. In both materials the strain localization was much more intense in the cold worked targets. Moreover, in the cold worked steel, a transformed band was observed to form at an impact velocity of 100 m/s (Fig. 9), although at 94 m/s only an intensely deformed band formed. The difference was established both from the appearance after etching and the high level of microhardness attained in the transformed band. These

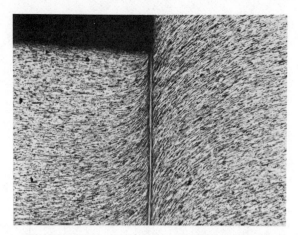

Fig. 9. Transformed shear band at corner of impact crater in 1018 steel, cold rolled 67% and impacted at 100 m/s. Etched in nital.

results confirm the role played by high strength and low strain-hardening capacity in increasing the susceptibility of metals to adiabatic strain localization independent of how these characteristics are obtained. The role of microstructural stability in susceptibility is still undetermined but must be secondary to that of the major material variables.

The above results also bear on the problem of whether or not a "critical strain rate" exists for adiabatic shearing in a given material. The above results and others obtained at Drexel appear to support this concept; the results of Timothy Hutchings[7] do not. Because all materials in fact conduct heat away from the zone of deformation and because no deformation is completely adiabatic, the rate of deformation must be a factor. Large strains can be achieved quasistatically in steels without transformed band formation; hence, provided that a minimum strain is achieved, there must exist a strain rate above which the removal of heat from the region of deformation is sufficiently limited that the temperature can rise above that needed for transformation to occur — a critical strain rate. At rates above this one would expect to see only limited effects of changes in strain rate. Thus, differences in the conditions of deformation may explain the apparently conflicting observations regarding the existence of a critical strain rate.

Fracture

The fracture of adiabatic shear bands frequently terminates the dynamic deformation process. The nature of the fracture or partial fracture can also provide information as to the stage of a complex loading situation in which fracture occurred. Although the fracture of these bands has been discussed in more detail elsewhere,[3,4] a few points will be made here. If the fracture is ductile in nature it almost certainly occurred during adiabatic deformation when the band was hot and weak. On the other hand, brittle fractures, particularly common in transformed bands in steels, occur subsequent to the termination of deformation and the quenching of the hot band by the adjacent matrix material to form the hard, brittle transformed structure. Brittle fracture in these fine-grain bands occurs normal to the tensile stress with little or no directionality attributable to their structure (Fig. 10). When ductile fracture occurs during deformation or immediately after deformation while the band is still hot and ductile, the type of fracture is determined by the stress state. The voids in the shear band in a U-2Mo alloy shown in Fig. 11 are equiaxed, indicating that the shearing deformation had

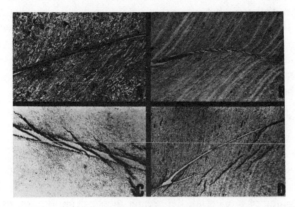

Fig. 10. Several types of brittle fractures of transformed adiabatic shear bands.[9]

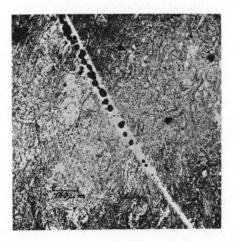

Fig. 11. Section of a fractured shear band in the nose portion of a U-2Mo alloy.[20]

ceased and that the hot band was subjected to normal tension. When bands fracture during shearing, the voids formed are extensively elongated in the shearing direction. Unfortunately, as in all fractures that take place by shearing under normal compressive loads, there is no tendency for the fracture surfaces to separate other than their possible nonplanarity, so that a large portion of the fracture is rubbed and scuffed and the original fracture surface completely obliterated in most areas.

Although it is not directly related to fracture during dynamic loading *per se,* an important practical point is frequently overlooked when considering the effects of adiabatic shear banding in steels. Once transformed bands have formed from one dynamic loading, although these bands may not be fractured, they remain as brittle fracture paths in what may be an otherwise ductile material. If the part or structure is then subjected to repeated loading, as is the case with hammers, ore-crushing equipment, and probably armor, chunks of the part can spall off as brittle fracture takes place through pre-existing brittle adiabatic shear bands. This may also occur during multiple-impact solid-particle erosion. In any case, the result may be catastrophic failure or rapid deterioration of the part containing such shear bands.

Discussion

The analyses and experimental evidence cited strongly support the argument that the smaller the rate of strain hardening and the greater the tendency for thermal softening of a material, the greater the susceptibility of that material to adiabatic strain localization. Considerable experimental evidence has also been accumulated to support the concept of a critical-shear-strain requirement for shear-band formation. The need for a minimum strain rate for transformed band formation is also fundamentally sound and is supported by the results on the cold worked 1018 steel. Once a minimum critical strain rate is exceeded, the further influences of strain rate and material strain-rate sensitivity are not clear. Additional study and evaluation are required to elucidate this aspect of adiabatic shearing in metals.

In addition to the specific material variables discussed, there are also a few general observations that must be kept in mind when considering the effect of adiabatic shearing on the response of materials to various dynamic loading conditions. From the preceding discussion of the influence of material strength on the tendency toward adiabatic strain localization, it is clear that the lower the strength of a material, all other material properties remaining the same, the more resistant it is to adiabatic shear-band formation. This is not necessarily equivalent to stating that, in a dynamic deformation situation, the stronger the material the more likely that adiabatic strain localization will occur. If the mode of loading is such that it is strain- or displacement-controlled, as in high-speed machining, this is true. However, if it is stress- or load-controlled, as in ballistic impact, the additional resistance of the high-strength material compared with a weaker material may prevent sufficient accumulation of deformation for such bands to develop, even though the bands may develop more easily for a given strain. Armor as well as hammers and other impact tooling should remain hard!

An aspect of the general problem that has not been adequately addressed is the temperature at which instability occurs if the deformation is initially uniform or, if not, when the initially inhomogeneous deformation begins to localize catastrophically. The analyses imply that there is little or no rise in the ambient matrix temperature prior to localization. There is nothing, however, that says that the adiabatic heating that occurs must be nonuniform *per se;* on the contrary, there may be a modest, essentially uniform temperature increase such as occurs during conventional deformation processes such as rolling. If instability does occur at temperatures above ambient, then obviously it is the material properties at that temperature that are the controlling factors rather than those at room temperature. The differences may or may not be minor depending on the temperature elevation prior to instability. At least the possibility should be kept in mind.

Of the many other possible topics for consideration that relate to adiabatic plastic deformation, there is one that is so basic that it cannot be ignored — the question of when and how frequently what is conventionally considered to be a thermally assisted catastrophic strain localization is in fact the result of frictional heating and rewelding of surfaces already fractured under highly compressive loading systems. Friction welding of two separate parts is a well-known and successful commercial joining process. It is also generally known that high-speed rubbing of two steel surfaces, such as occurs with aircraft carrier landing-arrest systems, can produce surface layers with characteristic properties very similar to those observed during adiabatic shearing. There is obviously no substantive difference between the frictional heating and welding of two separate bodies and a similar behavior of two adjacent fracture surfaces in the same body. That this does occur is graphically illustrated in Fig. 12. This is the etched cross section of a hardened O2 tool steel projectile that failed on impact as it struck a thick target. The projectile was circular in cross section and had a step in the center in the shape of a pillbox with a height of 0.94 mm. The "adiabatic shear band" that formed is strikingly revealed by etching. What in fact actually occurred is that a ring having a triangular cross section separated from the top of the step on impact or soon thereafter. Further penetration of the projectile into the target caused the ring to slide down the sloping fracture surface of the remainder of the step still attached to the projectile body. The increased axial and lateral pressure on the ring from the increasing penetration of the target caused the fracture surface of the ring to be forced heavily against the mating fracture surface as it slid toward the main body of the projectile. In addition to the compressive force normal to the sliding surfaces, there is an additional force required to expand the fractured

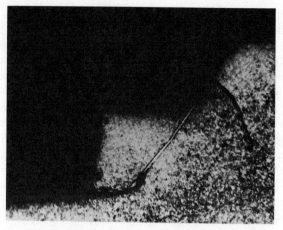

Fig. 12. Cross section of an O2 tool steel projectile fractured during impact. A white band appears between the fractured outer ring and the remainder of the original step. The white material does not, however, extend up to the original fracture surface.

ring as it moves toward the base of the truncated cone that comprises the remainder of the original step. The resulting frictional heating and welding of the two surfaces has produced what appears to be a classical transformed shear band.

Examination of the surface on the upper portion of the truncated conical remains of the original step shows, however, that fracture occurred by ductile rupture, with no evidence of a hard, brittle surface layer of adiabatically deformed and quenched steel. Despite the fact that much of the original fracture surface has been obliterated or smeared by sliding contact with the mating fracture surface of the ring, there remains a sufficient amount of the original fracture surface to clearly identify the fracture mode. Moreover, this mode of fracture is observed repeatedly in the failure of step-nose tool steel projectiles used in the impact studies.

These comments do not reflect the opinion that all observed transformed adiabatic shear bands are necessarily the result of the high-speed rubbing of two pre-existing fracture surfaces. Some probably are. In particular, some of the white bands observed at the surfaces of explosive fragments are of such a shape that it would appear to be difficult for them to have been generated by an adiabatic shearing process. Because all observations of this phenomenon are *a posteriori* as a result of the extreme brevity of the event, extra care must be taken in the interpretation of these observations.

The complexities of the problem do not permit broad generalizations to be made with a great degree of confidence. As discussed, it is even difficult to be certain what has occurred, let alone explain why it has occurred. With hypervelocity impact, pressure and inertial effects may be important additional perturbing factors for models developed at lower strain-rate ranges. Because of equipment and material limitations, there are few studies so comprehensive that they cover a broad range of deformation conditions, stress states, geometries, etc., for a wide variety of materials. Much work still remains before predictions of material behavior can be made with confidence from first principles and before practical materials can be developed that have maximum resistance to adiabatic strain localization while still being capable of carrying out their primary functions.

Acknowledgment

The ideas and original results presented herein have been generated as a result of a program at Drexel University that has been supported for several years by the U.S. Army Research Office. Dr. George Mayer's interest and stimulation are especially appreciated.

References

1. R. F. Recht, *J. Appl. Mech. Trans. ASME,* Vol 31E, 1964, p 189.
2. C. Zener and J. H. Holloman, *J. Appl. Phys.,* Vol 15, 1944, p 22.
3. H. C. Rogers, "Adiabatic Shearing: A Review," Drexel University Report prepared for U.S. Army Research Office, 1974.
4. H. C. Rogers, *Ann. Rev. Mater. Sci.,* Vol 9, 1979, p 283.
5. A. J. Bedford, A. L. Wingrove and K. R. L. Thompson, *J. Aust. Inst. Metals,* Vol 19, 1974, p 61.
6. H. C. Rogers, "Adiabatic Shearing–General Nature and Material Aspects," *Material Behavior Under High Stress and Ultra High Loading Rates,* in Proc. 29th Sagamore Army Res. Conf., Lake Placid, New York, 1982 (to be published).
7. S. P. Timothy and I. M. Hutchings, "Microstructural Features Associated with Ballistic Impact in Ti6Al4V," *Proc. 7th Int. Conf. on High Energy Rate Fabrication,* edited by T. Z. Blazynski, Univ. of Leeds, 1981, p 19.
8. R. E. Winter and I. M. Hutchings, *Wear,* Vol 34, 1975, p 141.
9. M. E. Backman and S. A. Finnegan, *Metallurgical Effects at High Strain Rates,* edited by R. W. Rohde et al., Plenum Press, New York, 1973, p 531.
10. H. C. Rogers and C. V. Shastry, in *Shock Waves and High-Strain-Rate Phenomena in Metals,* edited by M. A. Meyers and L. E. Murr, Plenum Press, New York, 1981, p 285.
11. S. L. Semiatin and G. D. Lahoti, *Met. Trans.,* Vol 12A, 1981, p 1705.
12. S. L. Semiatin, G. D. Lahoti and S. I. Oh, "The Occurrence of Shear Bands in Metalworking," *Material Behavior Under High Stress and Ultra High Loading Rates,* in Proc. 29th Sagamore Army Res. Conf. Lake Placid, New York, 1982 (to be published).
13. H. C. Rogers and C. V. Shastry, unpublished research.
14. R. J. Clifton, Adiabatic Shear Banding, Chap. 8 in *Materials Response to Ultra High Loading Rates,* National Materials Advisory Board Report No. NMAB-356, 1980.
15. M. R. Staker, *Acta Met.,* Vol 29, 1981, p 683.

16. G. B. Olson, J. F. Mescall and M. Azrin, in *Shock Waves and High-Strain-Rate Phenomena in Metals,* edited by M. A. Meyers and L. E. Murr, Plenum Press, New York, 1981, p 221.
17. R. S. Culver, *Metallurgical Effects at High Strain Rates,* edited by R. W. Rohde et al., Plenum Press, New York, 1973, p 519.
18. G. L. Wulf, *Int. J. Mech. Sci.,* Vol 21, 1979, p 713.
19. H.-D. Kunze, K.-H. Hartmann and J. Rickel, *Pract. Metallog.,* Vol 18, 1981, p 261.
20. C. J. Irwin, Metallographic Interpretation of Impacted Ogive Penetrators, DREV-R-652/72, Canada, 1972 (46 pages).

Characterization of Inhomogeneities in Complex Austenitic Stainless Steel Forgings

MARTIN C. MATAYA and MARTIN J. CARR
Department of Materials Technology
Rockwell International
Golden, CO 80401

Introduction

Forgings are by their nature inhomogeneous and anisotropic. Designs for forgings and their processing schemes have taken advantage of these qualities for many years now, either empirically (since the time of the Damascus sword),[1] or by science-based methodology (used more recently for jet engine components).[2] The selection of a particular forging design or forging process is generally made on the basis of the achievable properties at various places in the forging. The acceptability and ultimate performance of a forging are more commonly limited by localized departures from an intended structure or property. These limiting conditions may range from simple defects of one sort or another, to subtle variations in directional properties, or even to latent flaws which develop only when the finished product is exposed to the service environment. Most often, such problems are encountered during process development, but they occasionally crop up during a production run. In any case, whether the problem is a simple flaw or a subtle one encountered during development or production, productivity is best served when the problem is quickly recognized, traced to its source, and corrected. To do this efficiently, it is important to understand not only the sources of the strengths of forgings, but also the sources of their limitations.

In this paper, we will present case histories of investigations which illustrate practical methods for the recognition, elucidation and correction of three forging problems.

Overview

The case histories presented below were selected because they developed into in-depth studies of three basic sources of problems in complex, high-strength forgings.

445

The first problem arises from a compositional inhomogeneity that leads to the formation of a deleterious phase and poor properties. This general problem is common, but can take many other forms, such as gross banding (Fig. 1), locally high inclusion content (Fig. 2), or grain-size variations (Fig. 3). Treating any of these symptoms may lead to a temporary solution, but the root cause of the problem is best solved by improving mill procedures and homogenization.

The second case describes a particularly insidious problem wherein hemispherical parts which exhibited good microstructures, dimensional tolerances and tensile yield strengths were found to be very weak in compressive strength. The solution to this problem required a deep understanding of the origin of the Bauschinger effect in stainless steels and has led to a process capable of solving a class of similar problems.

The third case addresses a common problem in high-strength forgings, particularly forgings produced by high-rate forming processes — namely, the tendency of such forgings to develop shear bands. Shear bands are generally recognized as macroscopic defects which result from extreme localization of metal flow. Normally, metal-flow lines curve gently and follow the contours of a part (Fig. 4). Shear bands are regions where exceptional flow has occurred at the expense of uniform flow throughout the piece. Often, these areas recrystallize prematurely because of the extra deformation. While severe shear bands receive the most attention, nonuniform flow also occurs in more moderate degrees that result in kinked flow lines (Fig. 5) as well as grain-size variations in a forging cross section (Fig. 6, 7, and 8). The propensity for localized flow exhibited by a given

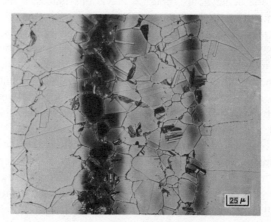

Fig. 1. Gross banding in aged JBK-75 (an alloy similar to A-286). The locally high concentration of titanium (3.0 wt % vs 2.0 wt % in the matrix) leads to the accelerated precipitation of γ' [$Ni_3(Ti,Al)$] and η (Ni_3Ti) phases during aging and the dark appearance of these areas on subsequent etching.

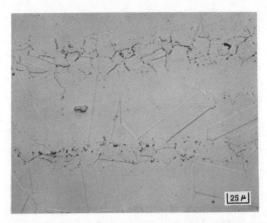

Fig. 2. High inclusion content caused by titanium segregation in JBK-75. The inclusions are titanium-rich and are in the high-titanium bands.

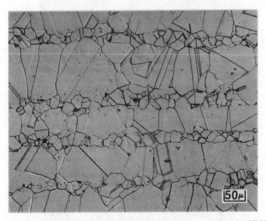

Fig. 3. Duplex grain size caused by titanium segregation in JBK-75. The fine grain size results from grain-boundary pinning by the titanium-rich inclusions (Fig. 2).

material is a function of a number of factors, including temperature and strain rate. Methods have been developed for quantifying the tendency for shear-band formation both empirically and analytically. When properly applied, these methods can be used to predict, and thus avoid, shear-band problems.

The case histories follow a chronological format. Unfolding as they do, they may touch on information that ultimately is shown to be superfluous or irrelevant. Such information is a common ingredient in investigative programs. One of the most useful skills is the ability to sort through all of the information and focus on

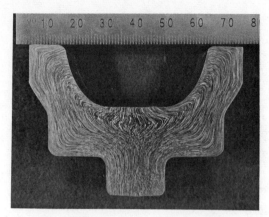

Fig. 4. Flow lines in a 21-6-9 forging.

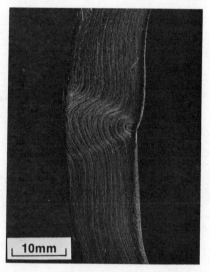

Fig. 5. Kinked flow lines in the wall of a spherical 304L forging.

the key facts. This skill is developed through experience and we hope that, by presenting our data in this manner, we impart this experience to the reader.

I.   Delta Ferrite in 21-6-9

I.A.   The Problem

Cracks were found in about ⅓ of a group of forgings. The forgings had been in production for some time with few problems. The forge shop personnel

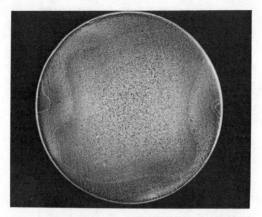

Fig. 6. Cross section of 2-in.-diam 304L bar with grain-size variation revealing square "halo" pattern. Processing history showed that the round bar was rolled from a 3-in.-square bar.

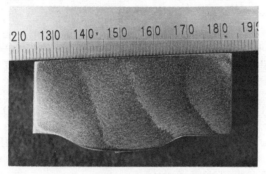

Fig. 7. Cross section of a JBK-75 forging which was formed in four blows. Note the four waves of grain-size variation.

suspected a die wear problem and were able to reduce the amount of cracking by changing some draft angles in the dies. In parallel with that activity, samples were submitted for metallographic analysis in order to document the problem more completely.

I.B. Background Information

The forging was high-energy-rate forged (HERF) from 21-6-9 austenitic stainless steel over a falling temperature range from 1040 to 815°C. The parts were water quenched after each step and reheated prior to the succeeding step.

The starting stock was received as 21-6-9 stainless steel plates, 1.1 m by 0.6 m by 4.45 cm, from which individual blanks were cut. A new supplier was chosen

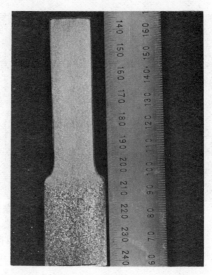

Fig. 8. JBK-75 stem forging showing coarse grains in body and fine grains in stem.

for this particular material, but the specifications for the steel remained the same. The plate was produced from electroslag remelted ingots and hot rolled to size. The newly supplied material differed from older material only in that the ferrite content of the incoming stock was now reported to be 1.5 to 3.0% (specified maximum, 2%) whereas previously it had been reported to be 0 to 0.5%.

The forging itself was of a complicated shape and was known to have an undesirable, transverse flow-line configuration at the flash line (Fig. 9). Extensive testing prior to production showed that no reduction in properties occurred due to this flow pattern.

The cracks that occurred in the forgings were detected by a dye-penetrant technique on rough (nonmachined) forgings. The location of the crack was the same in all cases: in the area of transverse flow at the flash line (shown by the dashed lines in Fig. 9).

I.C. The Investigation

Several typical cracked forgings were cut up and analyzed by optical and electron optical methods. Macroetching of the parts confirmed that the flow lines were nonuniform around the circumference. Specifically, the flow lines turn from circumferential to radial at two diametrically opposite locations. This flow pattern was previously noted and was attributed to excess material flow outward at the parting line of the two-piece die into the flash gutter. It is significant that the cracks, when they occurred, were found propagating along the direction of these radial flow lines.

Fig. 9. Macroetched section showing nonuniform flow lines at the parting line. View is into the bore of a complex, generally cylindrical 21-6-9 alloy forging. Dashed line shows parting line.

Optical metallographs showed that the cracks ran radially outward from the inside surface. The cracks ran along stringers and jumped at right angles between them. The stringers, which averaged 200 μm in length, were very common in the etched condition but were difficult to see in the as-polished condition. By comparison, manganese sulfides were clearly visible in the as-polished condition. A few sulfide inclusions were found, but they comprised only a very small fraction of the total inclusions, and did not appear to be associated with the cracks.

The ferrite contents of these sections of the forgings were nil, as measured both with the ferrite meter (<0.1%) and the Severn gage (<0.5%). This was notable because the starting plate was reported to have approximately 2% ferrite, and the stringer volume percentage from plate to forging did not change.

X-ray-energy-dispersive analyses in the SEM showed that the particles were rich in chromium relative to the 21-6-9 matrix. Furthermore, the stringers were seen to be made up of separate particles on the order of 3 to 5 μm in length (Fig. 10).

Optical and SEM fractography were done on the crack surface after it was pulled apart in a tensile tester. It was weak and broke easily. The fracture surface was blue to straw-colored, indicating that it was oxidized at an elevated temperature. The topography of the fracture surface was unusual (Fig. 11, top), having the form of elliptical, flat-top and flat-bottom projections and depressions with vertical sides. The typical size of the tops of these projections was about 400 by 200 μm. SEM examination showed that the flat tops and bottoms were generally covered with brittle, scalelike features (Fig. 11, bottom) that were higher in chromium content than the matrix. These were the same particles shown in Fig. 10, i.e., the stringerlike inclusions. The vertical sides in the optical

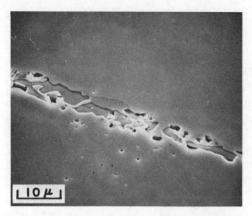

Fig. 10. Typical chromium-rich stringer found at the parting line in Fig. 9. Figures 12 and 13 identify the stringer as sigma phase.

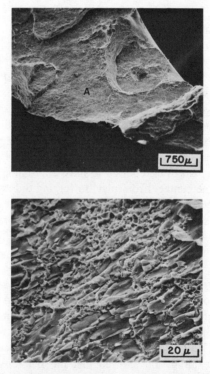

Fig. 11. (Top) Optical fractograph showing steplike features on cracked surface of a forging. (Bottom) SEM fractograph showing brittle fracture through high-chromium particles at area A in upper fractograph. Note that the stringers are present in the form of thin sheets, which is consistent with morphology of brittle areas in upper fractograph.

fractograph (Fig. 11, top) were covered with ductile dimples, and had the nominal composition of 21-6-9 stainless steel.

The inclusions were shown not to be carbides, nitrides or oxides by microprobe analysis. Quantitative analyses of several stringer particles showed an average composition of Fe-32Cr-2Ni-7.2Mn. The average matrix composition was Fe-19.6Cr-6.3Ni-8.1Mn. The complete data are given in Table 1.

Specimens for TEM were cut from near the crack. Particles were found in several foils and photographed with their associated diffraction patterns. Controlled tilting experiments allowed a series of related diffraction patterns to be obtained from a single particle (Fig. 12). When solved, these patterns (Table 2, Fig. 13) showed that these chromium-rich particles had the crystal structure reported for Fe-Cr sigma phase — namely, tetragonal (a = 8.80, c/a = 0.52). That these diffraction patterns were taken from the chromium-rich inclusions studied by the other methods described above was confirmed by re-examining the same foil that was photographed in the TEM with the SEM. The same area was relocated and shown by EDS to be chromium-rich with respect to the matrix (CR/Fe = 0.75, versus 0.48).

The results above indicated that the forgings cracked at high temperature (during forging) through stringers of chromium-rich sigma phase in the direction of the radial flow lines. Other analyses showed that sigma phase was present in similar amounts and distributions in both the cracked and uncracked forgings. Assumedly, minor variations during forging led to the actual production of cracks in this crack-sensitive microstructure.

Specimens taken from positions 90° to the flash line contained roughly the same volume fraction of chromium-rich stringers. However, the orientation of the stringers at these locations was circumferential, parallel to the flow lines in this area. Analysis of the stringers in these latter areas showed that both ferrite and sigma phase were present (Fig. 14). Figure 14 also shows that the transformation occurs by the nucleation of numerous sigma-phase particles at the ferrite-austenite interface. Subsequent growth of these particles transforms the initially

Table 1. Quantitative Microprobe Analyses of 21-6-9 (wt %)

Constituent	Cr	Ni	Mn	Si	Fe
Chromium-rich stringer in forging	32.0	2.0	7.2	0.6	58.1
Austenite matrix in forging ..	19.6	6.3	8.1	0.3	63.4
Low-ferrite plate (avg), γ matrix	19.2	7.2	9.6	0.5	63.6
High-ferrite plate (avg), γ matrix	20.9	6.9	8.7	0.5	63.0
High-ferrite plate, ferrite stringer	26.5	2.9	7.0	0.5	63.1

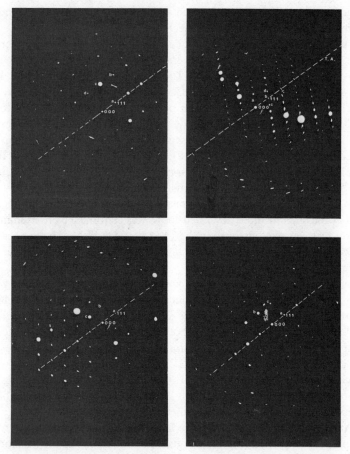

Fig. 12. Series of electron diffraction patterns obtained from a single particle in a stringer similar to that shown in Fig. 10 by controlled tilting about a common axis. Top left, $[01\bar{1}]$. Top right, $[\bar{1}4\bar{3}]$. Bottom left, $[\bar{1}3\bar{2}]$. Bottom right, $[\bar{1}2\bar{1}]$. See Table 2.

one-piece ferrite stringer into a similarly shaped stringer composed of many, discrete sigma particles.

Five specimens of 4.47-cm-thick plate from which forgings were made were examined in order to trace the formation of the sigma phase through the process history. Specimens were taken from lots of forgings having the highest and lowest frequencies of cracking (60% and 5%, respectively). A sample of old plate, considered to be typical of the type of material used to make uncracked forgings, was also examined. The chemical analyses of these specimens are listed in Table 1. Ferrite contents of these plate specimens were determined by both the ferrite meter and the Severn gage. The two crack-prone plates had essentially the

Table 2. Comparison of Measured Versus Calculated Distances and Angles in the Electron Diffraction Patterns in Fig. 12, Solved on the Basis of Fe-Cr Sigma Phase (Tetragonal, a = 8.80, c/a = 0.52)

Figure No.	Spot No.	Radius, cm, on negative	Calculated radius, cm	(hkl)	Angle between	Measured	Calculated	Zone of pattern
12 (top left)	a 0.26		0.27	(010)	(a-c)	90°	90°	[01$\bar{1}$]
	b 0.63		0.64	(111)				
	c 0.58		0.58	(101)	(b-c)	25.2°	24.6°	
	d 0.78		0.79	(121)	...	...	...	
					[01$\bar{1}$]-[$\bar{1}$43]	16.5°	14.5°	
12 (top right)	a 0.63		0.64	(111)				
	b 1.52		1.52	(5$\bar{2}$1)	(a-b)	37°	38.4°	[$\bar{1}$43]
	c 1.10		1.10	(410)	(a-c)	58°	59.6°	
	d 0.96		0.95	(30$\bar{1}$)	(a-d)	94.2°	95.0°	
					[$\bar{1}$43]-[$\bar{1}$32]	5°	4.8°	
12 (bottom left)	a 0.63		0.64	(111)				[$\bar{1}$32]
	b 0.73		0.74	(20$\bar{1}$)	(a-b)	104.3°	105.2°	
	c 0.83		0.84	(310)	(a-c)	57.5°	58.2°	
					[$\bar{1}$32]-[$\bar{1}$2$\bar{1}$]	9.5°	9.3°	
12 (bottom right)	a 0.63		0.64	(111)				[$\bar{1}$2$\bar{1}$]
	b 0.59		0.58	(210)	(a-b)	55°	56°	
	c 0.58		0.58	(10$\bar{1}$)	(a-c)	121.5°	121.8°	

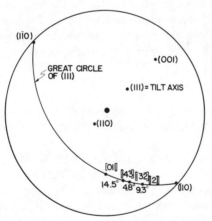

Fig. 13. A stereographic projection summarizing measured vs calculated angular relationships for a particle identified as sigma phase. See Table 2.

same ferrite content, varying from 1.5 to 2% near the top and bottom surfaces of the original plate to 3.5 to 4% at the centerline. The ferrite content of the old plate was very low, essentially zero by the ferrite meter and less than 0.5% by the Severn gage.

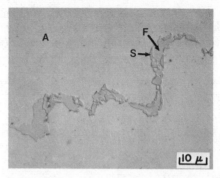

Fig. 14. A partially transformed stringer showing particles nucleated at the ferrite-austenite interface. The stringer was located 90° to the parting line in Fig. 9. A = austenite. F = ferrite. S = sigma.

Optical metallography showed that the ferrite was present in the form of stringers, elongated in the rolling direction. Unlike the sigma stringers, the ferrite stringers were individually continuous rather than broken into discrete particles. Re-etching of these specimens with Murikami's etch, which stains ferrite yellow and sigma phase blue, confirmed that essentially all of the stringers in the crack-prone plate material were ferrite. Quantitative microprobe measurements showed that the composition of the ferrite averaged Fe-26.5Cr-2.9Ni-7.0Mn-0.5Si.

These data showed that ferrite in the plate is chromium-rich to an extent which would require little solute redistribution during the transformation to sigma phase. The morphology of the ferrite in the plate was consistent with that of the resultant sigma phase, considering the additional metal flow which occurred during forging.

I.D. Discussion

The results of a number of investigations[3-13] of sigma formation in other stainless steels were helpful in our investigation; however, they will not be reviewed here.

The kinetics of the ferrite-to-sigma transformation in the material used in this investigation were studied by Packard-Ferrera[14] as a function of time, temperature and prior deformation. The results obtained are summarized in Fig. 15, which shows that prior working greatly accelerates the ferrite-to-sigma transformation, in the manner reported previously for other types of stainless steel.[5] The forgings in this investigation were not given prolonged high-temperature exposures, but, as a result of the multistep nature of the process, they were repeatedly heated through the temperature range in which sigma formation is most rapid. The precipitation and growth of sigma particles cause considerable changes in the properties of an alloy. The transformation of relatively ductile delta ferrite to the brittle sigma phase usually adversely affects the me-

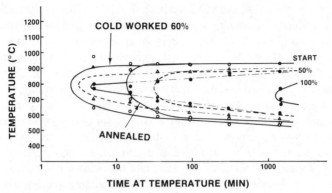

Fig. 15. Time-Temperature-Transformation (TTT) curves for sigma formation from ferrite.[14]

chanical properties.[15] The magnitude of the effect of sigma on the mechanical and corrosion properties of stainless steel is greatly dependent not only on the amount of sigma present, but also on the particle size and distribution.[16] The embrittling effect of sigma is more pronounced at room temperature than at elevated temperatures.[5]

The most sensitive test for the presence of sigma phase is the room-temperature impact test.[5] The effect of aging on the impact strength of the 21-6-9 stainless steel used to make the cracked forgings in this investigation was determined by Packard-Ferrera.[14] These data, presented in Fig. 16, show the dramatic drop in impact strength which occurs as ferrite is tranformed to sigma phase and also show that even relatively low amounts of prior ferrite have a profound effect.

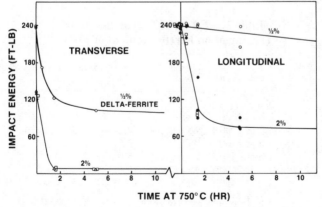

Fig. 16. Effect of aging on impact strength of 21-6-9 stainless steel containing ferrite.[14]

The ferrite in the plate was formed during the solidification of the ESR ingot. The ferrite content of austenitic stainless steels is controlled by composition and processing parameters.[17-19] The relative proportions of austenizing and ferritizing elements dictate the amount of ferrite which forms during actual solidification. The elimination of ferrite by the simple expedient of changing the composition is not a valid solution, because some ferrite-forming tendency is needed to prevent hot cracking in subsequent welds.[20,21] The phase diagram indicates that a fully austenitic structure should be formed on cooling. However, kinetic limitations may prevent this from occurring in economical processing times. The amount of retained ferrite is a function of the original amount of ferrite, its size and distribution, and the ingot breakdown and homogenization treatment. Table 3 shows the actual effects of temperature and hot work on the ferrite content of an experimental heat of 21-6-9.[14]

Summarizing the results of this investigation, it was found that the sigma phase in these forgings formed directly from ferrite that had a composition very near that required for sigma formation. While the transformation was accomplished by nucleation and growth, the need for long-range diffusion of large amounts of solutes was negated. Thus the sigma formation was much faster than would be expected for a homogeneous piece of 21-6-9. The strain imparted during forging further accelerated the transformation. The forging schedule was a complex one, and, although only the final blows were delivered at temperatures near the nose of the transformation curve, each furnace reheating of the forging between blows took the forging slowly through the critical 800°C range. Some of the forgings cracked as a result of decreased ductility due to the presence of sigma phase.

I.E. Solution

The solution to the problem was to adjust both the composition specification and the ingot breakdown and homogenization heat treatment of this material.

Table 3. Effect of Thermomechanical Treatment on Ferrite Content of 21-6-9[14]

Thermal treatment	Ferrite content, %	
	As cast(a)	Wrought(b)
None	5.4	0
3 h at 2000°F, WQ	3.7	0
3 h at 2200°F, WQ	2.2	0
3 h at 2300°F, WQ	3.9	0.4
3 h at 2350°F, WQ	15.0	3.2
3 h at 2400°F, WQ	9.6	4.9
3 h at 2450°F, WQ	10.5	13.5

(a) Ingot, 1 by 3 by 9 in. (b) As cast plus 2 h at 2000°F plus rolled at 2200°F to plate 0.8 by 3 by 11 in.

This compromise action resulted in low ferrite content in the plate used to make forgings while retaining sufficient ferrite-forming tendency to produce sound welds during assembly.

II. Yield-Strength Anisotropy in 21-6-9 Hemishells

II.A. The Problem

The aim of the investigation described below was to produce 150-mm-diam hemishells having a nominal tensile yield strength of 900 MPa. The alloy used was Nitronic 40 (Trademark of ARMCO Steel Corporation), often referred to as 21-6-9 in view of the nominal weight percentages of the alloying elements Cr, Ni and Mn, respectively. The composition of this alloy is given in Table 4. Since the pole area in a hemishell generally receives little strain during drawing, and therefore a negligible increase in strength via strain hardening, the plates from which forming blanks were obtained were cross rolled 20% at 25°C to bring the nominal starting tensile yield strength in the material before forming up to 900 MPa, as shown in Table 5. Round blanks, 203 mm in diameter by 7.9 mm thick, were trepanned from the cold rolled plates and formed at 25°C into hemishells. Figure 17 shows the orientations (longitudinal and circumferential) and locations of test specimens taken for mechanical-property determination. Table 6 gives a comparison of yield strength for the above orientations before and after forming. The initially high tensile yield strength in the starting blank (900 MPa) was retained in the longitudinal direction after forming; however, in the circumferential direction the tensile yield strength dropped dramatically to approxi-

Table 4. Chemical Composition of Alloy 21-6-9 (wt %)

Cr	Mn	Ni	N	C	Si	P	S	Fe
19.35	8.72	6.22	0.29	0.030	0.37	0.021	0.001	Rem

Table 5. Mechanical Properties of Annealed and Cold Rolled Plate

Rolling method	Yield stress (0.2% offset), MPa		Ultimate tensile stress, MPa	Tensile elongation, %	
	Tension	Compression		Uniform	Total
Annealed(a)......	406	420	787	44.3	54.2
Cold rolled 20% ..	889	655	978	14.6	28.9

(a) Processing schedule: 2 h at 1100°C, then roll in 7 passes; finishing temperature above 930°C; solution treat 1 h at 950°C, water quench. Microstructure: equiaxed grains, ASTM 7.

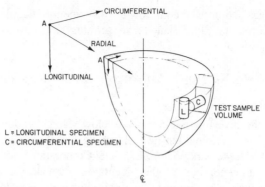

Fig. 17. Schematic cross section of hemishell, showing orientation nomenclature and specimen locations.

Table 6. Comparison of 0.2% Tensile Yield Strength (MPA) in 21-6-9 Before and After Hemishell Forming

Before forming	After forming at 25°C
889(a)	Longitudinal.............938
	Circumferential 586

(a) Uniform in the plane of the plate, after 20% cold cross rolling.

mately 600 MPa. As a result, the final tensile yield strength in the circumferential direction fell well below the minimum design limit of 900 MPa.

II.B. Background Information

Mechanical shaping of metal at temperatures well below the annealing range may very often result in a large yield-strength anisotropy for different orientations within the part, such as that observed for the 21-6-9 hemishells (statement of the problem) and large strength differentials between tension and compression for various orientations. Such phenomena are often related to the Bauschinger effect, as discussed below. Generally, when such a strength anisotropy occurs in formed parts, either the tensile or the compressive yield strength in a particular orientation will fall well below the yield strength of the material prior to shaping. If the reduced strength levels fall below design levels, the probability of part failure in service increases. Thus, it is important to assess the potential for yield-strength reduction via the Bauschinger effect during forming.

Bauschinger[22] found that, after a metal had been plastically deformed in one direction, re-straining in the opposite direction resulted in a lower elastic limit than would be experienced for re-straining in the same direction. Thus, anisotropic properties in a part may be explained by examining the sense of the strain

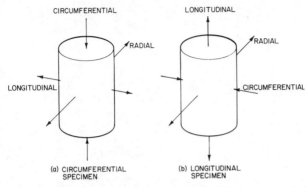

CIRCUMFERENTIAL

LONGITUDINAL

RADIAL

RADIAL

LONGITUDINAL

CIRCUMFERENTIAL

(a) CIRCUMFERENTIAL
SPECIMEN

(b) LONGITUDINAL
SPECIMEN

Fig. 18. Schematic illustration of the prestrain history imparted during forming for specimens machined from hemishells (Fig. 17) in the (a) circumferential and (b) longitudinal orientations. Arrows show sense of strain.

**Table 7. Forming Strain Imparted to Material
at the Hemishell Equator**

| Forming temperature | Forming strain | | |
	Circumferential	Longitudinal	Radial
25°C.	−26.5	21.3	12.1
480°C.	−26.5	22.0	11.5

imparted during forming and during testing. An examination of the prestrain history imparted during forming (Fig. 18 and Table 7) for the longitudinal and circumferential directions showed that the senses of the prestrain for the two orientations were different — tensile versus compressive, respectively. In lieu of the compressive prestrain imparted in the circumferential direction during forming and the subsequent determination of yield strength by re-straining the material in tension, the drop in yield strength from 900 to 600 MPa during hemishell formation could be explained by the existence of a significant Bauschinger effect in 21-6-9. However, a large Bauschinger effect is more characteristic of two-phase alloys[23,24] and would not necessarily be expected in a single-phase alloy such as 21-6-9. In order to clarify the importance of the Bauschinger phenomenon in 21-6-9 and to evaluate the role of this effect in the generation of a large yield-strength anisotropy in parts formed at room temperature, the following investigation was performed.

II.C. The Investigation

II.C.1. A New Bauschinger-Effect Parameter. Despite the importance of the Bauschinger effect in formed parts of complex shape and complex stress-

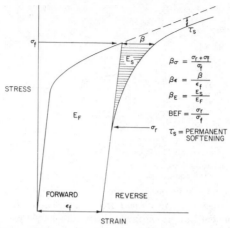

Fig. 19. Schematic illustration of various parameters used to measure the Bauschinger effect.

strain history, only a few investigations[25-27] have been made to characterize the role of the Bauschinger effect in the generation of mechanical-property anisotropy during forming. As a result, parameters for measuring the state of the Bauschinger effect in parts after forming have not been developed. The conventional Bauschinger parameters,[28-34] described schematically in Fig. 19, cannot be determined if prestraining is done at elevated temperature and if flow stress depends strongly on temperature,[26] or if the prestrain-stress history is unavailable, as is the case for formed parts. Therefore, for this study a new Bauschinger parameter, the Bauschinger Strength-Differential Factor (BSDF), shown schematically in Fig. 20, was formulated:

$$BSDF = \frac{\sigma_f' - \sigma_r}{\sigma_f'} \tag{1}$$

A distinct advantage of the BSDF parameter is that a measure of the Bauschinger effect can be obtained by testing without knowing the prestrain history, even though the Bauschinger effect is very much a function of prestrain history. Thus, the parameter can be used in tests which incorporate elevated-temperature prestraining and/or heat treatments given after prestrain deformation and prior to re-straining. Furthermore, the BSDF can be used for measuring the extent to which the Bauschinger effect has been developed in formed parts where the prior stress-strain behavior during forming is not readily available. Derivation of this parameter, and its limitations, are discussed elsewhere.[35]

II.C.2. Uniaxial Testing for the Bauschinger Effect. The Bauschinger effect in 21-6-9 was characterized by straining uniaxial specimens in tension and compression. The tensile and compressive yield strengths, TYS and CYS, mea-

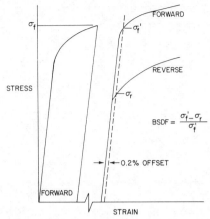

Fig. 20. Schematic illustration showing the method used to measure the Bauschinger effect via a parameter referred to as the Bauschinger Strength-Differential Factor (BSDF).

sured after a variety of strain and heat treatments applied to specimens machined from annealed plate are listed in Table 8, along with the resulting BSDF values. Examination of the values clearly shows that 21-6-9 does exhibit a large Bauschinger effect and that the effect can be reduced by prestraining at elevated temperature (480°C) and/or by heat treating at 480°C after prestraining and prior to re-straining. Examples of the flow curves which gave rise to the reduction in the BSDF are shown in Fig. 21.

The large increase in CYS shown in Fig. 21 and Table 8, which effectively reduces BSDF, may be attributed to several mechanisms. One might be a reduction in the elastic back stresses that accumulate during prestraining in tension at room temperature or a reduced rate of back-stress accumulation during 480°C prestraining and a concomitant loss in the level of assistance for dislocation movement in the opposite direction during re-straining in compression. A second mechanism, which is indicated by increases in TYS in Table 8 and the development of discontinuous yielding[35] after 480°C exposure, is strain aging. Strain aging may produce a significant increase in CYS because it effectively pins dislocations, which prior to aging have little resistance to motion in the reverse direction. In the 21-6-9 alloy, nitrogen is the major interstitial alloying addition (0.29 wt %) and may be the atom species that pins the dislocations.

II.C.3. Hemishell Properties and the Bauschinger Effect. Based on the data obtained from uniaxial tests (Table 8), the state of the Bauschinger effect in hemishells was determined after cold (25°C) and warm (480°C) forming, and after stress relieving at 480°C. The tensile and compressive yield strengths and associated BSDF values which were determined for the longitudinal and circum-

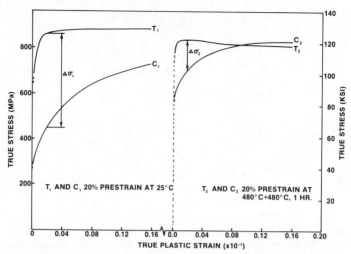

Fig. 21. True stress – true plastic strain curves for specimens uniaxially prestrained 20% in tension, showing the large reduction in strength differential ($\Delta\sigma_1$ vs $\Delta\sigma_2$) and elimination of permanent softening with exposure at 480°C. The material was solution heat treated for 1 h at 950°C and water quenched prior to testing.

Table 8. Uniaxial Test Results: Tensile and Compressive Yield Strength and BSDF After Different True Strains and Heating Treatments Applied to Annealed Plate (Table 5)

| History | Yield strength (0.2% offset), MPa | | BSDF |
	Tension	Compression	
No pretreatment	406(a)	420	· · ·
20% at 25°C	852	452	0.47
20% at 480°C	946	649	0.31
0.2% at 25°C + 1 h at 480°C . .	445	385	0.13
2.0% at 25°C + 1 h at 480°C . .	510	405	0.20
20% at 25°C + 1 h at 480°C . . .	887	695	0.22
20% at 480°C + 1 h at 480°C . .	845	709	0.16

(a) 195 MPa at 480°C.

ferential orientations at the hemishell equator (Fig. 17) are listed in Table 9. In the longitudinal direction, TYS > CYS, whereas CYS > TYS in the circumferential direction. The difference in behavior for the two orientations is due to a difference in the strain imparted during forming, i.e., the material was strained in tension in the longitudinal direction and in compression in the circumferential direction (see Fig. 18 and Table 7). Therefore, consistent with the Bauschinger

Table 9. Yield Strength and BSDF Values From Hemishell Equators

History		As formed			Heat treated 1 h at 480°C after forming		
Forming tempera- ture, °C	Specimen orien- tation(a)	Yield strength(b), MPa Tension	Compression	BSDF	Yield strength(b), MPa Tension	Compression	BSDF
25	L 938		660	0.30	1075	889	0.17
25	C 586		1062	0.45	820	1131	0.27
480	L 944		903	0.04	958	1048	−0.09
480	C 758		1060	0.28	800	1075	0.26

(a) L = longitudinal; C = circumferential. (b) At 0.2% offset.

effect, prestraining in tension results in a reduced CYS, while prestraining in compression results in a reduced TYS.

Consistent with the uniaxial test results presented in Table 8, the data in Table 9 clearly show that both stress relieving at 480°C after forming, and warm forming at 480°C, significantly reduce the Bauschinger effect and the associated yield-strength anisotropy. The reduction in the BSDF occurs mainly by increases in CYS in the longitudinal direction and in TYS in the circumferential direction. Note that a low TYS in the circumferential direction was highlighted in the statement of the problem and in Table 5. The increase in tensile strength in the longitudinal orientation (Fig. 22) is consistent with the strain aging that has taken place during the 480°C heat treatment. However, forming at 480°C, a process that requires only a few seconds, apparently leaves little time for aging, which explains the absence of an increase in flow stress on tensile loading of the longitudinal specimens. Thus, during 480°C forming, a reduction in the rate of the back-stress accumulation most probably accounts for the maintenance of high compressive flow stresses with a minimal contribution from aging. Prestraining of tensile specimens at 480°C did produce significant strain aging (Table 8), presumably because of the much longer duration (about 600 s) of prestraining compared with the time of hemishell formation (about 5 s).

Figure 23 shows the variation of BSDF for the longitudinal and circumferential directions with the relative exposure to 480°C received during forming and heat treating. As described elsewhere,[35] the BSDF is lower in the longitudinal direction than in the circumferential direction because of the tensile strain imparted to the longitudinal specimen in the off-axis radial direction (Fig. 18). More importantly, Fig. 23 shows that the BSDF value for the longitudinal specimen from the cold formed and heat treated hemishell does not lie on the expected curve, which roughly parallels the curve for the circumferential direction, but instead lies well above the expected curve. The reason for the seemingly anomalous BSDF value appears again to be associated with the radial strain direction in the

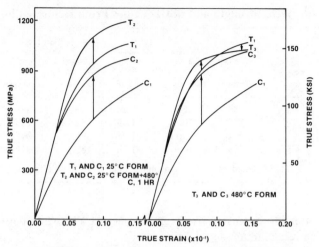

Fig. 22. True stress – true strain curves for longitudinal specimens (Fig. 17), showing large increases in both tensile and compressive flow curves (left) which occurred as a result of heat treating for 1 h at 480°C after forming at 25°C, and an increase in only the compressive flow curve (right) after forming at 480°C.

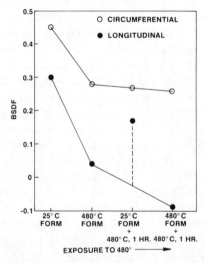

Fig. 23. Variation in BSDF with relative exposure at 480°C for the circumferential and longitudinal directions (Fig. 17) in the hemishells. Lines are drawn only to show trends. Note that one point lies well above the expected behavior.

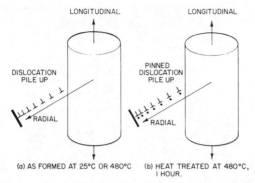

Fig. 24. Schematic representation of a model used to explain (a) the generally lower values of BSDF in the longitudinal orientation in Fig. 23, and (b) the unexpected increase in BSDF after forming at 25°C and heat treating for 1 h at 480°C. Arrows show sense of prestrain for the longitudinal specimen (Fig. 18b).

longitudinal specimen. In this case, the dislocation pile-ups which aid tensile strain and lower tensile stress in the as-formed condition (Fig. 24a) are pinned during heat treatment and cannot readily accommodate tensile strain (Fig. 24b). As a result, tensile flow stress is increased. Since solute pinning does not significantly increase the compressive flow stress because the dislocations are already relatively immobile in the forward radial direction, the difference between tensile and compressive flow stress increases, as does the BSDF. Thus, heat treating at 480°C increases the BSDF in the longitudinal direction after cold forming via strain aging. In this light, it appears that warm forming (480°C) offers a distinct advantage over cold forming (25°C) and heat treating with respect to minimizing the yield-strength differentials and the Bauschinger effect for the longitudinal direction in formed hemishells.

II.C.4. Microstructure. The microstructure of the annealed and 20% cold rolled plate was examined in order to search for possible correlations with the measured Bauschinger-effect behavior. The most striking microstructural feature is the presence of many sets of parallel striations within the deformed grains (Fig. 25). These linear features, observed by transmission electron microscopy, were identified as either deformation bands or deformation twins.[35] Analysis showed that the twins were not important features with respect to the Bauschinger effect.[35] However, the high density and the relatively uniform distribution of the twins show that twinning is an important deformation mechanism during room-temperature rolling of 21-6-9 stainless steel.

Figure 26 shows that the areas between the deformation twins contain numerous extended dislocations and associated stacking faults, the importance of which

Fig. 25. Microstructure at the equator of a hemishell formed from 20% cold rolled material at 25°C. Arrow shows longitudinal direction. Etched in oxalic acid.

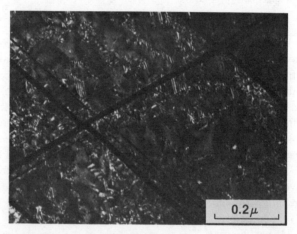

Fig. 26. TEM view of 20% cold rolled plate, showing the presence of stacking faults, loose dislocation cells, and deformation twins.

will be discussed later, and a loose or incipient dislocation cell structure. The 0.29 wt % N in the steel studied here places it in the low-stacking-fault-energy regime,[36,37] where extended dislocations and planar slip are expected. A low stacking-fault energy is also consistent with the deformation twinning observed in the room-temperature-deformed plate.

Figure 27 shows the substructure of a room-temperature-formed hemishell. Deformation twins residual from the room-temperature rolling are present in some of the austenite grains, whereas the remainder of the structure consists of a fine dislocation cell structure. Hemishells formed at 480°C showed similar deformation substructures; however, the cells were larger and the dislocations

Fig. 27. Transmission electron micrograph of a thin foil taken from the equator of a hemishell formed at 25°C, showing the inhomogeneous distribution of deformation twins which occurs on a fine scale and a relatively dense, cellular distribution of dislocations in the matrix.

were not as tightly packed in the cell walls.[35] The coarser cell structure observed after the 480°C deformation is consistent with the effect of temperature on cell size noted by Michel et al.[38] Stacking faults were not observed in the hemishells deformed at 480°C, perhaps because of an increase in stacking-fault energy with temperature.[39]

II.D. Discussion

II.D.1. General. Mechanisms for the Bauschinger effect have been addressed in a number of investigations.[23,24,26,40–45] Two generally accepted causes for the development of a Bauschinger effect during straining are (*a*) accumulation of long-range elastic back stresses during deformation that are not totally relieved by elastic unloading and that can subsequently assist dislocation motion when the sense of the deformation is reversed, and (*b*) generation of dislocations which have a reduced resistance to motion in the reverse direction because of the nature of the structure left behind the dislocations during forward motion. Long-range back stresses are a function of both dislocation arrangement[44] and dislocation density.[34,45–47] Because planar slip, which favors the formation of dislocation pile-ups with long-range back stresses,[44] has been observed in room-temperature-deformed 21-6-9,[36] the existence of a strong Bauschinger effect in this alloy may be expected.

Although the existence of significant back stresses in dislocation cell structures like those observed in the hemishells is less intuitive than in a linear dislocation pile-up, cell walls which are formed when only limited cross slip and climb occur can possess long-range elastic stresses.[44] The last dislocations to join the cell walls are not tightly bound and, under the influence of elastic stresses in the cell walls, can easily move under a reversed stress and therefore contribute to a Bauschinger effect.[44]

The occurrence of fine precipitates at 480°C could explain the observed changes in the Bauschinger effect, which so far have been attributed to dislocation pinning by solute atmospheres, but no TEM evidence for precipitation in 21-6-9 at 480°C could be found.[35]

II.D.2. Special Experiments: The Role of Epsilon Martensite and Stacking Faults. Because epsilon martensite was identified by Hills and Rack[48] in shock-loaded 21-6-9, some consideration was given to the possibility that a Bauschinger effect could arise from the reversion of epsilon martensite during reverse straining. A reduction in the Bauschinger effect should occur after heating to 480°C, because any martensite present would thermally revert to austenite, eliminating the possibility of mechanical reversion on re-straining. Furthermore, forming at 480°C would not result in any martensite formation at all. Thus, the Bauschinger effect would be reduced, as it was in this investigation, by stress relieving or warm forming.

In this portion of the investigation, the magnitude of the Bauschinger effect and the associated fine structure were determined after rolling the annealed material 15% at three different temperatures, -196, 25 and 480°C, and also after flash heating of the rolled material to 480°C. The heat treated specimens were up-quenched in a salt bath, held at 480°C for approximately 30 s and then water quenched.

Assuming that 480°C is above the epsilon M_d temperature, the flash heat treatment would allow reversion to take place but presumably would allow little time for solute pinning and/or stress relaxation. Thus, changes in the BSDF as a result of flash heating should be mainly associated with reversion of the epsilon martensite.

The mechanical properties of the rolled and rolled-plus-flash-heat-treated material are given in Table 10. Because the strain imparted to the mechanical test specimens, which were oriented in the rolling direction, is nonideal for a Bauschinger-effect measurement and similar to that shown in Fig. 18, the BSDF values underestimate the true Bauschinger effect.[35] In this light, the Bauschinger effect after rolling at -196°C is quite large (BSDF $= 0.21$). The flash heat treatment dramatically reduced the Bauschinger effect, as indicated by the change in BSDF (ΔBSDF $= -0.39$). Before heat treating, the fine structure consisted of many striations (Fig. 28, top), which were identified by electron diffraction[35] either as epsilon martensite or as deformation twins. After heat treating, deformation twins were the only linear structural features present. It should also be noted that the interstriation regions in the material rolled at -196°C were free of dislocations which, if present, could have contributed to the measured Bauschinger effect. Thus, it appears that a large Bauschinger effect was created by the formation of epsilon martensite during prestraining (rolling) and subse-

**Table 10.   Yield Strength and BSDF Values in Plate Rolled to 15%
Reduction in Thickness (Test Specimens Aligned With Rolling Direction)**

| Rolling tempera- ture, °C | As rolled | | | Heat treated 30 s at 480°C water quenched | | | ΔBSDF on heat treating |
| | Yield strength(a), MPa | | | Yield strength(a), MPa | | | |
	Tension	Compression	BSDF	Tension	Compression	BSDF	
−196	89.0	70.7	0.21	92.5	108.8	−0.18	−0.39
25	112.5	76.8	0.32	109.7	100.8	0.08	−0.24
480	118.6	125.0	−0.05	115.5	124.8	−0.08	0.03

(a) At 0.2% offset.

quently by its *mechanical reversion* during re-straining. It follows then, that the Bauschinger effect was reduced by *thermal reversion* of the martensite structure during the flash heat treating cycle.

Rolling 15% at 25°C produced a large Bauschinger effect (BSDF = 0.32) which was substantially reduced (ΔBSDF = −0.24) by the flash heat treatment. The fine structure prior to heat treating (Fig. 28, center) consisted of deformation twins with a dislocation structure in the intertwin regions, but no epsilon martensite. Assuming that little change in BSDF occurs via solute pinning and stress relaxation during the flash heat treatment, the relatively large change in BSDF (ΔBSDF = −0.24) must be accounted for by some other mechanism.

Figure 28 (top and center) show that stacking faults are also characteristic features in the structure of the material rolled at −196°C and 25°C. Abel and Muir[49] concluded that materials with low SFE have a large capacity to store energy associated with plastic deformation and can release the energy in a reversible manner. The energy is stored during prestraining (in our case, rolling) by stress-induced separation of partial dislocations and is recovered on unloading and during restraining by the return of the partials to an equilibrium separation distance. Figure 29 shows that the stacking faults present in 21-6-9 after rolling at 25°C are completely eliminated by the flash heat treatment. Closure of the partial dislocations is consistent with the increase in SFE that occurs with an increase in temperature. Energy which is stored via stress-induced partial dislocation separation during rolling should be released during heat treatment by closure of the partials. Because this energy would no longer be available for assistance of deformation on re-straining, the Bauschinger effect should be reduced by the flash heat treatment as it is in Table 10. Thus, a third mechanism — that associated with stacking faults — for the creation of the Bauschinger effect in 21-6-9 seems likely. The other two are associated with back stresses from general dislocation interactions and with epsilon martensite formation and reversion.

Rolling at 480°C did not produce a significant Bauschinger effect (BSDF = 0.05) and flash heat treating had little effect. Figure 28 (bottom) shows that a

Fig. 28. Transmission electron micrographs of 21-6-9 after a two-pass, 15% unidirectional rolling reduction at (top) −196°C, (center) 25°C and (bottom) 480°C. Prior to rolling, the material was solution treated for 1 h at 950°C and water quenched.

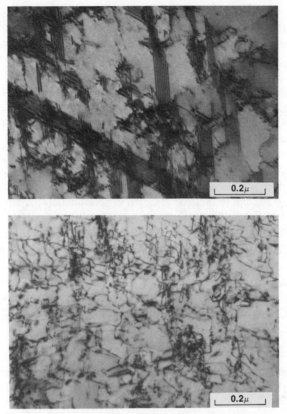

Fig. 29. Transmission electron micrographs of 21-6-9 which was reduced 15% by rolling at 25°C, showing (top) numerous stacking faults in the as-rolled condition and (bottom) no stacking faults after flash heating to 480°C.

well-developed dislocation cell structure was imparted during rolling. This cell structure did not change significantly during heat treatment. Consistent with the analysis of the Bauschinger effect in hemishells, the cell structure developed during 480°C forming must be relatively free of long-range back stresses.

II.E. Solution

The solution to the problem was to warm form and stress relieve the hemishells to provide the best combination of strength and isotropy.

III. Flow Localization and Shear-Band Formation

III.A. The Problem

Higher strengths in forgings can be attained by imparting and retaining a greater amount of strain hardening during the final stages of deformation.

This is often accomplished by lowering the final forging temperature to inhibit dynamic recrystallization and dynamic recovery, and by quenching the workpiece immediately after the forging operation to minimize static recrystallization and static recovery. Forging processes which impart higher strain rates during forming, i.e., high-energy-rate forging (HERF), often result in higher strengths because less time is available for recovery and recrystallization. This technique for increasing strength has been applied to various austenitic stainless steels (304L, 21-6-9, 22-13-5).[50] As forging temperature is lowered, the flow strength of the workpiece increases and a number of associated problems can be expected, i.e., excess die wear, increased tonnage required for deformation, increased number of blows to fill the die, the need to change lubricants, and a concomitant loss of workpiece malleability, which can eventually result in fracture.

However, an attempt to increase strength in precipitation-hardening austenitic stainless steel JBK-75 (Table 11 shows that JBK-75 is similar to A-286), by reducing the forging temperature from 900 to 825°C, resulted in an entirely new and unanticipated problem: the formation of shear bands which traversed the entire cross section of the forging (Fig. 30, left). Generally, shear bands represent discontinuities in required grain-flow orientation, and their presence is cause for part rejection as it was in this case. A secondary problem that was observed was cracking along the shear bands in some of the forgings (Fig. 30, top and bottom right).

III.B. Background Information

It is well known that flow instability and subsequent localization constitute a common occurrence during metal deformation. During sheet metal forming, the

Table 11. Elemental Analysis (wt %) of JBK-75 Studied in This Investigation Compared With A-286

	JBK-75	A-286
Cr.	14.6	13.5-16.5
Ni.	29.7	24.0-27.0
Al.	0.23	0.35 max
Ti.	2.3	1.9-2.35
Mo.	1.4	1.0-1.5
C	0.014	0.08 max
V	0.4	0.1-0.5
B	<5 ppm	0.001-0.010
S	0.003	0.03 max
P	0.006	0.04 max
Si.	0.016	0.04-1.0
O	0.004	$\cdots$
Mn.	0.003	1.0-2.0

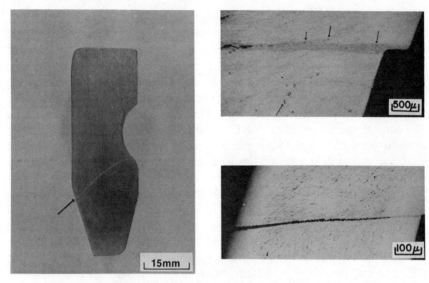

Fig. 30. (Left) Macroetched section of the subject JBK-75 alloy high-energy-rate-forged ring (forging temperature, 825°C), showing a shear band which perturbs the desired flow lines. Optical micrographs show (top right) that a high amount of localized plastic flow is concentrated in the shear band as evidenced by the large step in the inner surface of the forging, and (bottom right) that the shear bands often acted as sites for catastrophic fracture.

workpiece is primarily loaded in tension, and formability is limited by localized necking.[51] On the other hand, during primary or bulk forming operations, such as forging, extrusion or rolling, shaping is accomplished through compressive loading. In this case, regions of localized deformation are often revealed by highly curved flow lines, grain-size differences, property gradients, and/or shear bands. Shear bands are the most extreme examples of localized deformation, because they may span the entire cross section of a workpiece and in some cases may be associated with catastrophic shear failure or cracking. The occurrence of shear bands in both hot and cold forging has been reported for a number of materials, i.e., aluminum alloys;[52-56] steels;[57-63] titanium alloys;[64-68] austenitic stainless steels;[64,69,70] nickel-base alloys;[71] copper alloys, silver and gold;[52] and uranium and uranium alloys.[72,73]

Depending on the material, shear-band formation is a complex function of material parameters (strain-rate sensitivity of the flow stress, strain-hardening or flow-softening rate, temperature dependence of the flow stress, thermal conductivity, specific heat, and phase-transformation kinetics) and processing variables (deformation rate, workpiece temperature, die temperature, friction between workpiece and die, and workpiece and die geometry).[67] For example, examination of processing variables may explain shear-band formation during non-isothermal forging (dies colder than workpiece). In this case, the cold dies chill surface metal on contact, thus constraining flow in these areas, which are often referred to as "die locked," "die chilled" or "dead metal" zones. Subsequent flow of the softer bulk material past a stronger, chilled zone results in the generation of a locally steep metal velocity gradient and a condition which favors shear-band formation. On the other hand, during isothermal forging (dies and workpiece at the same temperature), shear-band formation is more directly a function of material parameters, i.e., strain-rate sensitivity and strain hardening.[74] With respect to strain-rate sensitivity, a material that hardens with increasing strain rate will deform more homogeneously because additional strain in areas deforming at a high strain rate requires a greater stress than that necessary for adjacent areas deforming at a relatively lower rate. Thus, subsequent deformation is distributed to adjacent areas. Similarly, with respect to strain hardening, a material that hardens rapidly with increasing strain will deform more homogeneously because additional strain will be shifted to adjacent areas which are less deformed and which require less stress to deform. Thus, for a material with low strain-rate sensitivity and a zero or negative strain-hardening rate, the concentration of deformation to that area which deforms first is an autocatalytic process and the propensity for shear-band formation is high.

Negative strain hardening is referred to as flow (or work) softening. The characterization of structural changes which occur during flow softening has been extensive, and a number of different mechanisms for the phenomenon have been reported—for example, dynamic recovery and recrystallization,[75-81] spheroidization of two-phase structures,[82,83] precipitate disturbance,[84] adiabatic heating,[67,85-88] and transformation from a hard to a soft texture.[83]

For materials with a large negative dependence of flow stress on temperature and poor thermal conductivity, adiabatic heating of material in a locally deforming area can reduce the local flow stress (flow softening) relative to adjacent, nondeforming material, causing additional deformation to be concentrated in the deformed area. Shear bands formed in this manner, referred to as adiabatic shear bands, are of concern in studies of ballistic penetration and high-rate machining and forming.[55,58,72,74,85,89]

Note that the above discussion of flow localization and shear-band formation does not circumscribe the most widely recognized concern during metalworking—the concern of fracture, whether it be associated with free surface cracks,[90] hot shortness,[91] central bursts,[92] grain-boundary and triple-point cracks,[93] or shear-band cracking.[74] Instead, it focuses on the subject of the last section of this chapter: prefracture defects or the intrinsic types of structure and property gradients caused by nonuniform flow.

III.C. The Investigation

III.C.1. Upset Forging. Early attempts to eliminate the shear bands from the forgings (statement of the problem) consisted of die design and lubricant modifications. Their failure, coupled with the sharp temperature dependence of shear-band formation in the 825 to 900°C range, suggested that shear-band formation was material- rather than process-dependent. Thus, an investigation of flow localization and shear-band formation in alloy JBK-75 was initiated. Because the shear bands were formed in the last stage of a multistage sequence, use of the forging itself as a test specimen in the study would have been inefficient. Therefore, a simple forging specimen which could be readily machined from bar stock and upset on existing forming equipment, i.e., HERF machines, hammers and presses, was used. The specimen was specifically designed for the purposes of measuring a material's propensity for localized flow and of examining the structure of shear bands, which were likely to form in the specimen during deformation.

The specimen is cylindrical and has a reduced gage section where deformation is concentrated (Fig. 31). The reduced gage section is used to calculate percent Distributed Gage Volume (DGV) by means of the following formula:

$$DGV\ (\%) = \frac{V_0 - V_f}{V_0} \times 100 \qquad (2)$$

where V_0 is the original gage volume and V_f is the final apparent gage volume between the specimen ends. Measurement, calculation and utilization of DGV are described in detail elsewhere.[64,70] Essentially, DGV is a measure of a material's propensity for localized flow during working. As shown in Fig. 32, a greater ability of a material to distribute deformation will result in greater penetration of the reduced gage section into the specimen ends and higher values of DGV. Low values of DGV are associated with localized plastic flow and a condition of material behavior which favors the formation of shear bands.

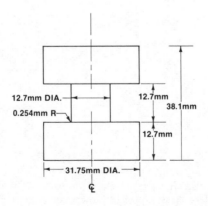

REDUCED GAGE SECTION UPSET SPECIMEN

Fig. 31. Shape and dimensions of the cylindrical compression specimen used to study localized flow and shear-band formation in JBK-75.

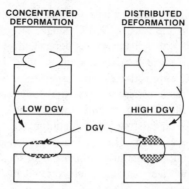

Fig. 32. Schematic illustration of the relationship between the parameter referred to as Distributed Gage Volume (DGV) and the type of flow which can occur during compression of the specimen shown in Fig. 31.

III.C.2. Nonprecipitation-Hardening Austenitic Stainless Steels. In order to generate a basis for comparing the flow behavior in JBK-75 with those in austenitic stainless steels which do not harden by precipitation, three other alloys were included in the study. Figure 33 shows the variation of percent DGV with upset temperature for the austenitic stainless steels, 304L, 21-6-9 and 22-13-5, after press forging (PF) and HERF. In general, percent DGV increased with strain

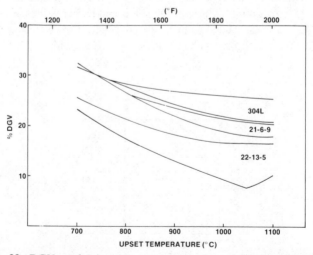

Fig. 33. DGV vs forging temperature for 304L, 21-6-9 and 22-13-5 austenitic stainless steels. Upper and lower curves are for high-energy-rate forging (1220-D Dynapak, $\dot{\varepsilon} \approx 1.4 \times 10^3$ s^{-1}) and for press forging (150-ton Erie, $\dot{\varepsilon} \approx 4$ s^{-1}), respectively. Prior to testing, the materials were solution treated for 1 h at 1000°C and water quenched.

rate and decreased as temperature was increased. Changes in percent DGV with temperature may be explained by the amount of hardening that occurs in the gage relative to the specimen ends. For example, at higher strain rates and lower temperatures, softening processes such as dynamic recovery and recrystallization are inhibited and the gage section hardens with deformation relative to the specimen ends. Subsequent deformation is, therefore, distributed to the specimen ends. Conversely, at lower strain rates and higher temperatures, softening processes are more competitive with hardening processes and the gage area does not harden as effectively. As a result, deformation is concentrated in the gage area and percent DGV is low. Note that even for the lowest DGV values, shear bands did not form in any of the alloys.

The high-temperature tensile properties for the three alloys are shown in Fig. 34 to 36. Comparing Fig. 35 and 36 with Fig. 33 shows that higher values of percent uniform elongation and work-hardening coefficient correspond to higher values of DGV and, thus, to an increased resistance to localized flow. In addition, the same temperature dependence of localized flow is predicted—that is, localized flow increases with increasing temperature.

III.C.3. A Precipitation-Hardening Austenitic Stainless Steel (JBK-75).
III.C.3(a) Effect of Temperature. Figure 37 shows DGV as a function of

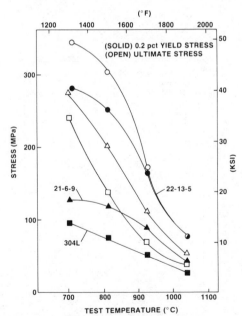

Fig. 34. Variation of elevated-temperature yield (0.2% offset) and ultimate stress with testing temperature for 304L, 21-6-9 and 22-13-5. Starting materials were solution treated as in Fig. 33. Strain rate $\approx 8 \times 10^{-4} \text{ s}^{-1}$.

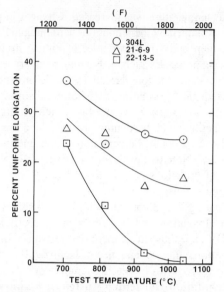

Fig. 35. Variation of elevated-temperature percent uniform elongation for 304L, 21-6-9 and 22-13-5 with testing temperature. See Fig. 34 for testing conditions.

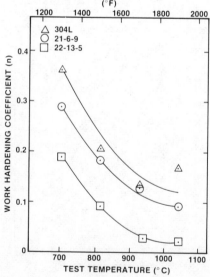

Fig. 36. Variation in elevated-temperature work-hardening coefficient for 304L, 21-6-9 and 22-13-5 with testing temperature. See Fig. 34 for testing conditions.

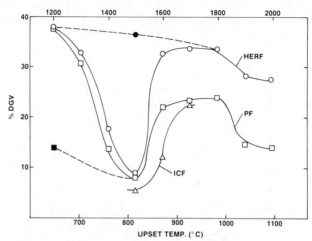

Fig. 37. DGV vs forging temperature for JBK-75 specimens deformed by three methods: High-Energy-Rate Forging (HERF), Press Forging (PF), and Isothermal Creep Forging (ICF), where $\dot{\varepsilon}_{ICF} < \dot{\varepsilon}_{PF} < \dot{\varepsilon}_{HERF}$. See text for discussion of extraordinary testing conditions for the solid data points and dashed curves. Prior to testing, the material was solution treated for 1 h at 1000°C and water quenched.

temperature for JBK-75 specimens deformed at three different strain rates that range over approximately six orders of magnitude (10^{-3} to 10^3 s^{-1}). In contrast to the austenitic stainless steels discussed above, JBK-75 can be strengthened by precipitation of Ni$_3$(Ti,Al), the γ' phase. Comparing Fig. 33 and 37 shows that JBK-75 exhibits an extreme drop in percent DGV between 649 and 816°C (1200 and 1500°F).

Consistent with the results for the three nonprecipitation-hardened alloys, percent uniform elongation and work-hardening coefficient for JBK-75 vary with temperature (Fig. 38) in a manner similar to DGV (Fig. 37). DGV, however, is the more sensitive parameter. Figure 39 shows that the flow stress of JBK-75 has a relatively large negative temperature dependence near the γ' solvus, compared with the relatively smaller dependences for the other alloys (Fig. 34). This indicates that the contribution of adiabatic heating to the flow-localization process will be more effective in JBK-75 than in the other alloys. It should be noted here that a major disadvantage of using only tensile properties to model behavior in compression is the lack of strict correspondence between the two modes of straining, i.e., flow is localized at lower strains in tension and also shear bands are more characteristic of compressive deformation.

Figure 40 shows macrographs of sectioned DGV specimens upset (via pressing) at 650, 815 and 870°C. The flow lines revealed by macroetching clearly show the severe strain localization that has occurred in the gage section of the

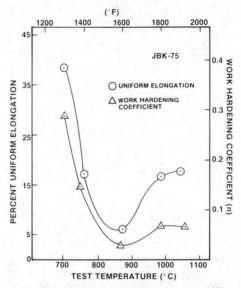

Fig. 38. Variation in elevated-temperature percent uniform elongation and work-hardening coefficient for JBK-75 with testing temperature. Starting material was solution treated as in Fig. 37. Strain rate $\approx 8 \times 10^{-4}$ s^{-1}.

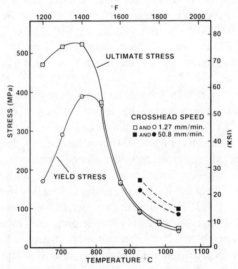

Fig. 39. Variation in elevated-temperature yield (0.2% offset) and ultimate stress with temperature for JBK-75. Starting material was solution treated as in Fig. 37.

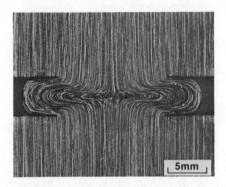

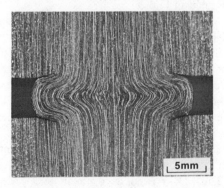

Fig. 40. Light micrographs of JBK-75, showing variations in flow-line contours and gage penetration into the specimen ends after press forging at (top) 650°C, (center) 815°C and (bottom) 870°C. Etched in oxalic acid.

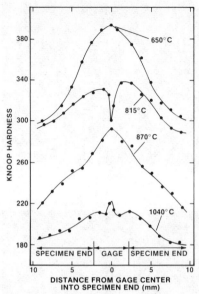

Fig. 41. Centerline microhardness traverses for press-forged JBK-75 specimens (Fig. 40) upset at different temperatures.

specimen deformed at 815°C. The other specimens show much less deformation of the gage section, deeper penetration of the gage section into the specimen ends, and greater over-all distribution of deformation in agreement with the measured values of DGV presented in Fig. 37.

Figure 41 shows centerline hardness traverses from specimens press forged between 650 and 1040°C. Work hardening of the gage section at 650 and 870°C corresponds to a high DGV and to the absence of recrystallization.[70] The lower over-all hardness at 870°C is attributed to the much greater role of dynamic recovery at higher deformation temperatures. At 815°C, the extreme softening of the gage section, attributed to extensive dynamic recrystallization and to dissolution of the γ' precipitate structure because of adiabatic heating during working,[70] is consistent with the low measured value of DGV. At 1040°C, deformation initiates dynamic recrystallization; however, softening with respect to adjacent material in the specimen ends is not extensive because of accelerated dynamic recovery. Thus, DGV is not as low as in the 815°C specimen, but is at an intermediate level. The small hardness peak shown in the 1040°C profile is due to work hardening of the dynamically recrystallized structure.

III.C.3(b)   Critical Experiments for JBK-75. In order to verify the effect of the presence or absence of fine γ' precipitation on strain localization, two experiments were specially designed. The first consisted of solution treating at 930°C, cooling within 30 s to 815°C, and immediately upsetting. No γ' precipitation could form within this time frame,[94] and, as a result, a high DGV (Fig. 37, solid circle) was measured for this specimen. Flow localization was minimized, shear bands did not form, and the JBK-75 deformed in a manner typical of single-phase austenitic stainless steel.

The second experiment consisted of aging at 815°C for one-half hour, water quenching, reheating to 650°C, and upsetting. In this case, γ' precipitation which was induced by the 815°C treatment resulted in a low DGV (Fig. 37, solid square) for this specimen, compared with that measured for a specimen upset normally at 650°C, without the pre-aging treatment. The flow was highly localized and took the form of well-defined, narrow shear bands which contained fine precipitate-free grains (Fig. 42). The results of these special tests confirm the

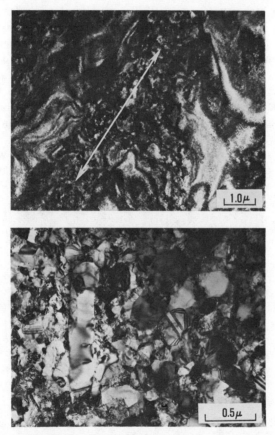

Fig. 42. Bright-field images of a shear band in a press-forged JBK-75 specimen which was heated for ½ h at 815°C, water quenched, reheated to 650°C, and upset. (Top) A shear band (arrow) in a heavily worked matrix. (Bottom) Fine, recrystallized, precipitate-free grains within the shear band in upper micrograph.

effect of γ' on deformation behavior—that is, the γ' precipitation sharply increases the propensity for flow localization and shear-band formation during forging of JBK-75.

III.C.3(c) Effects of Strain Rate. Contrary to the results for press forging of JBK-75 (Fig. 40, center), the extremely localized flow at 815°C (1500°F) resulted in macroscopic shear-band formation in HERF specimens. Figure 43 shows that the shear bands (arrows) are distinct macroscopic features which separate highly deformed from moderately deformed material. Shear strains in the bands, measured by dividing flow-line offset by band width, ranged from 5.7 to 9.6.

A microstructural characterization of shear-band formation as a function of strain rate for specimens deformed at 815°C is shown in Fig. 44. In the specimen deformed at the highest strain rate (HERF), a well-developed shear band is shown in Fig. 44 (top left). Figure 44 (top right) shows that the shear band is, in fact, composed of many very fine recrystallized grains on the order of 1 μm in diameter. At an intermediate rate (PF), deformation was more diffuse and resulted in many more and smaller shear bands distributed over a broader area (Fig. 44, center left). Also, within each band the shear strain is much smaller than in the shear band formed during the high-strain-rate deformation. Again, the shear bands consist of many fine recrystallized grains and span many of the original grains (Fig. 44, center right).

At the lowest strain rate (ICF), no macroscopic shear bands were observed (Fig. 44, bottom left). Figure 44 (bottom right) shows that the slow-rate defor-

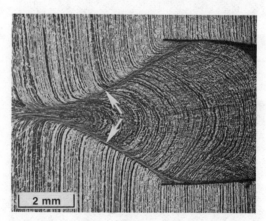

Fig. 43. Macroetched cross section of a JBK-75 specimen high-energy-rate forged at 815°C, showing that shear bands are distinct macroscopic features (arrows) and that they represent extremely sharp transitions or discontinuities in the flow-line pattern.

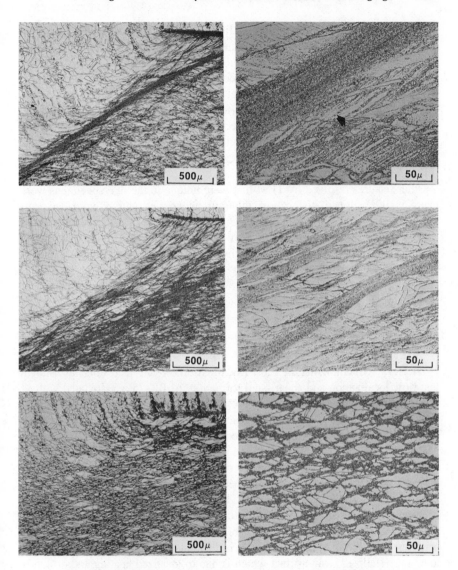

Fig. 44. Microstructures of JBK-75 specimens upset at 815°C by (top pair) high-energy-rate forging, (center pair) press forging, and (bottom pair) isothermal creep forging. The dark horizontal line is the separation between the bulged gage section (bottom) and the specimen end (top). Note the well-defined shear band in top pair, the diffuse shear bands in center pair, and the absence of shear bands in bottom pair. Oxalic acid etch. The arrow in micrograph at top right highlights fine, recrystallized, transgranular shear bands.

mation produced uniform networks of fine recrystallized grains along the grain boundaries of the starting grain structure. The branches in the network have a preferred orientation, apparently related to the direction of maximum resolved shear stress. The concentration of fine recrystallized grains at grain boundaries of the starting microstructure implies that the deformation was localized to the original grain boundaries. Figure 45 (top left) shows a precipitate-free zone (PFZ) in a thin foil taken from the specimen end. It is in the PFZ without the strengthening contribution of γ' precipitation where extensive deformation and subsequent recrystallization would be expected and were observed in the gage (Fig. 45, top right).

Other features resolved by TEM that were unique to the ICF specimens were many well-developed dislocation subgrain boundaries within deformed, un-recrystallized grains (Fig. 45, top right); partially precipitated recrystallized grains (Fig. 45, center left) and "willow-leaf" $\eta^{(70)}$ (Fig. 45, center right). The latter structure apparently was formed by the motion of a large-angle boundary which simultaneously made possible γ' solution (note bumps on upper boundary), rapid solute diffusion to the "willow-leaf" η plate, and recrystallization of the deformed matrix. Cellular eta colonies were also found in the ICF specimens (Fig. 45, bottom left and right).

III.D. Discussion

III.D.1. Mechanisms for Flow Localization.
So far, the results have shown that flow localization and shear-band formation in JBK-75 is associated with γ' precipitation; however, the mechanism for flow softening at the higher forming rates, whether it be localized work softening of the precipitate structure or localized dissolution of γ', has not been identified.

Closely spaced, fine precipitates can raise the flow stress of a microstructure to the point where dislocations cut through the precipitates, rather than bow between them.[95] The work of a number of investigators[96-102] emphasizes that particle cutting or shearing may initiate local softening and thus the strain localization that is observed in the 815°C-deformed JBK-75 specimens. Dissolution of the γ' via adiabatic heating above the solvus (850°C) could also account for local softening. Definition of the mechanism is difficult because shear bands are found to be recrystallized — an artifact which is due to localized heating and/or deformation and which obscures the earlier stages of shear-band formation. However, during an independent investigation of substructural development in JBK-75 HERF forward extrusions,[103] the early and late stages of shear-band formation were observed in two specimens that were given 45 and 60% reductions in area, respectively, at 803°C. With respect to the early stages, Fig. 46 (top left) shows an area near the intersection of two small shear bands lying parallel to {111} planes, with a visible offset in one band. When the right portion of this area was

Fig. 45. Bright-field images of the fine structure in an isothermally creep-forged JBK-75 specimen upset at 815°C and at $\dot{\varepsilon} \approx 8 \times 10^{-4}\ \mathrm{s}^{-1}$. (Top left) Precipitate-free zone associated with a high-angle grain boundary in a lightly deformed area in the specimen end. (Top right) Unrecrystallized area with coarsened γ' and sharp subgrain boundaries (left) adjacent to a band of precipitate-free, recrystallized grains (right) which formed on a high-angle grain boundary (see recrystallized grain necklaces in Fig. 44, bottom right). (Center left) Partially precipitated fine recrystallized grain. (Center right) "Willow-leaf eta" with γ' dissolution at the recrystallization front. (Bottom left) Cellular eta colony in specimen end (note, again, the γ' dissolution at the cell front). (Bottom right) Deformed cellular eta colony in the specimen gage section.

Fig. 46. Bright-field images from a foil of JBK-75 after 45% reduction in area by high-energy-rate forward extrusion at 803°C, showing early stages of shear-band development. Strain rate ≈ 1.4×10^3 s^{-1}. (Top left) Intersection of two small shear bands parallel to {111} planes. Note offset created in one band by the other. (Top right) Area 1 in micrograph at top left, showing a single slip trace (A), multiple slip traces (B) and sub-grains (C). (Bottom left) Area 2 in micrograph at top left, showing "ropey" or "braided" structure of subgrains in the band. (Bottom right) Area 3 in micrograph at top left, showing γ' dissolution (bumps) at the recrystallization front (see arrows), both in the matrix and in the shear band.

tilted slightly in the transmission electron microscope (Fig. 46, top right), it showed the sequence of shear-band development on a third {111} variant, from a slip trace (A) to paired slip traces (B) to multiple slip traces or subgrains (C). A fully developed shear band has a braided or "ropey" structure of subgrains (Fig. 46, bottom left). Note how recrystallization was accelerated along the shear band due to locally higher strain energy. Assuming little time for reprecipitation of the γ' after deformation, small bumps in the recrystallization front reveal the continued presence of γ' in and near the shear band (Fig. 46, bottom right). The original quality of a shear band in JBK-75 is then found to be localized work softening of the precipitation-hardened structure, rather than localized dissolution of γ'. At a slightly later stage in development, the small shear bands are fully recrystallized either statically or dynamically (i.e., see arrow in Fig. 44, top right), and the γ' is no longer present. If recrystallization occurs dynamically, the flow-localization process continues in these regions because of localized γ' dissolution and dislocation-density reduction.

At a still later stage in development, the shear bands grow and propagate across grain boundaries (Fig. 47) to become transgranular features. Finally, these microscopic shear bands link together to form macroscopic shear bands, which can act as crack-initiation sites and crack-propagation paths. An example of shear-band cracking was observed in the 60%-reduced HERF extrusion (Fig. 48).

At low strain rates the mechanism for flow localization is substantially different. The unrecrystallized areas in the ICF DGV specimens deformed at 815°C contained widely spaced γ' particles, which coarsened from 2 to 12 nm to 20 to 30 nm in diameter during deformation, in contrast to the dissolution of particles in the DGV specimens deformed at higher strain rates. These observations are consistent with the fact that there was no measured adiabatic heating during ICF and that the long deformation times in ICF allowed the precipitates to coarsen. Note, a 50°C temperature rise was measured in the gage section during press forging. Flow localization in the ICF specimens deformed at 815°C was due to the dynamic microstructural changes, i.e., dynamic recovery, dynamic recrystallization and particle coarsening, which become more prevalent at low strain rates. The dominant effect is dynamic recrystallization along the high-angle grain boundaries associated with localized deformation of the PFZ's which occurred vis-a-vis limited deformation in the γ'-strengthened grain interiors. A corroborating observation is that annealing-twin boundaries where there were no PFZ's exhibited neither strain concentration nor localized dynamic recrystallization.

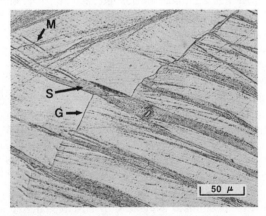

Fig. 47. Optical micrograph of a JBK-75 specimen press forged at 710°C, showing propagation of small shear band (S) across a high-angle grain boundary (G). Note small crossing bands (M) similar to those described in Fig. 46. Etched in oxalic acid.

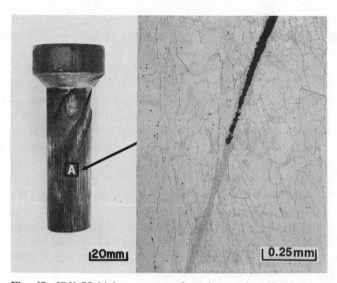

Fig. 48. JBK-75 high-energy-rate-forged extrusion (forging temperature, 815°C; 65% reduction in area; $\dot{\varepsilon} \approx 1.4 \times 10^3$ s^{-1}. (Left) View of extrusion showing spiral cracks. (Right) Optical micrograph (oxalic acid) showing the microstructure at the tip of one of the cracks in the extrusion (area A in view at left). Note that the crack initiated in a macroscopic shear band which formed first at the lead end of the extrusion.

III.D.2. An Analytical Analysis of Flow Localization. The percent DGV test described above has a number of advantages, i.e., it is intuitively easy to understand, the specimen can be deformed in available forming equipment, and the data can be quickly obtained and analyzed, which make it attractive to fabricators with immediate, applied needs. However, it has been the mathematical characterization of the process of metal flow which has led to the interpretation of flow localization in terms of material properties[72,104–108] and which has permitted computer simulation of metalworking processes via rigid-plastic (and rigid-viscoplastic) finite-element methods.[109–112] This section is included to give the reader a brief exposure to a more sophisticated analysis of flow localization as it is applied to JBK-75, and to show that the results and conclusions drawn from both methods of analysis, simple and complex, are consistent with each other.

Analysis of uniaxial isothermal compression for a cylindrical workpiece has revealed that flow localization as evidenced by unstable bulging in the specimen contour is dependent on two material properties: γ' (referred to as γ^* hereafter, to avoid confusion with the precipitate γ'), the normalized flow-softening rate; and m, the strain-rate sensitivity. The following analysis, employed by Semiatin and Lahoti,[67] by Jonas et al.[107] and by Dadras and Thomas,[113] reveals the associated mathematical relationship.

For a uniaxial stress state the axial force can be expressed as

$$F = \sigma A \tag{3}$$

where σ is the axial stress and A is the cross-sectional area of the workpiece. Taking the differential of Eq 3, and assuming that at equilibrium F is the same throughout the specimen, we obtain

$$dF = \sigma dA + A d\sigma = 0 \tag{4}$$

Because flow stress is a function of strain, strain rate and temperature, $\sigma = f(\varepsilon, \dot{\varepsilon}, T)$, Eq 4 becomes

$$0 = \sigma dA + A \left[\left(\frac{\partial \sigma}{\partial \varepsilon} \right) \Big|_{\dot{\varepsilon},T} d\varepsilon + \left(\frac{\partial \sigma}{\partial \dot{\varepsilon}} \right) \Big|_{\varepsilon,T} d\dot{\varepsilon} + \left(\frac{\partial \sigma}{\partial T} \right) \Big|_{\varepsilon,\dot{\varepsilon}} dT \right] \tag{5}$$

where σdA is a geometric hardening term — that is, bulging is stemmed by an increase in dA, which in turn causes σ to decrease. The terms within the brackets are: first, a strain-hardening (or softening) term; second, a strain-rate-hardening term; and third, the dependence of flow stress on temperature.

Noting that in compression $-d\varepsilon = dA/A$, dividing by $\sigma A d\varepsilon$ and rearranging, Eq 5 becomes

$$0 = -1 + m \frac{1}{\dot{\varepsilon}} \frac{d\dot{\varepsilon}}{d\varepsilon} + \left[\left(\frac{\partial \sigma}{\partial \varepsilon} \right) \Big|_{\dot{\varepsilon},T} d\varepsilon + \left(\frac{\partial \sigma}{\partial T} \right) \Big|_{\varepsilon,\dot{\varepsilon}} dT \right] \Big/ \sigma d\varepsilon \tag{6}$$

where m is defined as $(\dot{\varepsilon}/\sigma)(\partial\sigma/\partial\dot{\varepsilon})_{\varepsilon,T}$. The term $-(1/\dot{\varepsilon})(d\dot{\varepsilon}/d\varepsilon)$, the fractional change in strain rate with strain, is referred to as α (a flow-localization parameter) and is used to rate the propensity of a material to flow locally or inhomogeneously. The last term in Eq 6 is the normalized strain-hardening or softening rate, γ^*, which is determined at a constant strain rate. In this analysis, γ^* includes the effect of deformation heating and is defined as

$$\gamma^* = \frac{1}{\sigma} \frac{d\sigma}{d\varepsilon} \Big|_{\dot{\varepsilon}} \tag{7}$$

Note that $d\sigma|\dot{\varepsilon}$ evaluates the term in brackets in Eq 6. Substituting γ^* and α in Eq 6 yields an expression for the tendency for localized flow in terms of the material properties γ^* and m. The expression is

$$\alpha = \frac{\gamma^* - 1}{m} \tag{8}$$

The value of γ^* can be determined from constant-strain-rate compression tests and the value of m from compression tests with incrementally different constant strain rates or from "step-strain rate change" tests. The parameter α will generally vary with strain and so should be evaluated as a function of strain. Following the suggestion of Jonas et al.,[107] Semiatin and Lahoti[67] showed that, although flow localization theoretically starts when $\alpha = 0$, it is practically significant for a number of materials only when $\alpha \geq 5$ approximately.

Under plane-strain conditions, localization occurs along a direction of pure shear and $dA = 0$.[72] If we set $\sigma dA = 0$ in Eq 5, Eq 8 becomes

$$\alpha = \frac{\gamma^*}{m} \qquad (9)$$

for plane-strain deformation. A comparison of Eq 8 and 9 shows that the tendency for flow localization and shear-band formation in plane-strain deformation, which occurs in forming operations such as lateral side pressing and rolling, is greater than in uniaxial compression (i.e., upsetting), where the strain is triaxial.

The correlation between α and the occurrence of shear bands in JBK-75 was examined by isothermally upsetting right circular cylinders to obtain the material parameters γ^* and m, and by observing the flow behavior in plane-strain lateral side pressings. Test temperatures of 816 and 982°C (1500 and 1800°F) were selected because of the differing tendencies for localized flow (high and low, respectively) determined from the DGV results (Fig. 37). The true stress–true strain curves (Fig. 49) show that JBK-75 exhibits significant flow softening at 815°C and little softening at 980°C. Thus, γ^*_{max}, the maximum softening rate after peak stress, was also significantly different (Table 12). The values of strain-

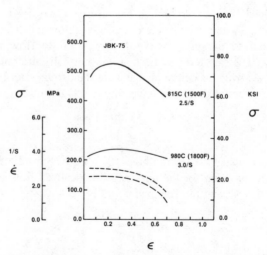

Fig. 49. True stress–true strain compression flow curves for JBK-75 under isothermal conditions at 980 and 815°C and the associated strain rate vs strain curves (dashed lines).

**Table 12.   Parameters Obtained From
JBK-75 Isothermal Compression Tests
(Fig. 50) for Calculation of the Flow-
Localization Parameter, α_{max}**

Temperature					
°C	°F	$\dot{\epsilon}$, s^{-1}	γ^* max	m	α_{max}
815	1500.....2.5		0.70	0.04	17.5
980	1800.....3.0		0.33	0.14	2.4

rate sensitivity (m) were estimated by comparing the flow strength at each temperature for different strain rates, and α was calculated for plane-strain conditions using Eq 9. Considering the rule of thumb that $\alpha \geq 5$ for the occurrence of significant localization, shear bands would be expected to occur during lateral side pressing at 815°C ($\alpha = 17.5$) but not at 980°C ($\alpha = 2.4$). Examination of macroetched, transverse cross sections of the isothermally side-pressed cylinders (Fig. 50) corroborates these predictions and confirms the strength of this analytical analysis technique.

III.D.3.   Computer Simulation. Recently, a rigid-plastic finite-element method has been used to develop a computer program called ALPID (Analysis of Large Plastic Incremental Deformation) for analyzing and simulating forming operations.[114] Among its applications, this program has been used to study the conditions for flow localization and shear-band formation.[72] In particular, it can investigate the effects of the material properties γ^* and m on flow localization.

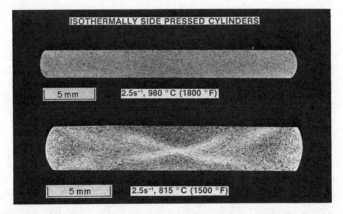

Fig. 50. Macroetched cross sections (axial view) of isothermally side-pressed JBK-75 cylinders, showing shear bands in the 815°C but not in the 980°C specimen. Note, from Table 12, that $\alpha = 17.5$ and 2.4, respectively.

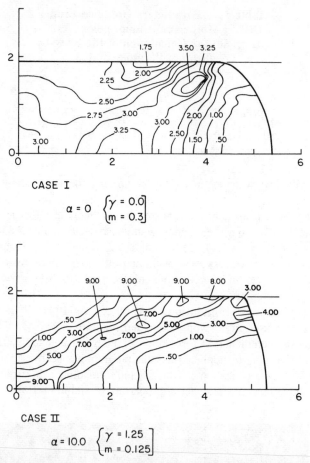

Fig. 51. Computer simulation (Ref 80, Fig. 11) of effective isostrain-rate contours developed in side-pressed cylinders (upper right quarter section, axial view) after 47.5% reduction in height for two combinations of flow-softening rate, γ^*, and strain-rate sensitivity, m. Strain rates are in units of s^{-1}. Note the consistency with the experimental results shown in Fig. 50.

For example, predicted concentrations of effective strain rate in lateral side pressings for two different sets of γ^* and m values are shown in Fig. 51.[72] For these two cases, deformation is relatively uniform when $\alpha = 0$. However, marked flow localization is predicted when $\alpha = 10$. Note that only the upper right quadrant of the total cross section is shown in Fig. 51, because of symmetry. Thus, a qualitative agreement between computer simulation and the α-parameter method for assessment of flow localization has been demonstrated. The advantage of the former is that an extensive evaluation of metal flow in complex shapes is possible. Predictions of mechanical states ($\dot{\varepsilon}$, ε, σ, velocity, flow lines, temperature profiles) at any time during a shaping cycle and at any point in the workpiece can now be obtained, and the ability to predict metallurgical states (microstructure and mechanical properties) throughout the workpiece cross section will be the focal point for future advancements in computer-simulation technology.

III.E. Solution

The solution to the problem was to HERF well above the γ' solvus, where the propensity for shear-band formation was quite low, and to accept and design for the associated loss in strength. Later it was found that strength could be increased by press forging the part. Evidently, the lower relative strain rate resulted in less adiabatic heating, less dynamic softening and static recrystallization, and thus higher retained strength.

Acknowledgments

The authors would like to acknowledge the technical support received from Tom Wilford (TEM), from Angelo Cossa and Gary Robinson (mechanical testing), and from Dick Rhoads, Carlo Pepper, George Snyder, Jack Steele and Gene Parson (optical metallography), as well as the engineering support of Carol Packard-Ferrera in the sigma studies and Cathy Perkins in the flow-localization investigation. The technical contribution of Dr. George Krauss, AMAX Professor, Colorado School of Mines, Golden, Colorado, was integral to the development of the Bauschinger study in 21-6-9 and the flow-localization study in JBK-75. The work on shear-band occurrence in JBK-75 side-pressed cylinders and on the isothermal compression testing of JBK-75 was part of a joint effort by the authors and by Dr. S. L. Semiatin and Tom Nikander of Battelle-Columbus Laboratories, Columbus, Ohio.

References

1. C. S. Smith, "The 1963 ASTM Lecture on Outstanding Research," ASTM, Philadelphia, 1964.
2. G. D. Lahoti, S. L. Semiatin, S. I. Oh, T. Altan and H. L. Gegel, "Development of Process Models to Produce a Dual-Property Titanium Alloy Compressor Disk," Battelle-Columbus Laboratories, Columbus, OH, 1982.
3. C. T. Sims and W. C. Hagel, *The Superalloys,* John Wiley & Sons, New York, 1972.
4. J. H. Westbrook, *Intermetallic Compounds,* John Wiley & Sons, New York, 1972, p 197.
5. E. D. Hall and S. H. Algie, "The Sigma Phase," *Met. Rev.,* Vol 11, 1966.
6. E. J. Dulis and G. V. Smith, "Identification and Mode of Formation and Resolution of Sigma Phase in Austenite Cr-Ni Steels," in "Symposium on the Nature, Occurrence, and Effects of Sigma Phase," STP No. 110, ASTM, Philadelphia, 1951, p 3.

7. H. Thielsch, "Physical Metallurgy of Austenitic Stainless Steels," *Welding J.,* Dec 1950, p 584-5.

8. B. Weiss and R. Stickler, "Phase Instabilities During High Temperature Exposure of 316 Austenitic Stainless Steel," *Met. Trans.,* Vol 3, 1972, p 865.

9. P. Duhaj, J. Ivan and E. Makovicky, "Sigma-Phase Precipitation in Austenitic Steels," *J. Iron Steel Inst.,* Dec 1968, p 1245.

10. J. E. Spruiell, J. A. Scott, C. S. Ary and R. L. Hardin, "Microstructural Stability of Thermal-Mechanically Pretreated Type 316 Austenitic Stainless Steel," *Met. Trans.,* Vol 4, 1973, p 1533.

11. A. S. Grot and J. E. Spruiell, "Microstructural Stability of Titanium-Modified Type 316 and Type 321 Stainless Steel," *Met. Trans. A,* Vol 6A, 1975, p 2025.

12. A. J. Lena, "Sigma Phase–a Review," *Metal Prog.,* July 1, 1954, p 88.

13. L. K. Poole, "Sigma – an Unwanted Constituent in Stainless Weld Metal," *Metal Prog.,* June 1954, p 109.

14. C. L. Packard-Ferrera, M. C. Mataya and C. M. Edstrom, "Study of the Formation and Effects of Sigma Phase in 21-6-9 Stainless Steel," Rockwell International, Rocky Flats Plant, Golden, CO, RFP-3031, 1981.

15. R. J. Gray et al., "A Metallographic Study of Ferrite-Sigma Transformation Using Ferromagnetic Colloid, Microprobe Analysis, and Color Etching," *Microstructural Sci.,* Elsevier North-Holland,Inc., 1977, Vol 5, p 81.

16. A. J. Lena, "Effect of Sigma Phase on Properties of Alloys," *Metal Prog.,* August 1954, p 94.

17. S. A. David, G. M. Goodwin and D. N. Braski, "Solidification Behavior of Austenitic Stainless Steel Filler Metals," *Welding J.,* November 1979, p 335-S.

18. H. Fredricksson, "The Solidification Sequence in an 18-8 Stainless Steel, Investigated by Directional Solidification," *Met. Trans.,* Vol 3, November 1972, p 2989.

19. J. C. Lippold and W. F. Savage, "Solidification of Austenitic Stainless Steel Weldments," *Welding J.,* December 1979, p 362-S.

20. N. Suutala, T. Takalo and T. Moiso, "The Relationship Between Solidification and Microstructure in Austenitic and Austenitic-Ferritic Stainless Steel Welds," *Met. Trans. A,* Vol 10A, 1979, p 512.

21. N. Suutala, T. Takalo and T. Moiso, "Ferritic-Austenitic Solidification Mode in Austenitic Stainless Steel Welds," *Met. Trans. A,* Vol 11A, 1980, p 717.

22. J. Bauschinger, *Civilingenieur,* Vol 27, 1881, p 289.

23. D. V. Wilson, *Acta Met.,* Vol 13, 1965, p 807.

24. D. V. Wilson, *Metals Tech.,* Vol 2, 1975, p 8.

25. S. T. Rolfe, R. P. Haak and J. H. Gross, *J. Basic Eng.,* 1968, p 408.

26. C. C. Li, J. D. Flasck, J. A. Yaker and W. C. Leslie, *Met. Trans. A,* Vol 9A, 1978, p 85.

27. D. Uko, R. Sowerby and J. D. Embury, *Metals Tech.,* September 1980, p 359.

28. N. H. Polakowski, *Research,* Vol 5, 1952, p 143.

29. C. F. Tipper, *J. Iron Steel Inst.,* Vol 172, 1952, p 143.

30. E. H. Edwards, J. Washburn and E. R. Parker, *Trans. AIME,* Vol 197, 1953, p 1526.

31. E. H. Edwards and J. Washburn, *J. Metals,* Vol 200, 1954, p 1239.

32. R. L. Woolley, *Phil Mag.,* Vol 44, 1953, p 597.

33. S. N. Buckley and K. M. Entwistel, *Acta Met.,* Vol 4, 1956, p 352.

34. A. Abel and H. Muir, *Phil. Mag.,* Vol 26, 1972, p 489.

35. M. C. Mataya, M. J. Carr and G. Krauss, *Mater. Sci. Eng.,* Vol 57, No. 2, 1983, p 205-222.

36. B. C. Odegard, A. J. West and J. A. Brooks, *Effects of Hydrogen on Behavior of Materials*, edited by A. W. Thompson and I. M. Bernstein, AIME, New York, 1976, p 189.
37. R. E. Stoltz and J. B. Vander Sande, *Met. Trans. A.*, Vol 11A, 1980, p 1033.
38. D. J. Michel, J. Moteff and A. J. Lovell, *Acta Met.*, Vol 21, 1973, p 1269.
39. L. Remy, *Acta Met.*, Vol 25, 1977, p 173.
40. R. Orowon, *Internal Stresss and Fatigue in Metals*, Elsevier, New York, 1959, p 59.
41. F. A. McClintock and A. S. Argon, *Mechanical Behavior of Materials*, Addison-Wesley, 1966, p 185.
42. R. V. Milligan, W. H. Koo and T. E. Davidson, *J. Basic Eng.*, Vol 88(D), 1966, p 480.
43. R. E. Stoltz and R. M. Pelloux, *Met. Trans. A.*, Vol 7A, 1976, p 1295.
44. J. D. Embury, Strengthening Methods in Crystals, Applied Science Publishers, Ltd., England, 1971, p 348-362.
45. N. Ibraham and J. D. Embury, *Mater. Sci. Eng.*, Vol 19, 1975, p 147.
46. J. H. Hausselt and W. D. Nix, *Acta Met.*, Vol 25, 1977, p 595.
47. M. C. Mataya and R. A. Fournelle, *Met. Trans. A*, Vol 9A, 1978, p 917.
48. C. R. Hills and H. J. Rack, *Mater. Sci. Eng.*, Vol 51, 1981, p 231.
49. A. Abel and H. Muir, *Phil. Mag.*, Vol 26, No. 2, 1972, p 489.
50. M. C. Mataya, M. J. Carr, R. W. Krenzer and G. Krauss, "Processing and Structure of High Energy Rate Forged 21-6-9 and 304L Forgings," Rockwell International, Rocky Flats Plant, Golden, CO, RFP-3020, 1981.
51. S. S. Hecker, A. K. Ghosh and H. L. Gegel, *Formability*, AIME, New York, 1978.
52. J. Nutall and J. Nutting, *Metal Sci.*, September 1978, p 430.
53. S. Nourbakhst and J. Nutting, *Acta Met.*, Vol 28, 1980, p 357.
54. K. Brown, *J. Inst. Metals*, Vol 100, 1972, p 341.
55. M. N. Janardhana and S. K. Biswas, *Int. J. Mech. Sci.*, Vol 21, 1979, p 699.
56. Y. W. Chang and R. J. Asaro, *Acta Met.*, Vol 24, 1981, p 241.
57. A. J. Bedford, A. L. Wingrove and K. R. L. Thompson, *J. Aust. Inst. Met.*, Vol 19, No. 1, 1974, p 61.
58. M. R. Staker, *Scripta Met.*, Vol 14, 1980, p 677.
59. V. Osina, *Met. Treat.*, Vol 33, May 1966, p 193.
60. W. Johnson, G. L. Baraya and R. A. C. Slater, *Int. J. Mech. Sci.*, Vol 6, 1964, p 409.
61. R. H. Ernst and J. W. Spretnak, *Trans. Iron Steel Inst. Japan*, Vol 9, 1969, p 361.
62. G. L. Moss, Tech. Report ARBRL-TR-02242, BRL., Aberdeen Proving Ground, MD, May 1980.
63. P. S. Mathur and W. A. Backofen, *Met. Trans.*, Vol 4, 1973, p 643.
64. M. C. Mataya and G. Krauss, *J. Appl. Metalwork.*, Vol 2, No. 1, 1981, p 28.
65. G. L. Wulf, *Inst. J. Mech. Sci.*, Vol 21, 1979, p 713.
66. S. L. Semiatin and G. D. Lahoti, *Met. Trans. A.*, Vol 14A, 1982, p 275.
67. S. L. Semiatin and G. D. Lahoti, *Met. Trans. A.*, Vol 12A, 1981, p 1705.
68. S. L. Semiatin and G. D. Lahoti, ibid., p 1719.
69. M. Blicharski and S. Gorczyca, *Metal Sci.*, July 1978, p 303.
70. M. C. Mataya, M. C. Carr and G. Krauss, *Met. Trans. A*, Vol 13A, 1982, p 1263.
71. F. N. Lake and D. J. Moracz, Technical Report AFML-TR-69-174, TRW, Inc., Cleveland, OH, June 1969.
72. S. L. Semiatin, G. D. Lahoti and S. I. Oh, "The Occurrence of Shear Bands in Metalworking": 29th Sagamore Army Materials Research Conference, *Materials Response to Large Plastic Deformation*, edited by M. G. Stout and S. S. Hecker,

Lake Placid, NY, July 19-23, 1982.
73. M. C. Mataya and M. J. Weis, unpublished data, Rockwell International, 1982.
74. H. C. Rogers, *Ann. Rev. Mater. Sci.*, Vol 9, 1979, p 283.
75. J. J. Jonas and H. J. McQueen, *Treatise on Materials Science and Technology*, Vol 6, "Plastic Deformation of Metals," edited by R. J. Arsenault, Academic Press, New York, 1975, p 394.
76. D. J. Abson and J. J. Jonas, *Metals Tech.*, October 1977, p 462.
77. B. Brenna and A. Luft, *Mater. Sci. Eng.*, Vol 52, 1982, p 229.
78. J. P. Imarigeon and J. J. Jonas, "Flow Stress and Substructural Change During the Transient Deformation of ARMCO Iron and Silicon Steel," *Acta Met.*, Vol 19, 1971, p 1053-1061.
79. M. J. Luton and C. M. Sellars, "Dynamic Recrystallization in Nickel and Nickel-Iron Alloys During High Temperature Deformation," *Acta Met.*, Vol 17, 1969, p 1033-1043.
80. H. J. McQueen and J. J. Jonas, "Recovery and Recrystallization During High Temperature Deformation," in *Plastic Deformation of Materials*, edited by R. J. Arsenault, Academic Press, New York, 1975, p 393-493.
81. J. J. Jonas, C. M. Sellars and W. J. McG. Tegart, "Strength and Structure Under Hot Working Conditions," *Met. Rev.*, Vol 14, No. 130, 1969, p 1-24; *Metals Mater.*, Vol 3, 1969, p 33.
82. J. L. Robbins, O. C. Shepard and O. D. Sherby, *J. Iron Steel Inst.*, October, 1964, p 804.
83. J. J. Jonas and M. J. Luton, *Advances in Deformation Processing*, Plenum Press, New York, 1978, p 215-243.
84. J. J. Jonas, B. Heritier and M. J. Luton, *Met. Trans. A*, Vol 10A, 1979, p 611.
85. G. B. Olson, J. F. Mescall and M. Azrin, *Proc. Int. Conf. Metal Effects of High Strain-Rate Deformation and Fabrication*, Plenum Press, 1980, p 221-247.
86. C. M. Young and O. D. Sherby, *Metal Forming – Interrelation Between Theory and Practice*, edited by A. L. Hoffmanner, Plenum Press, New York, 1971, p 429.
87. P. Dadras and J. F. Thomas, *Met. Trans. A*, Vol 12A, 1981, p 1867.
88. A. A. Guimaraes and J. J. Jonas, *Met. Trans. A*, Vol 12A, 1981, p 1655.
89. M. E. Backman and S. A. Finnegan, *Metallurgical Effects at High Strain Rates*, edited by R. W. Rhode, B. M. Butcher, J. R. Holland and C. H. Karnes, Plenum Press, New York, 1973, p 531-43.
90. P. W. Lee and H. Kuhn, *Met. Trans. A*, Vol 4A, 1973, p 969.
91. K. A. Reynolds, *Deformation Under Hot Working Conditions*, edited by C. M. Sellars and W. J. McG. Tegart, Iron and Steel Institute, London, 1968, p 107.
92. A. L. Hoffmanner, "The Use of Workability Results to Predict Processing Limits," in *Metalforming – Interrelation Between Theory and Practice*, edited by A. L. Hoffmanner, Plenum Press, New York, 1971.
93. R. C. Koeller and R. Raj, *Acta Met.*, Vol 26, 1978, p 1551.
94. T. J. Headley, M. M. Karnowsky and W. R. Sorenson, *Met. Trans. A*, Vol 13A, 1982, p 345.
95. E. Hornbogen, in AMAX Symposium: *Steel Strengthening Mechanisms*, American Metal Climax, Inc., 1970, p 1-15.
96. E. Hornbogen and G. Staniek, *J. Mater. Sci.*, Vol 9, 1974, p 879-886.
97. C. Calabrese and C. Laird, *Mater. Sci. Eng.*, Vol 13, 1974, p 141-157.
98. S. Nourbakhsh and J. Nutting, *Acta Met.*, Vol 28, 1980, p 357-365.
99. S. P. Lynch, *Metal Sci.*, Vol 9, 1975, p 401-410.
100. J. J. Jonas, B. Heritier and M. J. Luton, *Met. Trans. A*, Vol 10A, 1979, p 611-620.

101. W. J. Plumbridge and D. A. Ryder, *Met. Rev.*, Review No. 136, 1970, p 119-141.
102. C. H. Wells and C. P. Sullivan, *Trans. ASM,* Vol 57, 1974, p 841.
103. M. C. Mataya, M. J. Carr and G. Krauss, *Met. Trans. A* (in press).
104. M. Considere, *Annls. Ponts Chauss.,* Vol 9, 1885, p 574.
105. W. A. Backofen, in *Fracture of Engineering Materials,* ASM, Metals Park, OH, 1964, p 107.
106. E. W. Hart, *Acta Met.,* Vol 15, 1967, p 351.
107. J. J. Jonas, R. A. Holt and C. E. Coleman, *Acta Met.,* Vol 24, 1976, p 911.
108. M. R. Staker, *Acta Met.,* Vol 29, 1981, p 683.
109. C. H. Lee and S. Kobayashi, *Trans. ASME, J. Eng. Ind.,* Vol 95, 1973, p 865.
110. S. I. Oh, N. Rebelo and S. Kobayashi, "Finite Element Formulation for the Analysis of Plastic Deformation of Rate-Sensitive Materials for Metalforming," *Metal Forming Plasticity,* IUTAM Symposium, Tutzing, Germany, 1978, p 273.
111. S. I. Oh, *Inter. J. Mech. Sci.,* Vol 24, 1982, p 479.
112. S. Kobayashi, *J. Appl. Metalwork.,* Vol 2, No. 3, July 1982, p 163.
113. P. Dadras and J. F. Thomas, Jr., *Res. Mechanica Letters,* Vol 1, 1981, p 97.
114. S. I. Oh, G. D. Lahoti and T. Altan, *J. Metalwork. Tech.,* Vol 6, 1982, p 277.

Index

NOTE: The symbol (F) or (T) following a page number in this index indicates that information on the subject is presented in a figure or a table.

503